Lecture Notes in Artificial Intelligence   11288

Subseries of Lecture Notes in Computer Science

LNAI Series Editors

Randy Goebel
University of Alberta, Edmonton, Canada
Yuzuru Tanaka
Hokkaido University, Sapporo, Japan
Wolfgang Wahlster
DFKI and Saarland University, Saarbrücken, Germany

LNAI Founding Series Editor

Joerg Siekmann
DFKI and Saarland University, Saarbrücken, Germany

More information about this series at http://www.springer.com/series/1244

Ildar Batyrshin
María de Lourdes Martínez-Villaseñor
Hiram Eredín Ponce Espinosa (Eds.)

Advances in Soft Computing

17th Mexican International Conference
on Artificial Intelligence, MICAI 2018
Guadalajara, Mexico, October 22–27, 2018
Proceedings, Part I

 Springer

Editors
Ildar Batyrshin
Instituto Politécnico Nacional
Mexico City, Mexico

María de Lourdes Martínez-Villaseñor
Universidad Panamericana
Mexico City, Mexico

Hiram Eredín Ponce Espinosa
Faculty of Engineering
Universidad Panamericana
Mexico City, Mexico

ISSN 0302-9743   ISSN 1611-3349   (electronic)
Lecture Notes in Artificial Intelligence
ISBN 978-3-030-04490-9   ISBN 978-3-030-04491-6   (eBook)
https://doi.org/10.1007/978-3-030-04491-6

Library of Congress Control Number: 2018958468

LNCS Sublibrary: SL7 – Artificial Intelligence

This Springer imprint is published by the registered company Springer Nature Switzerland AG
The registered company address is: Gewerbestrasse 11, 6330 Cham, Switzerland

Preface

The Mexican International Conference on Artificial Intelligence (MICAI) is a yearly international conference series that has been organized by the Mexican Society of Artificial Intelligence (SMIA) since 2000. MICAI is a major international artificial intelligence forum and the main event in the academic life of the country's growing artificial intelligence community.

MICAI conferences publish high-quality papers in all areas of artificial intelligence and its applications. The proceedings of the previous MICAI events have been published by Springer in its *Lecture Notes in Artificial Intelligence* series, vol. 1793, 2313, 2972, 3789, 4293, 4827, 5317, 5845, 6437, 6438, 7094, 7095, 7629, 7630, 8265, 8266, 8856, 8857, 9413, 9414, 10061, 10062, 10632, and 10633. Since its foundation in 2000, the conference has been growing in popularity and improving in quality.

The proceedings of MICAI 2018 are published in two volumes. The first volume, *Advances in Soft Computing*, contains 33 papers structured into three sections:

– Evolutionary and nature-inspired intelligence
– Machine learning
– Fuzzy logic and uncertainty management

The second volume, *Advances in Computational Intelligence*, contains 29 papers structured into three sections:

– Knowledge representation, reasoning, and optimization
– Natural language processing
– Robotics and computer vision

This two-volume set will be of interest for researchers in all areas of artificial intelligence, students specializing in related topics, and for the public in general interested in recent developments in artificial intelligence.

The conference received 149 submissions for evaluation from 23 countries: Argentina, Australia, Brazil, Canada, Colombia, Costa Rica, Cuba, Czech Republic, Finland, France, Hungary, Iran, Italy, Mexico, Morocco, Pakistan, Peru, Poland, Russia, Spain, Thailand, Turkey, and USA. Of these submissions, 62 papers were selected for publication in these two volumes after a peer-reviewing process carried out by the international Program Committee. Therefore, the acceptance rate was 41%.

The international Program Committee consisted of 113 experts from 17 countries: Azerbaijan, Brazil, Canada, Colombia, Cuba, France, Greece, India, Israel, Italy, Japan, Mexico, Portugal, Singapore, Spain, UK, and USA.

MICAI 2018 was honored by the presence of renowned experts who gave excellent keynote lectures:

– Alexander S. Poznyak Gorbatch, CINVESTAV-IPN Campus Mexico City, Mexico
– Jeff Clune, Uber AI Labs, University of Wyoming, USA

- David J. Atkinson, Silicon Valley Research & Development Center Continental AG, USA
- Gregory O'Hare, School of Computer Science, University College Dublin, Ireland
- Srinivas V. Chitiveli, Offering Manager & Master Inventor, IBM PowerAI Vision, USA

The technical program of the conference also featured 11 tutorials:

- Smart Applications with FIWARE, by Miguel Gonzalez Mendoza
- Intelligent Management of Digital Data for Law Enforcement Purposes, by Jesus Manuel Niebla Zatarain
- Introduction to Data Science: Similarity, Correlation, and Association Measures, by Ildar Batyrshin
- Brain–Computer Interface (BCI) and Machine Learning, by Javier M. Antelis, Juan Humberto Sossa Azuela, Luis G. Hernandez, and Carlos D. Virgilio
- New Models and Training Algorithms for Artificial Neural Networks, by Juan Humberto Sossa Azuela
- Intelligent Chatbots Using Google DialogFlow, by Leonardo Garrido
- Spiking Neural Models and Their Applications in Pattern Recognition: A beginner's Tutorial, by Roberto A. Vazquez
- Introduction to Quantum Computing, by Salvador Venegas
- Deep-Learning Principles and Their Applications in Facial Expression Recognition with TensorFlow-Keras, by Luis Eduardo Falcón Morales and Juan Humberto Sossa Azuela
- Introduction to Natural Language Human–Robot Interaction, by Grigori Sidorov
- Knowledge Extraction from Fuzzy Predictive Models, by Félix A. Castro Espinoza

Three workshops were held jointly with the conference:

- HIS 2018: 11th Workshop of Hybrid Intelligent Systems
- WIDSSI 2018: 4th International Workshop on Intelligent Decision Support Systems for Industry
- WILE 2018: 11th Workshop on Intelligent Learning Environments

The authors of the following papers received the Best Paper Awards based on the paper's overall quality, significance, and originality of the reported results:

- First place: "Universal Swarm Optimizer for Multi-Objective Functions," by Luis Marquez and Luis Torres Treviño, Mexico
- Second place: "Topic-Focus Articulation: A Third Pillar of Automatic Evaluation of Text Coherence," by Michal Novák, Jiří Mírovský, Kateřina Rysová and Magdaléna Rysová, Czechia
- Third place: "Combining Deep Learning and RGBD SLAM for Monocular Indoor Autonomous Flight," by José Martínez Carranza, L. Oyuki Rojas Pérez, Aldrich A. Cabrera Ponce and Roberto Munguia Silva, Mexico

The cultural program of the conference included a tour of Guadalajara and the Tequila Experience tour.

We want to thank everyone involved in the organization of this conference. In the first place, the authors of the papers published in this book: It is their research effort that gives value to the book and to the work of the organizers. We thank the track chairs for their hard work, the Program Committee members, and additional reviewers for their great effort in reviewing the submissions.

We would like to thank the Tecnológico de Monterrey Campus Guadalajara for hosting the workshops and tutorials of MICAI 2018, with special thanks to Dr. Mario Adrián Flores, Vice President of the University, and Dr. Ricardo Swain, Dean of the Engineering and Sciences School, for their generous support. We also thank Dr. José Antonio Rentería, Divisional Director of the Engineering and Sciences School, for his kind support. We also want to thank Erik Peterson from Oracle, Rodolfo Lepe and Leobardo Morales from IBM, Oscar Reyes from SinergiaSys, and Luis Carlos Garza Tamez from Grupo ABSA, for their support in organization of this conference. The entire submission, reviewing, and selection process, as well as the preparation of the proceedings, were supported free of charge by the EasyChair system (www.easychair. org). Finally, yet importantly, we are very grateful to the staff at Springer for their patience and help in the preparation of this volume.

October 2018   Ildar Batyrshin
                                        María de Lourdes Martínez-Villaseñor
                                               Hiram Eredín Ponce Espinosa

Conference Organization

MICAI 2018 was organized by the Mexican Society of Artificial Intelligence (SMIA, Sociedad Mexicana de Inteligencia Artificial) in collaboration with the Tecnológico de Monterrey Campus Guadalajara, the Tecnológico de Monterrey CEM, the Centro de Investigación en Computación of the Instituto Politécnico Nacional, the Facultad de Ingeniería or the Universidad Panamericana, and the Universidad Autónoma del Estado de Hidalgo.

The MICAI series website is www.MICAI.org. The website of the Mexican Society of Artificial Intelligence, SMIA, is www.SMIA.org.mx. Contact options and additional information can be found on these websites.

Conference Committee

General Chair

Miguel González Mendoza	Tecnológico de Monterrey CEM, Mexico

Program Chairs

Ildar Batyrshin	Instituto Politécnico Nacional, Mexico
María de Lourdes Martínez Villaseñor	Universidad Panamericana, Mexico
Hiram Eredín Ponce Espinosa	Universidad Panamericana, Mexico

Workshop Chairs

Obdulia Pichardo Lagunas	Instituto Politécnico Nacional, Mexico
Noé Alejandro Castro Sánchez	Centro Nacional de Investigación y Desarrollo Tecnológico, Mexico
Félix Castro Espinoza	Universidad Autónoma del Estado de Hidalgo, Mexico

Tutorials Chair

Félix Castro Espinoza	Universidad Autónoma del Estado de Hidalgo, Mexico

Doctoral Consortium Chairs

Miguel Gonzalez Mendoza	Tecnológico de Monterrey CEM, Mexico
Antonio Marín Hernandez	Universidad Veracruzana, Mexico

Keynote Talks Chair

Sabino Miranda Jiménez	INFOTEC, Mexico

Publication Chair

Miguel Gonzalez Mendoza   Tecnológico de Monterrey CEM, Mexico

Financial Chair

Ildar Batyrshin   Instituto Politécnico Nacional, Mexico

Grant Chairs

Grigori Sidorov   Instituto Politécnico Nacional, Mexico
Miguel Gonzalez Mendoza   Tecnológico de Monterrey CEM, Mexico

Organizing Committee Chairs

Luis Eduardo Falcón   Tecnológico de Monterrey, Campus Guadalajara
  Morales
Javier Mauricio Antelis   Tecnológico de Monterrey, Campus Guadalajara
  Ortíz

Area Chairs

Machine Learning

Felix Castro Espinoza   Universidad Autónoma del Estado de Hidalgo, Mexico

Natural Language Processing

Sabino Miranda Jiménez   INFOTEC, Mexico
Esaú Villatoro   Universidad Autónoma Metropolitana Cuajimalpa,
  Mexico

Evolutionary and Evolutive Algorithms

Hugo Jair Escalante   Instituto Nacional de Astrofísica, Óptica y Electrónica,
  Balderas   Mexico
Hugo Terashima Marín   Tecnológico de Monterrey CM, Mexico

Fuzzy Logic

Ildar Batyrshin   Instituto Politécnico Nacional, Mexico
Oscar Castillo   Instituto Tecnológico de Tijuana, Mexico

Neural Networks

María de Lourdes Martínez   Universidad Panamericana, Mexico
  Villaseñor

Hybrid Intelligent Systems

Juan Jose Flores   Universidad Michoacana, Mexico

Intelligent Applications

Gustavo Arroyo   Instituto Nacional de Electricidad y Energias Limpias, Mexico

Computer Vision and Robotics

José Martínez Carranza   Instituto Nacional de Astrofísica, Óptica y Electrónica, Mexico

Daniela Moctezuma   Centro de Investigación en Ciencias de Información Geoespacial, Mexico

Program Committee

Rocío Abascal-Mena   Universidad Autonoma Metropolitana – Cuajimalpa, Mexico

Giner Alor Hernandez   Instituto Tecnologico de Orizaba, Mexico

Matias Alvarado   Centro de Investigación y de Estudios Avanzados del IPN, Mexico

Nohemi Alvarez   Centro de Investigación en Geografía y Geomática Ing. Jorge L. Tamayo A.C., Mexico

Gustavo Arechavaleta   Centro de Investigación y de Estudios Avanzados del IPN, Mexico

Gustavo Arroyo-Figueroa   Instituto Nacional de Electricidad y Energías Limpias, Mexico

Maria Lucia Barrón-Estrada   Instituto Tecnológico de Culiacán, Mexico

Rafael Batres   Tecnológico de Monterrey, Mexico

Ildar Batyrshin   CIC, Instituto Politécnico Nacional, Mexico

Davide Buscaldi   LIPN, Université Paris 13, Sorbonne Paris Cité, France

Hiram Calvo   CIC, Instituto Politécnico Nacional, Mexico

Nicoletta Calzolari   Istituto di Linguistica Computazionale, CNR, Italy

Jesus Ariel Carrasco-Ochoa   Instituto Nacional de Astrofísica, Óptica y Electrónica, Mexico

Oscar Castillo   Instituto Tecnológico de Tijuana, Mexico

Felix Castro Espinoza   CITIS, Universidad Autónoma del Estado de Hidalgo, Mexico

Noé Alejandro   Centro Nacional de Investigación y Desarrollo
Castro-Sánchez   Tecnológico, Mexico

Jaime Cerda Jacobo   Universidad Michoacana de San Nicolás de Hidalgo, Mexico

Ulises Cortés   Universitat Politècnica de Catalunya, Spain

Paulo Cortez   University of Minho, Portugal

Laura Cruz   Instituto Tecnologico de Cd. Madero, Mexico

Israel Cruz Vega   Instituto Nacional de Astrofísica, Óptica y Electrónica, Mexico

Andre de Carvalho   University of São Paulo, Brazil

Jorge De La Calleja	Universidad Politécnica de Puebla, Mexico
Omar Arturo Domínguez-Ramírez	CITIS, Universidad Autónoma del Estado de Hidalgo, Mexico
Leon Dozal	CentroGEO, Mexico
Hugo Jair Escalante	Instituto Nacional de Astrofísica, Óptica y Electrónica, Mexico
Bárbaro Ferro	Universidad Panamericana, Mexico
Denis Filatov	Sceptica Scientific Ltd., UK
Juan José Flores	Universidad Michoacana de San Nicolás de Hidalgo, Mexico
Anilu Franco-Árcega	CITIS, Universidad Autónoma del Estado de Hidalgo, Mexico
Sofia N. Galicia-Haro	Universidad Nacional Autónoma de México, Mexico
Cesar Garcia Jacas	Universidad Panamericana, Mexico
Milton García-Borroto	Universidad Tecnológica de la Habana José Antonio Echeverría (CUJAE), Cuba
Alexander Gelbukh	CIC, Instituto Politécnico Nacional, Mexico
Carlos Gershenson	Universidad Nacional Autónoma de México, Mexico
Eduardo Gomez-Ramirez	Dirección de Posgrado e Investigación, Universidad La Salle, Mexico
Enrique González	Pontificia Universidad Javeriana de la Compañía de Jesús, Colombia
Luis-Carlos González-Gurrola	Universidad Autónoma de Chihuahua, Mexico
Miguel Gonzalez-Mendoza	Tecnológico de Monterrey Campus Estado de México, Mexico
Mario Graff	Infotec, Centro de Investigación e Innovación en Tecnologías de la Información y Comunicación, Mexico
Fernando Gudiño	FES Cuautitlán, Universidad Nacional Autónoma de México, Mexico
Miguel Angel Guevara Lopez	Computer Graphics Center, Portugal
Andres Gutierrez	Tecnológico de Monterrey, Mexico
J. Octavio Gutierrez-Garcia	Instituto Tecnológico Autónomo de México, Mexico
Rafael Guzman Cabrera	Universidad de Guanajuato, Mexico
Yasunari Harada	Waseda University, Japan
Jorge Hermosillo	Universidad Autónoma del Estado de Hidalgo, Mexico
Yasmin Hernandez	Instituto Nacional de Electricidad y Energías Limpias, Mexico
José Alberto Hernández	Universidad Autónoma del Estado de Morelos, Mexico
Oscar Herrera	Universidad Autónoma Metropolitana – Azcapotzalco, Mexico
Pablo H. Ibarguengoytia	Instituto Nacional de Electricidad y Energías Limpias, Mexico

Sergio Gonzalo Jiménez Vargas	Instituto Caro y Cuervo, Colombia
Angel Kuri-Morales	Instituto Tecnológico Autónomo de México, Mexico
Carlos Lara-Alvarez	Centro de Investigación en Matemáticas (CIMAT), Mexico
Eugene Levner	Ashkelon Academic College, Israel
Fernando Lezama	Instituto Nacional de Astrofísica, Óptica y Electrónica, Mexico
Rodrigo Lopez Farias	CONACYT, Mexico; Consorcio CENTROMET, Mexico
Omar Jehovani López Orozco	Instituto Tecnológico Superior de Apatzingán, Mexico
Omar López-Ortega	CITIS, Universidad Autónoma del Estado de Hidalgo, Mexico
Octavio Loyola-González	Escuela de Ingeniería y Ciencias, Tecnológico de Monterrey, Mexico
Yazmin Maldonado	Instituto Tecnológico de Tijuana, Mexico
Cesar Martinez Torres	Universidad de las Américas Puebla, Mexico
María De Lourdes Martínez Villaseñor	Universidad Panamericana, Mexico
José Martínez-Carranza	Instituto Nacional de Astrofísica, Óptica y Electrónica, Mexico
José Fco. Martínez-Trinidad	Instituto Nacional de Astrofísica, Óptica y Electrónica, Mexico
Antonio Matus-Vargas	Instituto Nacional de Astrofísica, Óptica y Electrónica, Mexico
Patricia Melin	Instituto Tecnológico de Tijuana, Mexico
Ivan Vladimir Meza Ruiz	Instituto de Investigaciones en Matemáticas Aplicadas y en Sistemas, Universidad Nacional Autónoma de México, Mexico
Efrén Mezura-Montes	Universidad Veracruzana, Mexico
Sabino Miranda-Jiménez	Infotec, Centro de Investigación e Innovación en Tecnologías de la Información y Comunicación, Mexico
Daniela Moctezuma	CONACyT, Mexico; Centro de Investigación en Ciencias de Información Geoespacial, Mexico
Raul Monroy	Tecnológico de Monterrey Campus Estado de México, Mexico
Marco Morales	Instituto Tecnológico Autónomo de México, Mexico
Annette Morales-González	CENATAV, Cuba
Masaki Murata	Tottori University, Japan
Antonio Neme	Universidad Autónoma de la Ciudad de México, Mexico
C. Alberto Ochoa-Zezatti	Universidad Autónoma de Ciudad Juárez, Mexico
José Luis Oliveira	University of Aveiro, Portugal

Jose Ortiz Bejar	Universidad Michoacana de San Nicolás de Hidalgo, Mexico
José Carlos Ortiz-Bayliss	Tecnológico de Monterrey, Mexico
Juan Antonio Osuna Coutiño	Instituto Tecnológico de Tuxtla Gutiérrez, Mexico
Partha Pakray	National Institute of Technology Silchar, India
Leon Palafox	Universidad Panamericana, Mexico
Ivandre Paraboni	University of São Paulo, Brazil
Obdulia Pichardo-Lagunas	Unidad Profesional Interdisciplinaria en Ingeniería y Tecnologías Avanzadas, Instituto Politécnico Nacional, Mexico
Garibaldi Pineda García	Universidad Michoacana de San Nicolás de Hidalgo, Mexico; University of Manchester, UK
Hiram Eredin Ponce Espinosa	Universidad Panamericana, Mexico
Soujanya Poria	Nanyang Technological University, Singapore
Belem Priego-Sanchez	Benemérita Universidad Autónoma de Puebla, Mexico; Université Paris 13, France; Universidad Autónoma Metropolitana – Azcapotzalco, Mexico
Luis Puig	Universidad de Zaragoza, Spain
Vicenç Puig	Universitat Politècnica de Catalunya, Spain
Juan Ramirez-Quintana	Instituto Tecnológico de Chihuahua, Mexico
Patricia Rayón	Universidad Panamericana, Mexico
Juan Manuel Rendon-Mancha	Universidad Autónoma del Estado de Morelos, Mexico
Orion Reyes	University of Alberta Edmonton, Canada
José A. Reyes-Ortiz	Universidad Autónoma Metropolitana, Mexico
Noel Enrique Rodriguez Maya	Instituto Tecnológico de Zitácuaro, Mexico
Hector Rodriguez Rangel	University of Oregon, USA
Alejandro Rosales	Tecnológico de Monterrey, Mexico
Christian Sánchez-Sánchez	Universidad Autónoma Metropolitana, Mexico
Ángel Serrano	Universidad Rey Juan Carlos, Spain
Shahnaz Shahbazova	Azerbaijan Technical University, Azerbaijan
Grigori Sidorov	CIC, Instituto Politécnico Nacional, Mexico
Juan Humberto Sossa Azuela	CIC, Instituto Politécnico Nacional, Mexico
Efstathios Stamatatos	University of the Aegean, Greece
Eric S. Tellez	CONACyT, Mexico; Infotec, Mexico
Esteban Tlelo-Cuautle	Instituto Nacional de Astrofísica, Óptica y Electrónica, Mexico
Nestor Velasco-Bermeo	Tecnológico de Monterrey Campus Estado de México, Mexico
Francisco Viveros Jiménez	Efinfo, Mexico
Carlos Mario Zapata Jaramillo	Universidad Nacional de Colombia, Colombia

Saúl Zapotecas Martínez   Universidad Autónoma Metropolitana – Cuajimalpa,
                                Mexico
Ramón Zatarain   Instituto Tecnológico de Culiacán, Mexico

Additional Reviewers

David Tinoco   Kazuhiro Takeuchi
Atsushi Ito   Rafael Rivera López
Ryo Otoguro   Adan Enrique Aguilar-Justo

Organizing Committee

Local Chairs

Luis Eduardo Falcón   Tecnológico de Monterrey, Campus Guadalajara
  Morales
Javier Mauricio Antelis   Tecnológico de Monterrey, Campus Guadalajara
  Ortíz

Logistics Chairs

Olga Cecilia García Rosique   Tecnológico de Monterrey, Campus Guadalajara
Edgar Gerardo Salinas   Tecnológico de Monterrey, Campus Guadalajara
  Gurrión
Omar Alejandro Robledo   Tecnológico de Monterrey, Campus Guadalajara
  Galván

Finance Chair

Mónica González Frías   Tecnológico de Monterrey, Campus Guadalajara

Contents – Part I

Machine Learning

Fuzzy Logic and Uncertainty Management

Contents – Part II

Robotics and Computer Vision

Evolutionary and Nature-Inspired Intelligence

A Genetic Algorithm to Solve Power System Expansion Planning with Renewable Energy

Lourdes Martínez-Villaseñor[1], Hiram Ponce[1(✉)], José Antonio Marmolejo[1],
Juan Manuel Ramírez[2], and Agustina Hernández[2]

[1] Facultad de Ingeniería, Universidad Panamericana,
Augusto Rodin 498, 03920 Mexico City, Mexico
{lmartine,hponce,jmarmolejo}@up.edu.mx
[2] Centro de Investigación y de Estudios Avanzados Instituto Politécnico Nacional,
Zapopan, Mexico
{jramirez,ahernandez}@gdl.cinvestav.mx

Abstract. In this paper, a deterministic dynamic mixed-integer programming model for solving the generation and transmission expansion-planning problem is addressed. The proposed model integrates conventional generation with renewable energy sources and it is based on a centralized planned transmission expansion. Due a growing demand over time, it is necessary to generate expansion plans that can meet the future requirements of energy systems. Nowadays, in most systems a public entity develops both the short and long of electricity-grid expansion planning and mainly deterministic methods are employed. In this study, an heuristic optimization approach based on genetic algorithms is presented. Numerical results show the performance of the proposed algorithm.

Keywords: Genetic algorithms
Generation and transmission problem · Power system planning

1 Introduction

Electric power systems, as well as any consumer system, present dynamic behaviors that require efficient planning for their operation over time. The dynamics of their behavior are due to a growing demand over time that requires reliable supply and as economically as possible. The elements that constitute an energy system are summarized in elements of generation, operation, transport and distribution mainly. Regarding the operation of power systems, two important factors can be mentioned: the capacity of their transmission lines and the capacity of the generation mix. These factors should be planned in time horizons that consider the growth of energy demand. An optimal reliability level should be achieved that will guarantee a continuous power flow with a reasonable cost. In the literature, this problem has been addressed through the generation and

© Springer Nature Switzerland AG 2018
I. Batyrshin et al. (Eds.): MICAI 2018, LNAI 11288, pp. 3–17, 2018.
https://doi.org/10.1007/978-3-030-04491-6_1

transmission (G&TEP) optimization problem. The mathematical modeling of this problem allows the decision makers to determine the optimal transimisión and generation-capacity expansion plan. Real-world G&TEP problems are nonlinear and complex, given that it entails handling high number of constraints, high dimensionality, and uncertainty related to the demand, fuel price, market price, interest rate, among other factors. Typically, the problem of expansion of generation and transmission have been addressed separately, therefore their results are considered suboptimal for the joint problem. For this reason, in this work both expansion problems are addressed in the same mathematical model. The model presented is based on [8], but considering the inclusion of renewable generation sources.

Regarding the solution methods, metaheuristics find high quality solutions at reasonable computational costs for complex problems like the combined G&TEP, hence optimality cannot be guaranteed. Nevertheless, heuristic and metaheuristic approaches generally perform better than the classical ones for very large problems [23]. Reference [2] mentioned two difficulties when implementing evolutionary optimization methods for G&TEP problem: handling the highly constraints in large and medium G&TEP problems, and large computational time consuming algorithms that do not permit online applications. There are several works reported in literature for generation expansion planning problem (GEP) [33], for transmission expansion planning problem (TEP) [15], and for the combined G&TEP [16]. Pure metaheuristic techniques are manly used to solve GEP and TEP problems. Since G&TEP real-world problems are nonlinear, handle great number of constraints, imply high dimensionality, and deal with many uncertainties, they are resolved with high cost computational methods [22,26].

The aim of this paper is to determine if a pure metaheuristic method, i.e. genetic algorithms can deal with this complex problem. It is a trend to use hybrid metaheuristic approaches for G&TEP problem at high computational cost. Reference [4] stated that a pure metaheuristic works well, there is no need to use more complex processes. In this work we explore the possibility of avoiding the computational costs of hybrid metaheuristics.

The paper is organized as follows. In Sect. 2 we present an overview of the state-of-art in metaheuristics for the G & TEP problem. The mathematical model is described in Sect. 3. Our proposed genetic algorithm approach is detailed in Sect. 4. Experiments and results are shown in Sect. 5. Finally, we conclude and present future work in Sect. 6.

2   Metaheuristics for G&TEP Problem

It is frequent that in real-world problems with high dimensionality, deterministic methods do not reach optimal or good solutions in reasonable computing time. It is common knowledge nowadays that metaheuristic approaches are robust and efficient for complex problems, delivering near to optimal solutions. Hence, there is no guarantee that the solutions found in reasonable amount of time are optimal. According to [39], the term metaheuristic has been used with two different meanings. The first meaning conceives metaheuristic as a framework, a set of concept and strategies that guide the development of optimization algorithms. The second meaning refers to a specific implementation or algorithm based on certain strategies. Metaheuristic concepts and strategies imitate nature, social culture, biology or laws of physics to guide the search in an optimization problem [34]. The implementations of metaheuristic algorithms have the ability to conduct global searches avoiding local optimal solutions [36].

GEP as well as TEP, and the combined G&TEP have been solved with deterministic and stochastic approaches for decades. These problems are nonlinear and very complex given that they usually handle a high number of constraints and are of high dimensionality. A summary of the most relevant surveys and works that solve the problems mentioned above with metaheuristic approaches is presented next.

The GEP problem has been addressed with many so called pure metaheuristics [33]. Genetic algorithms (GA) [28] and particle swarm optimization (PSO) [27] are the most commonly used techniques. Other metaheuristic methods are also reported in literature namely: Tabu search (TS) [41], simulated annealing (SA), modified honey bee mating optimization (MHBMO) algorithm [1], artificial immune systems (AIS) [7], modified shuffled frog leaping algorithm (SFL) [17], NSGA [21,25], differential evolution algorithm (DE) [29], and others.

TEP problem has been also solved with many metaheuristic optimization methods [15], such as genetic algorithms [6,18,38], simulated annealing [9,31], tabu search [12,13], ant colony optimization [37], artificial immune system [30], harmony search [40], particle swarm optimization [20,35], and hybrid metaheuristic methods [10,32]. Reference [23] proposed to solve TEP problem using an imperialist competitive algorithm (ICA) comparing the results with other evolutionary methods.

The most relevant works to this paper are those that present metaheuristic-based solutions for the combined G&TEP problem. Reference [16] presented a review of generation and transmission planning solution and stated that metaheuristics are well suited to address this complex problem. Among these works in this regard is [19] that presented a multi-objective framework to evaluate the integration of distant wind farm. Reference [3] proposed a multi-period integrated framework based in genetic algorithms for GEP and TEP, and natural gas grid expansion planning (NGGEP) for large-scale systems. It was applied to the Iranian power system proving that the proposed framework can be applicable for large-scale real-world problems.

In [24], authors proposed a framework for transmission planning considering also generation expansion. In order to solve this interrelated multilevel optimization problem, the authors present an iterative solution linking agent-based and search-based algorithms.

Murgan and Kannan published works using the NSGA-II to solve generation expansion planning problem [21,25]. In [26], the authors applied the elitist NSGA-II to the combined transmission and generation expansion-planning multi-objective problem.

3   Generation and Transmission Expansion Planning Problem

To solve the G&TEP problem, in the following a linearized deterministic integer-mixed multi-period problem is proposed, assuming the insertion of renewable energies. The problem minimizes an objective function of costs (1)–(2), subject to a set of constraints (3)–(24), which include: (i) investment constraints; and (ii) operating constraints. The latter refer to both conventional and renewable generation costs, non-supplied demand, and costs for failing with a percentage of clean energies; such costs are multiplied by a weighting factor associated with the number of hours of operation. The investment and operation costs are added along the entire planning horizon. The final investment integrates the sum of all period times.

The set of constraints corresponding to investment (3)–(7) takes into account the physical limitations of the generating units and the budget limit. Within the group of operating constraints (3)–(24); power balance constraints, thermal capacity limits in existing lines and those proposed by the model, constraints of DC power flows for existing lines and candidates, are represented in expressions (12) and (13), respectively. The maximum generation capacity for existing and proposed units (conventional and clean) is guaranteed by (14)–(19). The limits on not-supplied demand and the phase angle are described in (20)–(21). The clean energy deficit for each operating point is represented by (22); such a constraint implies that a fixed percentage of the loads must be supplied from renewable energies.

Thus, the objective function (1)–(2) and constraints (3)–(24) are the following:

$$
\min SD = \sum_t \left[\sum_o \rho_o \left(\sum_g C_g^E P_{got}^E + \sum_c C_c^C P_{cot}^C + \sum_d C_d^{LS} P_{dot}^{LS} \right. \right.
$$

$$
\left. + \sum_{gw} C_{gw}^{EW} P_{gwot}^{EW} + \sum_{cw} C_{cw}^{CW} P_{cwot}^{CW} + \gamma_o TT_{ot} \right) \tag{1}
$$

$$
\left. + \alpha_t \left(\sum_c \tilde{I}_c P_{ct}^{C^{max}} + \sum_{cw} \tilde{I}_{cw} P_{ct}^{CW^{max}} + \sum_{l \in \Omega^{L+}} \tilde{I}_l x_{lt}^L \right) \right]
$$

Where:

$$
SD = \left\{ P_{got}^E, P_{cot}^C, P_{gwot}^{EW}, P_{cwot}^{CW}, P_{dot}^{LS}, P_{lot}^L, \theta_{not}, P_{ct}^{CW^{max}}, P_{ct}^{C^{max}}, x_{lt}^L, TT_{ot} \right\} \tag{2}
$$

Subject to constraints:

$$
0 \leq \sum_t P_{cwt}^{CW^{max}} \leq \overline{P}_{cw}^{CW^{max}} , \quad \forall cw \tag{3}
$$

$$0 \leq \sum_t P_{ct}^{C^{max}} \leq \overline{P}_c^{C^{max}}, \quad \forall c \tag{4}$$

$$x_{lt}^L \in \{0,1\}, \quad \forall l \in \Omega^{L^+}, t \tag{5}$$

$$\sum_t x_{lt}^L \leq 1, \quad \forall l \in \Omega^{L^+} \tag{6}$$

$$\sum_{cw} \tilde{I}_{cw} P_{ct}^{CW^{max}} \leq I_t^{CW,max}, \quad \forall t \tag{7}$$

$$\sum_c \tilde{I}_c P_{ct}^{C^{max}} \leq I_t^{C,max}, \quad \forall t \tag{8}$$

$$\sum_{l \in \Omega^{L^+}} \tilde{I}_l x_{lt}^L \leq I_t^{L,max}, \quad \forall t \tag{9}$$

$$\sum_{g \in \Omega_n^E} P_{got}^E + \sum_{c \in \Omega_n^C} P_{cot}^C + \sum_{gw \in \Omega_n^{EW}} P_{gwot}^{EW} + \sum_{cw \in \Omega_n^{CW}} P_{cwot}^{CW} - \sum_{l|s(l)=n} P_{lot}^L$$
$$+ \sum_{l|r(l)=n} P_{lot}^L = \sum_{d \in \Omega_n^D} (P_{dot}^{D^{max}} - P_{dot}^{LS}), \forall n, o, t \tag{10}$$

$$P_{lot}^L = B_l(\theta_{s(l)ot} - \theta_{r(l)ot}), \forall l \backslash l \in \Omega^{L^+}, o, t \tag{11}$$

$$-\sum_{\tau \leq t} x_{l\tau}^L F_l^{max} \leq P_{lot}^L \leq \sum_{\tau \leq t} x_{l\tau}^L F_l^{max}, \forall l \in \Omega^{L^+}, o, t \tag{12}$$

$$-(1-\sum_{\tau \leq t} x_{l\tau}^L)M \leq P_{lot}^L - B_l(\theta_{s(l)ot} - \theta_{r(l)ot}) \leq (1-\sum_{\tau \leq t} x_{l\tau}^L)M, \forall l \in \Omega^{L^+}, o, t \tag{13}$$

$$-F_l^{max} \leq P_{lot}^L \leq F_l^{max}, \forall l, o, t \tag{14}$$

$$0 \leq P_{got}^E \leq P_g^{E^{max}}, \forall g, o, t \tag{15}$$

$$0 \leq P_{gwot}^{EW} \leq P_{gw}^{EW^{max}}, \forall gw, o, t \tag{16}$$

$$0 \leq P_{cwot}^{CW} \leq \sum_{\tau \leq t} P_{cw\tau}^{CW^{max}}, \forall cw, o, t \tag{17}$$

$$0 \leq P_{cot}^C \leq \sum_{\tau \leq t} P_{c\tau}^{C^{max}}, \forall c, o, t \tag{18}$$

$$0 \leq P_{dot}^{LS} \leq P_{dot}^{D^{max}}, \forall d, o, t \tag{19}$$

$$-\pi \leq \theta_{not} \leq \pi, \forall n, o, t \tag{20}$$

$$\theta_{not} = 0, \quad n : ref \quad \forall o, t \tag{21}$$

$$\sum_{gw \in \Omega_n^{EW}} P_{gwot}^{EW} + \sum_{cw \in \Omega_n^C} P_{cwot}^{CW} + TT_{ot} \geq r(t) \tag{22}$$

$$\left\{ \sum_{g \in \Omega_n^E} P_{got}^E + \sum_{gw \in \Omega_n^{EW}} P_{gwot}^{EW} + \sum_{c \in \Omega_n^C} P_{cot}^C + \sum_{cw \in \Omega_n^{CW}} P_{cwot}^{CW} \right\} \quad \forall g, o, t \tag{23}$$

$$TT_{ot} \geq 0, \quad \forall o, t \tag{24}$$

4   Genetic Algorithm for G&TEP

Hybrid metaheuristic approaches are used to solve complex real-world optimization problems. In this work, we propose a Genetic Algorithm approach to avoid high computational costs if possible.

Firstly, we present the description of the chromosome or the array of variables to be optimized. The chromosome that defines and individual solution for the G&TEP proposed problem consists in an array of 88 real variables that represent the variables to be optimized of the case study described in Sect. 5. The description of a chromosome is presented in Table 1. Each gene (element of the array) represents a decision variable of the G&TEP model. The possible values for gene are determined according to the description of the possible values for the decision variables.

Table 1. Description of the chromosome.

Variable	Number of genes	Positions in chromosome	Lower value	Upper value
P_{lot}^L	23	0–23	-40	40(for 0–3 and 20–23 genes)
			-100	100(for 4–7 genes)
			-140	140(for 8–11 genes)
			-105	105(for 12–15 genes)
			-200	200(for 16–19 genes)
θ_{not}	16	24–39	-3.141592	3.141592
P_{got}^E	8	40–47	0	400(for 480–527)
				150(for 528–575)
P_{gwot}^{EW}	4	48–51	0	100
P_{cwot}^{CW}	4	52–55	0	50
P_{cot}^C	4	56–59	0	190
$P_{cwt}^{CW\,max}$	2	60–61	0	50
$P_{ct}^{C\,max}$	2	62–63	0	190
P_{dot}^{LS}	16	64–79	0	According to demand for each operating, condition and period along the planning horizon
TT_{ot}	4	80–83	0	100
X_{lt}^L	4	84–87	0	1

The evaluation function ff(individual) is based on the objective function (SD) described in (1) and is presented in Algorithm 1. In order to deal with the constraints of the G&TEP model, we present a fitness function using a penalization P. The constrained problem is then transformed to an unconstrained problem using a penalization in the fitness function to decrease the aptitude of those individuals that violate one or more constraints.

A strategy to handle infeasibility of the chromosomes is proposed to guide the evolution of the generations improving the performance of the genetic algorithm. First, we count the number of constraints that are not fulfilled by each individual (n). The fitness function is designed based in the consideration of this number n multiplied by a penalty factor. In order to guide the performance of the genetic algorithm we determined three heuristic variables ($penalty, threshold, bias$) that influence the evaluation of ff(individual), as shown in Algorithm 1.

Algorithm 1. Definition of the fitness function.

1: **procedure** $ff(individual)$
2:      $z \leftarrow$ evaluation of the individual in the optimization problem
3:      $n = \sum_i c_i$ is the number of constraints (c_i) not fulfilled by the individual
4:      **if** $n > threshold$ **then**
5:          $f = penalty * n + bias$
6:      **else**
7:          $f = penalty * n + z$
8:      **end if**
9: **return** f
10: **end procedure**

The variables *penalty*, *threshold* and *bias* are better explained in Sect. 5. The parents of each generation are selected by the tournament selection method, which randomly picks a small subset of mating pool and the lowest cost chromosome becomes a parent. This method avoids sorting as in elitisms providing a good choice of selection for large population sizes [14].

Single-point crossover operator that picks two selected individuals with probability p_c and randomly determines a crossover point to form two new individuals. Uniform random mutation operator selects each gene of new individuals and changes its value with probability p_m.

5   Experiments and Results

Three experiments were designed in this work to evaluate the performance of the optimization of the G&TEP problem using the genetic algorithm described above. These experiments aim to refine the heuristic variables and identify if the proposed strategy to handle the feasibility of the chromosomes is fitted for the proposed problem. For all experiments the penalty was considered as the heuristic number 10,000,000, crossover probability p_c was set to 0.8 and mutation probability p_m was equal to 0.1. The bias in Experiments 2 and 3 was set as 1000,000,000.

The genetic algorithm was implemented using the Distributed Evolutionary Algorithms in Python (DEAP) [11], while the problem was model in GAMS [5]. Then, we interface both applications using the proper interface API from GAMS [5]. With the purpose of validating our proposal, we evaluated this same model in GAMS (23.6.5) program and obtained the optimal solution with CPLEX method, which is also presented in Figs. 2, 3 and 4 for comparison.

The power test system analyzed is showed in Fig. 1. We consider two operating condition o, four existing lines l, two candidate lines $l \in \Omega_L^+$, four demand buses n, two existing conventional generators g, one existing clean energy generators gw, one candidate conventional generators c, one candidate clean energy generators cw and two time period t.

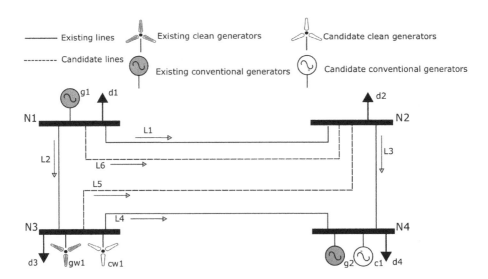

Fig. 1. Four-buses test power system

The parameters in the case study (see Table 2) are:

$B_l = 500\,[\$/\text{MWh}]$.
$C_c^C = 35\,[\$/\text{MWh}]$.
$C_{cw}^{CW} = 25\,[\$/\text{MWh}]$.
$C_d^{LS} = 80\,[\$/\text{MWh}]$.
$C_g^E = 35\,[\$/\text{MWh}]$.
$C_g^{EW} = 25\,[\$/\text{MWh}]$.
$F_l^{max} = [40, 100, 140, 105, 200, 40]\,[\text{MW}]$ for each line l respectively.
$I^{CW,max} = 200000000\,[\$]$.
$I^{C,max} = 200000000\,[\$]$.
$I^{L,max} = 2000000\,[\$]$.

$\tilde{I}_{cw} = 700000\,[\$/\text{MWh}]$.

$\tilde{I}_{c} = 700000\,[\$/\text{MWh}]$.

$\tilde{I}_{l} = 1000000$ for each candidate line l [$\$$].

$\overline{P}_{cw}^{CW^{max}} = 50\,[\text{MW}]$.

$\overline{P}_{c}^{C^{max}} = 190\,[\text{MW}]$.

$P_{dot}^{D^{max}}$ See Table 2.

$P_{g}^{E^{max}} = [400, 150]$ for each unit g respectively [MW].

$P_{gw}^{E^{max}} = 100\,[\text{MW}]$.

$\rho_{o} = [6000, 2760]$ Weight of the operating condition o [h].

$\gamma_{o} = [100, 100]$ for each operating condition o [$\$/\text{MWh}$].

$\alpha_{t} = [20\%, 10\%]$ for each time period t.

Table 2. Maximum and minimum demand by bus [MWh/h].

Demand	t_1		t_2	
	Low	High	Low	High
d_1	40	52	50	65
d_2	136	176.8	170	221
d_3	160	208	200	260
d_4	64	83.2	80	104

In the initial experiment (Experiment 1), we do not use the condition described in previous section. In this experiment the finalization condition was 1000 epochs. Each individual is evaluated only using:

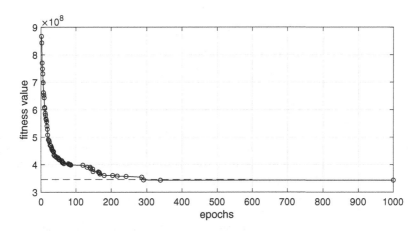

Fig. 2. Performance of Experiment 1 (no condition strategy) showing the evaluation of the best individual at each epoch.

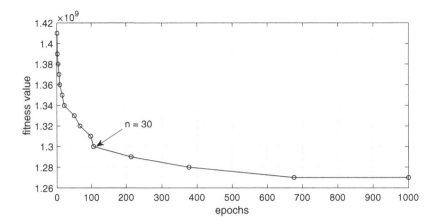

Fig. 3. Performance of Experiment 2 (condition strategy with threshold 5) showing the evaluation of the best individual at each epoch.

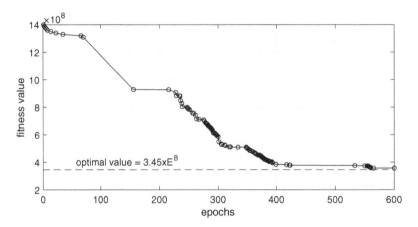

Fig. 4. Performance of Experiment 3 (condition strategy with threshold 30) showing the evaluation of the best individual at each epoch.

$$ff(individual) = penalty * n + z \qquad (25)$$

The results of this initial experiment are shown in Fig. 2. The genetic algorithm converges toward the optimal solution starting from about 300 generation. Nevertheless, given that the fitness functions is evaluated based on the number of constraints not fulfilled but is really unconstrained, the GA kept searching for a minimum value increasing the infeasibility of the solution.

In the second experiment (Experiment 2) the proposed strategy was used evaluating of individuals according to the condition described in previous section. In this experiment the finalization condition was also 1000 epochs. A threshold of 5 constraints not fulfilled was used with the aim of minimizing the evaluation of individual that violate many constraints and avoid increasing the infeasibility. The results are presented in Fig. 3. We can observe that it never reached the breaking point to start the evaluation of the fitness function considering z.

Based on the results of Experiment 2, we determined a threshold of 30 constraints not fulfilled for the third experiment, and 600 epochs were performed. Figure 4 shows the results of the third experiment. in which the performance of GA is improved reaching a solution near to the optimal obtained with GAMS.

6   Conclusions

In this paper, we presented a genetic algorithm approach to solve the complex G&TEP problem to avoid high computational costs. A strategy to handle infeasibility of the chromosomes is proposed to guide the evolution of the generations improving the performance of the genetic algorithm. The proposed method was proven in a determinsitic dynamic G&TEP model. The results are promising given that the solutions found by the GA were near the optimum; nevertheless infeasibility is still an issue to be further investivated.

For future work, this method should be proven in real-world scenarios, and hybrid metaheuristics are also considered.

7   Notation

Indices:

n	Buses.
g	Existing conventional generators.
gw	Existing clean energy generators.
c	Candidate conventional generators.
cw	Candidate clean energy generators.
d	Demand.
l	Transmission lines.
O	Operating conditions.
t	Time periods.

Sets:

Ω_L^+	Candidate transmission lines.
Ω_n^E	Conventional existing generators located at bus n.
Ω_n^{EW}	Existing wind generators located at bus n.
Ω_n^C	Candidate conventional generators located at bus n.
Ω_n^{CW}	Candidate clean energy generators located at bus n.
Ω_n^D	Demand at bus n.
$s(l)$	Sending bus for transmission line l.
$r(l)$	Ending bus for transmission line l.

Parameters:

B_l	Susceptance of transmission line l.
C_c^C	Production cost of the conventional candidate generating unit [\$/MWh].
C_{cw}^{CW}	Production cost of the clean candidate generating unit [\$/MWh].

C_d^{LS} Cost of the not supplied demand not supplied [\$/MWh].

C_g^E Production cost for the existing conventional generating unit [\$/MWh].

C_g^{EW} Production cost of the existing clean generating unit [\$/MWh].

F_l^{max} Rating of the transmission line l [MW].

$I^{CW,max}$ Investment budget to build the clean generation candidate unit [\$].

$I^{C,max}$ Investment budget to build the conventional generating candidate unit [\$].

$I^{L,max}$ Investment budget to build the candidate transmission line l [\$].

$\tilde{I}_{cw}$ Investment cost for the candidate clean generation unit cw [\$/MWh].

$\tilde{I}_c$ Investment cost for the candidate conventional generating unit c [\$/MWh].

$\tilde{I}_l$ Investment cost for the candidate transmission line l [\$].

$\overline{P}_{cw}^{CW^{max}}$ Maximum generation capacity of the candidate clean generation unit cw [MW].

$\overline{P}_c^{C^{max}}$ Maximum generation capacity of the candidate conventional generation unit c [MW].

$P_{dot}^{D^{max}}$ Maximum demand d [MW].

$P_g^{E^{max}}$ Maximum production capacity of the existing conventional generation unit g [MW].

$P_{gw}^{E^{max}}$ Maximum production capacity of the existing clean energy generation unit gw [MW].

ρ_o Weight of the operating condition o [h].

γ_o Penalization cost for breaching the renewable portfolio requirement [\$/MWh].

α_t Amortization rate [%].

Binary Variables:

X_{lt}^L Binary variable: equal to 1 if the candidate transmission line is constructed, and 0 otherwise.

Continuous Variables:

P_{got}^E Power supplied by the existing conventional generating unit [MW].

P_{gwot}^{EW} Power supplied by the existing wind generating unit [MW].

P_{cwot}^{CW} Power supplied by the candidate wind generator unit [MW].

P_{cot}^C Power supplied by the conventional candidate generating unit [MW].

P_{dot}^{LS} Not supplied demand [MW].

P_{lot}^L Power flow through the transmission line l [MW].

θ_{not} Angle of voltage at bus n [rad].

$P_{cwt}^{CW^{max}}$ Rating of the candidate clean generation unit [MW].

$P_{ct}^{C^{max}}$ Rating of the candidate conventional generating unit [MW].

TT_{ot} Deficiency of the renewable goal [MW].

References

1. Abbasi, A.R., Seifi, A.R.: Energy expansion planning by considering electrical and thermal expansion simultaneously. Energy Convers. Manag. **83**, 9–18 (2014)
2. Alizadeh, B., Jadid, S.: Reliability constrained coordination of generation and transmission expansion planning in power systems using mixed integer programming. IET Gener., Transm. Distrib. **5**(9), 948–960 (2011)
3. Barati, F., Seifi, H., Sepasian, M.S., Nateghi, A., Shafie-khah, M., Catalão, J.P.: Multi-period integrated framework of generation, transmission, and natural gas grid expansion planning for large-scale systems. IEEE Trans. Power Syst. **30**(5), 2527–2537 (2015)
4. Blum, C., Puchinger, J., Raidl, G.R., Roli, A.: Hybrid metaheuristics in combinatorial optimization: a survey. Appl. Soft Comput. **11**(6), 4135–4151 (2011)

5. Bussieck, M.R., Meeraus, A.: General algebraic modeling system (GAMS). In: Kall-rath, J. (ed.) Modeling Languages in Mathematical Optimization. APOP, vol. 88, pp. 137–157. Springer, Boston (2004). https://doi.org/10.1007/978-1-4613-0215-5_8

6. Cadini, F., Zio, E., Petrescu, C.A.: Optimal expansion of an existing electrical power transmission network by multi-objective genetic algorithms. Reliab. Eng. Syst. Saf. **95**(3), 173–181 (2010)

7. Chen, S.L., Zhan, T.S., Tsay, M.T.: Generation expansion planning of the utility with refined immune algorithm. Electr. Power Syst. Res. **76**(4), 251–258 (2006)

8. Conejo, A., Baringo, L., Kazempour, S., Siddiqui, A.: Investment in Electric-ity Generation and Transmission: Decision Making under Uncertainty, 1st edn. Springer, Heidelberg (2016). https://doi.org/10.1007/978-3-319-29501-5

9. Cortes-Carmona, M., Palma-Behnke, R., Moya, O.: Transmission network expan-sion planning by a hybrid simulated annealing algorithm. In: 2009 15th Interna-tional Conference on Intelligent System Applications to Power Systems, ISAP 2009, pp. 1–7. IEEE (2009)

10. Faria, H., Binato, S., Resende, M.G., Falcão, D.M.: Power transmission network design by greedy randomized adaptive path relinking. IEEE Trans. Power Syst. **20**(1), 43–49 (2005)

11. Fortin, F.A., De Rainville, F.M., Gardner, M.A., Parizeau, M., Gagné, C.: DEAP: evolutionary algorithms made easy. J. Mach. Learn. Res. **13**, 2171–2175 (2012)

12. Gallego, R., Monticelli, A., Romero, R.: Transmision system expansion planning by an extended genetic algorithm. IEE Proc.-Gener., Transm. Distrib. **145**(3), 329–335 (1998)

13. Gallego, R.A., Romero, R., Monticelli, A.J.: Tabu search algorithm for network synthesis. IEEE Trans. Power Syst. **15**(2), 490–495 (2000)

14. Haupt, R.L., Haupt, S.E., Haupt, S.E.: Practical Genetic Algorithms, vol. 2. Wiley, New York (1998)

15. Hemmati, R., Hooshmand, R.A., Khodabakhshian, A.: State-of-the-art of trans-mission expansion planning: comprehensive review. Renew. Sustain. Energy Rev. **23**, 312–319 (2013)

16. Hemmati, R., Hooshmandd, R.A., Khodabakhshian, A.: Comprehensive review of generation and transmission expansion planning. IET Gener., Transm. Distrib. **7**(9), 955–964 (2013)

17. Jadidoleslam, M., Ebrahimi, A.: Reliability constrained generation expansion plan-ning by a modified shuffled frog leaping algorithm. Int. J. Electr. Power Energy Syst. **64**, 743–751 (2015)

18. Jalilzadeh, S., Shabani, A., Azadru, A.: Multi-period generation expansion plan-ning using genetic algorithm. In: 2010 International Congress on Ultra Modern Telecommunications and Control Systems and Workshops (ICUMT), pp. 358–363. IEEE (2010)

19. Javadi, M.S., Saniei, M., Mashhadi, H.R., Gutiérrez-Alcaraz, G.: Multi-objective expansion planning approach: distant wind farms and limited energy resources integration. IET Renew. Power Gener. **7**(6), 652–668 (2013)

20. Jin, Y.X., Cheng, H.Z., Yan, J., Zhang, L.: New discrete method for particle swarm optimization and its application in transmission network expansion plan-ning. Electr. Power Syst. Res. **77**(3), 227–233 (2007)

21. Kannan, S., Baskar, S., McCalley, J.D., Murugan, P.: Application of NSGA-II algorithm to generation expansion planning. IEEE Trans. Power Syst. **24**(1), 454–461 (2009)

22. Khakpoor, M., Jafari-Nokandi, M., Akbar Abdoos, A.: A new hybrid GA-fuzzy optimization algorithm for security-constrained based generation and transmission expansion planning in the deregulated environment. J. Intell. Fuzzy Syst. **33**(6), 3789–3803 (2017)
23. Moradi, M., Abdi, H., Lumbreras, S., Ramos, A., Karimi, S.: Transmission expansion planning in the presence of wind farms with a mixed AC and DC power flow model using an imperialist competitive algorithm. Electr. Power Syst. Res. **140**, 493–506 (2016)
24. Motamedi, A., Zareipour, H., Buygi, M.O., Rosehart, W.D.: A transmission planning framework considering future generation expansions in electricity markets. IEEE Trans. Power Syst. **25**(4), 1987–1995 (2010)
25. Murugan, P., Kannan, S., Baskar, S.: Application of NSGA-II algorithm to single-objective transmission constrained generation expansion planning. IEEE Trans. Power Syst. **24**(4), 1790–1797 (2009)
26. Murugan, P., Kannan, S., Baskar, S.: NSGA-II algorithm for multi-objective generation expansion planning problem. Electr. Power Syst. Res. **79**(4), 622–628 (2009)
27. Neshat, N., Amin-Naseri, M.: Cleaner power generation through market-driven generation expansion planning: an agent-based hybrid framework of game theory and particle swarm optimization. J. Clean. Prod. **105**, 206–217 (2015)
28. Pereira, A.J., Saraiva, J.T.: Generation expansion planning (GEP)-a long-term approach using system dynamics and genetic algorithms (GAs). Energy **36**(8), 5180–5199 (2011)
29. Rajesh, K., Bhuvanesh, A., Kannan, S., Thangaraj, C.: Least cost generation expansion planning with solar power plant using differential evolution algorithm. Renew. Energy **85**, 677–686 (2016)
30. Rezende, L.S., Leite da Silva, A.M., de Mello Honório, L.: Artificial immune system applied to the multi-stage transmission expansion planning. In: Andrews, P.S., et al. (eds.) ICARIS 2009. LNCS, vol. 5666, pp. 178–191. Springer, Heidelberg (2009). https://doi.org/10.1007/978-3-642-03246-2_19
31. Romero, R., Gallego, R., Monticelli, A.: Transmission system expansion planning by simulated annealing. In: 1995 IEEE Proceedings of the Conference on Power Industry Computer Application, pp. 278–283. IEEE (1995)
32. Sadegheih, A., Drake, P.: System network planning expansion using mathematical programming, genetic algorithms and tabu search. Energy Convers. Manag. **49**(6), 1557–1566 (2008)
33. Sadeghi, H., Rashidinejad, M., Abdollahi, A.: A comprehensive sequential review study through the generation expansion planning. Renew. Sustain. Energy Rev. **67**, 1369–1394 (2017)
34. Saka, M.P., Hasançebi, O., Geem, Z.W.: Metaheuristics in structural optimization and discussions on harmony search algorithm. Swarm Evol. Comput. **28**, 88–97 (2016)
35. Shayeghi, H., Mahdavi, M., Bagheri, A.: Discrete PSO algorithm based optimization of transmission lines loading in TNEP problem. Energy Convers. Manag. **51**(1), 112–121 (2010)
36. da Silva, A.M.L., Freire, M.R., Honório, L.M.: Transmission expansion planning optimization by adaptive multi-operator evolutionary algorithms. Electr. Power Syst. Res. **133**, 173–181 (2016)
37. da Silva, A.M.L., Rezende, L.S., da Fonseca Manso, L.A., de Resende, L.C.: Reliability worth applied to transmission expansion planning based on ant colony system. Int. J. Electr. Power Energy Syst. **32**(10), 1077–1084 (2010)

38. da Silva, E.L., Gil, H.A., Areiza, J.M.: Transmission network expansion planning under an improved genetic algorithm. In: Proceedings of the 21st 1999 IEEE International Conference on Power Industry Computer Applications, PICA 1999, pp. 315–321. IEEE (1999)
39. Sorensen, K., Sevaux, M., Glover, F.: A history of metaheuristics. arXiv preprint arXiv:1704.00853 (2017)
40. Verma, A., Panigrahi, B., Bijwe, P.: Harmony search algorithm for transmission network expansion planning. IET Gener., Transm. Distrib. 4(6), 663–673 (2010)
41. Yoza, A., Yona, A., Senjyu, T., Funabashi, T.: Optimal capacity and expansion planning methodology of PV and battery in smart house. Renew. Energy 69, 25–33 (2014)

Memetic Algorithm for Constructing Covering Arrays of Variable Strength Based on Global-Best Harmony Search and Simulated Annealing

Jimena Timaná[1], Carlos Cobos[1(✉)], and Jose Torres-Jimenez[2]

[1] Information Technology Research Group (GTI), Universidad del Cauca,
Popayán, Colombia
{jtimana, ccobos}@unicauca.edu.co
[2] Center for Research and Advanced Studies of the National Polytechnic
Institute, Ciudad Victoria, Tamaulipas, Mexico
jtj@cinvestav.mx

Abstract. Covering Arrays (CA) are mathematical objects widely used in the design of experiments in several areas of knowledge and of most recent application in hardware and software testing. CA construction is a complex task that entails a high run time and high computational load. To date, research has been carried out for constructing optimal CAs using exact methods, algebraic methods, Greedy methods, and metaheuristic-based methods. These latter, including among them Simulated Annealing and Tabu Search, have reported the best results in the literature. Their effectiveness is largely due to the use of local optimization techniques with different neighborhood schemes. Given the excellent results of Global-best Harmony Search (GHS) algorithm in various optimization problems and given that it has not been explored in CA construction, this paper presents a memetic algorithm (GHSSA) using GHS for global search, SA for local search and two neighborhood schemes for the construction of uniform and mixed CAs of different strengths. GHSSA achieved competitive results on comparison with the state of the art and in experimentation did not require the use of supercomputers.

Keywords: Covering array · Metaheuristics · Global-best Harmony Search
Simulated annealing

1 Introduction

A Covering Array (CA) is a mathematical object used widely in the design of experiments and today is a key tool for testing software and hardware by helping to reduce their cost and time, achieving the greatest possible coverage of fault detection.

A covering array denoted by CA $(N; k, v, t)$ is a matrix of size $N \times k$, where N indicates the number of rows (test cases) of the CA and k the number of variables or columns; v indicates the number of possible symbols (alphabet) that each of the k variables takes and t indicates the strength of interaction between the variables.

© Springer Nature Switzerland AG 2018
I. Batyrshin et al. (Eds.): MICAI 2018, LNAI 11288, pp. 18–32, 2018.
https://doi.org/10.1007/978-3-030-04491-6_2

Construction of a CA is a complex task entailing a high run time and involving a high computational load. The aim of the methods and algorithms used in constructing CAs is to satisfy the coverage properties with the least possible number of rows, i.e. looking for optimal CAs.

Much research has been done on different methods for constructing CAs, including exact methods [1], algebraic methods [2], Greedy methods [3] and those based on metaheuristics [3, 4], the latter having reported the best results to date.

Among prominent metaheuristic algorithms in CA construction are Simulated Annealing (SA) [5], Tabu Search [6], genetic algorithms [7] and ant colony optimization [8]. Most owe their effectiveness to the additional use that they make of local optimization techniques with different structures for exploiting the neighborhood.

The metaheuristic algorithm Global-best Harmony Search (GHS) meanwhile has reported excellent results in resolving complex problems of a continuous, discrete and binary nature in various areas of knowledge [9], but never in the context of CA construction.

This work presents an algorithm for constructing uniform (same alphabet for each of k variables) and mixed CAs (different alphabet in at least one of the k variables of the CA) of different strengths, optimal or close to optimal, based on GHS for global search, SA for local search, and two schemes for exploiting the neighborhood.

The paper is structured as follows: Sect. 2 presents and details the proposed algorithm called GHSSA; Sect. 3 describes the three experiments carried out and analyzes the results; and Sect. 4 presents research conclusions and future work that the group hopes to carry out.

2   GHSSA Algorithm

GHSSA is a memetic algorithm that constructs Covering Arrays (CAs) with different strength levels ($2 \leq t \leq 6$), based on the Global-best Harmony Search metaheuristic, as a global optimization strategy, Simulated Annealing as a local optimization strategy and two schemes for the construction of the neighborhood. The GHSSA algorithm is summarized in Fig. 1 and explained below.

Step 1: The algorithm starts defining values for the following input parameters: *NI*, which represents the maximum number of iterations or improvisations; *ParMin*, which represents the minimum pitch adjustment rate; *ParMax*, which represents the maximum pitch adjustment rate and that must be greater than ParMin; *HMCR*, which represents the harmony memory consideration rate; *HMS* that represents harmony memory size (or population size), *MCAConfiguration* that indicates the configuration of the Mixed Covering Array (MCA) or Covering Array (homogeneous CA) to be constructed, and *MaxIterationSA* that corresponds to the maximum number of iterations that the local optimizer based on Simulated Annealing can perform in each execution.

Step 2: Initialize the Harmony Memory (*HM*). The harmonies that fill the *HM* are created one by one. Each harmony stores three elements. The first element, called CA, has an integer value N that indicates the number of rows of the CA, an integer value k that indicates the number of variables/columns of the CA, a vector of integers V of size k, which stores for each column the set of values or alphabet that each variable of

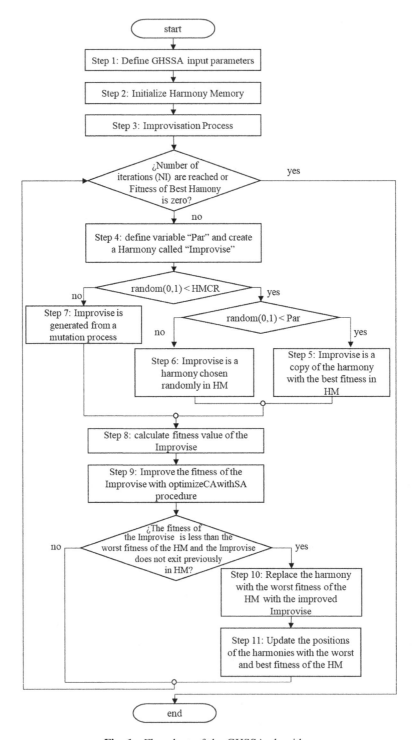

Fig. 1. Flowchart of the GHSSA algorithm

the MCA/CA takes, an integer value t that represents the strength or degree of interaction between the variables, and a matrix where the MCA/CA is constructed and stored. This matrix is of integers with N rows and k columns. The values of N, k, V, and t are assigned from the provided configuration of the MCA/CA being constructed. The second element of a harmony is an integer value that represents the *Fitness* of the harmony for the *CA matrix* that has been constructed. The third element of the harmony is an auxiliary matrix of integer values called *matrixP* that makes it possible to calculate and update (incrementally) the value of the fitness of the CA.

For each of the harmonies in the *HM* do the following:

- Step 2.1: Fill the *CA matrix* by means of a Greedy method. This method seeks that each of the rows added to the *CA matrix* are as different from each other as possible. This is done by selecting the row most different from those already included in the *CA matrix*, from a set of randomly generated candidate rows. The most different row is the one with the greatest Hamming distance with respect to the rows already stored, which means that it has less similarity with respect to these.
- Step 2.2: Calculate the *Fitness* value. It stores the number of missing alphabet combinations with the level of strength required between the variables so that the matrix stored within the harmony is the MCA/CA that is being sought to construct. A fitness of zero indicates that the desired MCA/CA has already been constructed in the harmony.
- Step 2.3: Optimize the *CA matrix* in the harmony with the Simulated Annealing algorithm that is explained below and summarized in Fig. 2. This algorithm is based on the proposal presented in [5].

The previous process (step 2.1 to 2.3) is repeated *HMS* times. Once the harmony memory is full, the position in the harmony memory where the harmony that registers the worst fitness is found, which in this case is the highest value (the harmony that is lacking the greatest number of missing alphabet combinations with the level of strength required between the variables to become the sought MCA/CA). This is called the *Worst Harmony*. Finally, the position in the harmony memory where the harmony that registers the best fitness is found, which in this case is the lowest value. This is called the *Best Harmony*.

Step 3: The iterative process of improvisation is carried out, framed in a specific number of improvisations defined by the *NI* parameter. The fitness of the best harmony in the harmony memory is evaluated to discover if it is zero. If so, the cycle of improvisations is terminated since the MCA/CA has already been constructed for the configuration sought. If the required MCA/CA has not yet been found, the step 4 is continued with, in which the variable *par* is defined, and a harmony type variable called *improvise* is created. The variable *par* represents the pitch adjustment rate, whose value is assigned through an expression defined within the original algorithm of Global-best Harmony Search [9] which uses *ParMin* and *ParMax* input parameters. The variable *par* changes dynamically with each iteration or improvisation.

Then, it is asked if the value of a random number generated uniformly between 0 and 1 is less than the value of the *HMCR* parameter. If so, it is asked if the value of a new random number uniformly generated between 0 and 1 is less than the value of the

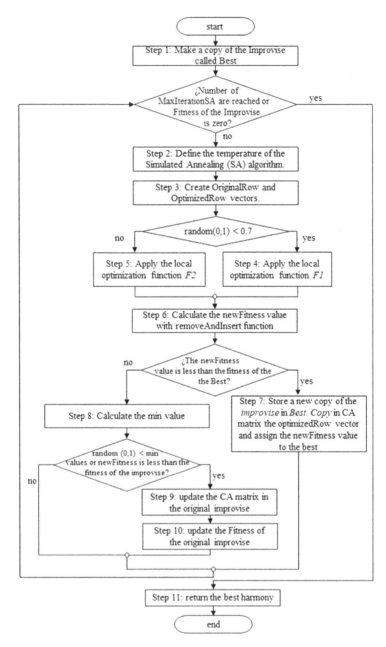

Fig. 2. optimizeCAwithSA procedure

variable *par*. If so, in step 5, a copy of the harmony with the best fitness recorded in the harmony memory is stored in the *improvise*. If not, in step 6, a copy of a harmony randomly chosen from the harmony memory is stored in the *improvise*.

If the random value generated was not less than the value of the *HMCR* parameter, in step 7, the *improvise* is generated using a mutation process, which modifies some values in the rows (10%) of the *CA matrix* of a harmony that has been randomly selected from the harmony memory.

In step 8, the *Fitness* value for the *improvise* is calculated. Then, in step 9, the optimizeCAwithSA procedure is invoked to optimize or improve the *Fitness* for the *improvise*.

Then, two questions are asked. The first, if the *Fitness* of the *improvise* (improved in step 9) is less than the *Fitness* of the harmony with the worst fitness of the harmony memory and the second, if the *improvise* that was generated is not previously in the harmony memory. If both conditions are fulfilled, then in step 10, the "improved" *improvise* enters to replace the harmony with the worst fitness of the harmony memory. In step 11, the positions of the harmonies with the worst and best fitness of the harmony memory are updated. The iterative process ends when the number of iterations defined in the *NI* parameter are reached or until the MCA/CA that reports fitness equal to zero is found.

The *optimizeCAwithSA* procedure is summarized in Fig. 2. In step 1, a copy of the original *improvise* that arrives as a parameter to the procedure is made and is stored in a variable called *Best*, which will be optimized.

The optimization process is carried out, framed by a maximum number of optimization iterations defined by the *MaxIterationSA* parameter. It is asked if the *Fitness* of the *CA* within the original *improvise* is zero, if so, the cycle ends, as does the optimization process. In step 2, the temperature of the SA algorithm is defined, which begins at 1.0 for the first iteration and decreases linearly to a value close to 0.0 in the last iteration.

In step 3, the *originalRow* and *optimizedRow* vectors, both of size k, are created, which will store a row of the CA. Then, a uniform random number between 0 and 1 is generated, and depending on the value that number takes, one of the two local optimization functions *F1* (with a probability of choice of P (*F1*)) or *F2* (with a probability of choice of $1 - P$ (*F1*)) is chosen. In the experimentation it was found that better results are obtained when P (*F1*) = 0.7. The purpose of local optimization functions is to generate a neighbor that is represented by the result of interchanging in the CA the vector *originalRow* by the vector *optimizedRow*.

If the random number is less than 0.7, in step 4, the *optimizedRow* vector is assigned to the row of the matrix within the CA that has been optimized through the local optimization function *F1*. This optimization function first randomly selects one row of the MCA/CA matrix that is being constructed and then iteratively takes each zero of the *matrixP* (that is, those combinations of columns and alphabets that have not been fulfilled to build the desired MCA/CA) and make a change of values in the selected row to eliminate the zero, then evaluate how much the fitness improves with the change and store that information.

After having made the evaluation of possible changes, it then selects the one that best improves the fitness and returns the number of the selected row called *rowCA* and the way it should be to obtain this change, that is, the *optimizedRow*. In the *originalRow* vector is stored that row of the matrix inside the CA that is found in the *rowCA* position.

Otherwise, if the random number is greater than or equal to 0.7, in step 5, *rowCA* stores the row position of a randomly selected row of the *CA matrix*. In the *originalRow* vector, that row of the matrix inside the CA that is in the *rowCA* position is stored. In the *optimizedRow* vector the row that has been optimized is saved through the *F2* optimization function. The *F2* function takes the original row previously selected in a random way and iteratively changes the values of each cell for the possible values of its alphabet, evaluates the improvement on the fitness and before changing the next cell, leaves the previous cell with its original value. Finally, the change of value is selected in a cell that made the greatest improvement in fitness and it is returned as an optimized row (*optimizedRow*).

In step 6, the *newFitness* value stores the fitness or number of zeros registered in the *matrixP* after applying the function named *removeAndInsert*. The function removes from the *matrixP* the alphabet's combinations by column's combinations given their place by the removal of the *originalRow* vector in the CA of the original *improvise*. It then adds to the *matrixP* the alphabet's combinations by column's combinations given their place by putting the vector *optimizedRow* in the CA of the *improvise*. Next, the number of zeros in *matrixP* is calculated. That new fitness is that returned by the *removeAndInsert* function. This function receives a Boolean parameter with a false value and that indicates that once the number of zeros in the *matrixP* has been recalculated after applying the function, the changes on the *matrixP* must be undone and leaving it as it was before applying the function (incremental calculation of the change in *Fitness* value).

Then, it is asked if the *newFitness* values is less than the fitness of *Best* (copy of Improvise). If it is less, in step 7 a new copy of the *improvise* is stored in *Best*. Then in the *CA matrix* of the best solution, the *optimizedRow* vector is copied in the *rowCA* position. Subsequently the fitness of the best solution is assigned the value of the *newFitness*.

In step 8, the *min* value is calculated. Which it a number that indicates the probability of acceptance that a harmony would have with a new fitness not as good in relation to the fitness of the original *improvise* (simulated annealing mechanism, with which solutions are accepted that are not better than the current one, as a strategy to avoid being trapped in local optima).

The acceptance threshold for the harmony with the new fitness is established, based on fulfilling either of the following conditions. The first condition evaluates whether a uniform random value between 0 and 1 is less than the *min* value. With this question it is desired to accept a harmony with a greater fitness (poor) with respect to the fitness of the original *improvise*. The second condition asks if the value of *newFitness* is less than the fitness of the original improvise. If either condition is met, in step 9, it is begun to update the *CA matrix* in the original *improvise*, with the optimized row obtained with the application of one of the two local optimization functions. In the step 10, the *Fitness* of the original *improvise* is updated with the value returned by the *removeAndInsert* function. On this occasion, the Boolean parameter takes the value of true, which indicates that once the number of zeros in the *matrixP* has been recalculated, after making the row change, the changes made to the *matrixP* must persist (incremental fitness value calculation). Finally, in step 11 the best harmony found in the optimization process (*Best*) is returned.

3   Experiments

To measure the performance of the proposed algorithm, three experimental comparisons were made.

- In the first, the proposed algorithm, GHSSA, was compared with the original versions of Global-best Harmony Search (GHS) and Simulated Annealing (SA) algorithms adapted for the construction of uniform and mixed CAs of variable strength. This experiment was carried out seeking to define the contribution of each main component of the proposed algorithm.
- In the second, GHSSA was compared with state-of-the-art algorithms that construct uniform, binary Covering Arrays of variable strength.
- In the third, GHSSA was compared with state-of-the-art algorithms that construct mixed Covering Arrays of variable strength.

3.1   First Experiment

The initial implementations of the proposed GHSSA algorithm, which did not use local optimization techniques, managed to construct a few simple, mixed CAs characterized by having few variables and strengths and small alphabets. However, they failed to construct complex, mixed CA configurations with a considerable number of variables and alphabets and where interaction strength was high. The performance of the GHSSA algorithm improved considerably by including both Simulated Annealing and a local optimization technique such as the two schemes for the construction of the neighborhood (functions $F1$ and $F2$), achieving the construction of configurations not possible with the initial versions of the algorithm.

This first experiment was carried out on a desktop computer with the following description: Dell Vostro 460 with 4 GB of RAM memory, Intel Core i5-2400 3.1 GHz processor and Windows 7 Home Premium operating system.

The GHS, SA and GHSSA algorithms were run on the same machine, to guarantee a fair comparison. After running each algorithm, a record was taken of the MCA/CA found, the number of zeros in matrix (fitness values) if the solution sought was not found, and the time in seconds that it used to construct the MCA/CA.

Each algorithm (GHS, SA and GHSSA) was run in parallel on four and eight threads, each with a different seed (initial number for the generation of random numbers) to have different initial solutions in the search space, trying to increase the chance of finding the solution for the MCA/CA sought. The process that obtained the shortest construction time for a specific MCA/CA configuration is the one that is reported for each of the algorithms.

Table 1 shows the results of this first experiment. Thirty CAs were constructed, featuring both uniform and mixed. The first column of this table presents the identifier of the CA to be constructed. The second column indicates the configuration of the MCA/CA to be constructed. The third column indicates the algorithm with which it was constructed. The fourth column indicates the status reached when executing the algorithm. This state can be (S) indicating that the construction of the CA was successful and (P) indicating that it was able to construct a CA, but with more rows (test cases).

The fifth column indicates the fitness achieved by each algorithm within a predefined number of iterations whose value was 25 million. The sixth column indicates the time in seconds that construction of the MCA/CA took for each algorithm. The seventh column called N' indicates the number of test cases with which the algorithm was able to find the desired MCA/CA, usually greater than the N defined for the MCA/CA sought.

Analyzing the results in Table 1 it is observed that the three algorithms managed to construct several configurations of the desired MCA/CA, achieving a fitness of zero, which indicates a successful construction, as for example for Id 14.

In 20 of the 30 tests performed, GHSSA managed to be faster than the other two algorithms. In the remaining 10 tests, SA was fastest. However, the difference between SA and GHSSA in most cases is minimal since it occurs in milliseconds.

On analyzing the CA with Id 17, it is observed that the GHSSA and SA algorithms managed to construct the proposed CA configuration. However, when running the GHS algorithm, it fell into a local optimum, thus remaining stagnant within the pre-defined number of iterations, which caused the process to be intentionally halted. On stopping the process and observing that it was not possible to construct the indicated CA, the value of N was increased, and the algorithm was run again. For the example in question, the search for a CA with $N = 15$ became one with N' = 16. In case the algorithm is stuck in a local optimum again, the process is stopped, and the N value is increased once more. The GHS algorithm managed to construct the configuration with N' = 18.

Of the 30 tests performed, GHS was able to successfully construct only six (6) MCA/CA configurations within the predefined number of iterations, while in the remaining 24 configurations it did not manage to do so. It thus proceeded to find the respective N'. From the above it can be clearly seen that the performance of the GHS algorithm for the construction of MCA/CAs is poor.

Analyzing the MCA with Id = 5, the proposed GHSSA algorithm constructed the configuration indicated, while the SA algorithm failed in the construction of the MCA for the predefined number of iterations, since the final fitness was equal to 2. This was the only case in which SA could not find the required MCA/CA. It is also observed that the GHS algorithm, within the number of predefined iterations, fell into a local optimum, therefore the process was intentionally stopped. GHS managed to construct the CA for an N' = 50.

3.2   Second Experiment

The second and third experiments were carried out on different computers with different hardware and software configurations. For these experiments the time spent in the construction of the CAs was not considered, because the algorithms of the state of the art used in the comparisons were run in several cases on supercomputers. Therefore, only the value of N obtained in the construction of the MCA/CAs was considered. The lower the value of N, the better the MCA/CA generated. The desktops and laptops used in the second and third experiments were:

Table 1. Comparison between the GHS, GHSSA and SA algorithms

Id	MCA	Algorithm	State	Fitness reached	T (seconds)	N'
1	N5K4V2^4t2	GHS	S	0	0,020	–
		GHSSA	S	0	0,011	–
		SA	S	0	0,020	–
2	N15K11V5^1-3^8-2^2t2	GHS	P	–	–	20
		GHSSA	S	0	11,752	–
		SA	S	0	36,568	–
3	N19K9V4^5-3^4t2	GHS	P	–	–	21
		GHSSA	S	0	37,393	–
		SA	S	0	186,513	–
4	N36K20V6^2-4^9-2^9t2	GHS	P	–	–	44
		GHSSA	S	0	98,248	–
		SA	S	0	15,886	–
5	N37K16V6^4-4^5-2^7t2	GHS	P	–	–	50
		GHSSA	S	0	25373,737	–
		SA	F	2	22090,729	–
6	N42K19V7^1-6^1-5^1-4^5-3^8-2^3t2	GHS	P	–	–	47
		GHSSA	S	0	5,148	–
		SA	S	0	6,580	–
7	N30K61V4^15-3^17-2^29t2	GHS	P	–	–	37
		GHSSA	S	0	3190,504	–
		SA	S	0	1723,872	–
8	N36K10V6^2-5^2-4^2-3^2-2^2t2	GHS	P	–	–	41
		GHSSA	S	0	0,156	–
		SA	S	0	0,30	–
9	N64K8V8^2-7^2-6^2-5^2t2	GHS	P	–	–	90
		GHSSA	S	0	968,846	–
		SA	S	0	445,215	–
10	N49K12V7^2-6^2-5^2-4^2-3^2-2^2t2	GHS	P	–	–	57
		GHSSA	S	0	81,222	–
		SA	S	0	13,540	–
11	N100K5V10^2-8^3t2	GHS	P	–	–	160
		GHSSA	S	0	11,656	–
		SA	S	0	9,480	–
12	N8K4V2^4t3	GHS	P	–	–	10
		GHSSA	S	0	0,038	–
		SA	S	0	0,040	–
13	N10K5V2^5t3	GHS	S	0	5,539	–
		GHSSA	S	0	0,052	–
		SA	S	0	0,010	–
14	N12K6V2^6t3	GHS	S	0	1,833	–
		GHSSA	S	0	0,040	–
		SA	S	0	0,090	–
15	N12K7V2^7t3	GHS	P	–	–	13
		GHSSA	S	0	0,01	–
		SA	S	0	0,125	–

(*continued*)

Table 1. (*continued*)

Id	MCA	Algorithm	State	Fitness reached	T (seconds)	N'
16	N12K11V2^11t3	GHS	S	0	47,158	–
		GHSSA	S	0	0,230	–
		SA	S	0	0,281	–
17	N15K12V2^12t3	GHS	P	–	–	18
		GHSSA	S	0	0,320	–
		SA	S	0	0,370	–
18	N100K6V5^2-4^2-3^2t3	GHS	P	–	–	130
		GHSSA	S	0	205,101	–
		SA	S	0	2398,327	–
19	N16K14V2^14t3	GHS	P	–	–	19
		GHSSA	S	0	207,501	–
		SA	S	0	199,888	–
20	N13K11V2^11t3	GHS	S	0	26,012	
		GHSSA	S	0	0,028	–
		SA	S	0	0,042	–
21	N360K7V10^1-6^2-4^3-3^1t3	GHS	P	–	–	800
		GHSSA	S	0	350,699	–
		SA	S	0	75,169	–
22	N400K12V10^2-4^1-3^2-2^7t3	GHS	P	–	–	710
		GHSSA	S	0	35,305	–
		SA	S	0	34,659	–
23	N16K5V2^5t4	GHS	S	0	17,050	–
		GHSSA	S	0	0,162	–
		SA	S	0	0,032	–
24	N21K6V2^6t4	GHS	P	–	–	22
		GHSSA	S	0	1,817	–
		SA	S	0	10,546	–
25	N24K12V2^12t4	GHS	P	–	–	48
		GHSSA	S	0	0,848	–
		SA	S	0	1,040	–
26	N32K6V2^6t5	GHS	P	–	–	33
		GHSSA	S	0	0,071	–
		SA	S	0	0,436	–
27	N42K7V2^7t5	GHS	P	–	–	61
		GHSSA	S	0	0,034	–
		SA	S	0	0,423	–
28	N52K8V2^8t5	GHS	P	–	–	60
		GHSSA	S	0	29,875	–
		SA	S	0	371,133	–
29	N64K7V2^7t6	GHS	P	–	–	97
		GHSSA	S	0	0,043	–
		SA	S	0	0,084	–
30	N85K8V2^8t6	GHS	P	–	–	160
		GHSSA	S	0	5,340	–
		SA	S	0	61,932	–

- Dell Vostro 460 with 4 GB of RAM, Intel Core i5-2400 3.1 GHz processor, 256 GB DD and Windows 7 Home Premium 64-bit operating system.
- Dell Latitude 3470 with 16 GB of RAM, Intel Core i7 2.7 GHz processor, 1 TB DD and Windows 10 Pro 64-bit operating system.
- Lenovo Z50 with 16 GB of RAM, Intel Core i7-4510U 2.6 GHz. 1 TB DD and Windows 10 64-bit operating system.
- Mac Book Air with 8 GB of RAM, Intel Core i5 1.6 GHz processor. 256 SSD and Windows 10 64-bit operating system.

This second experimental comparison was carried out for a total of 12 configurations of uniform, binary CAs, with the best results reported by the algorithms of the state of the art, namely: DDA [10], TS [11], IPOG-F [12] and an improved SA algorithm [13]. The results obtained are shown in Table 2.

Table 2. Comparison between DDA, TS, IPOG-F, SA and GHSSA

Id	Binary CA	DDA	TS	IPOG-F	SA	Best (β)	GHSSA (θ)	Δ
1	N8K4V2^4t3	8	8	8	8	8	8	0
2	N10K5V2^5t3	10	10	11	10	10	10	0
3	N12K11V2^11t3	20	12	18	12	12	12	0
4	N15K12V2^12t3	21	15	19	15	15	15	0
5	N16K14V2^14t3	27	16	21	16	16	16	0
6	N17K16V2^16t3	27	17	22	17	17	19	+2
7	N16K5V2^5t4	16	16	22	16	16	16	0
8	N21K6V2^6t4	26	21	26	21	21	21	0
9	N24K12V2^12t4	52	48	47	24	24	24	0
10	N32K6V2^6t5	32	32	42	32	32	32	0
11	N42K7V2^7t5	52	56	57	42	42	42	0
12	N52K8V2^8t5	76	56	68	52	52	52	0

The first column of this table shows the ID of the CA, the second column shows the configuration of the uniform, binary CA to be constructed. From column 3 to 6, the algorithms with which the proposed algorithm was compared are shown. The seventh column of the table denoted as Best (β) presents the best level (test cases) reported in the literature. The eighth column shows the level obtained by the proposed algorithm GHSSA (θ) and the ninth column presents the difference between the result obtained by GHSSA and the best solution reported in the state of the art $\Delta = \theta - \beta$.

In 11 of the 12 tests conducted in this second experiment, the GHSSA algorithm managed to construct the established configurations. Only for the configuration with Id 6 was it not possible to construct the CA and two additional rows were required to do that, which is quite close to the best reported.

3.3 Third Experiment

This last experiment was performed on 19 configurations of mixed CAs of variable strength, compared to the best results reported in the state of the art by the following algorithms: IPOG [14], IPOG-F [12], SA-VNS [15], MiTS [16] y SA-H [17]. Table 3 shows the results obtained from this experiment. The first column shows the ID of the MCA. The second column shows the configuration of the MCA to be constructed. From column 3 to 7, the algorithms with which the proposed algorithm was compared are shown. The eighth column of the table denoted as Best (β) presents the best level reported in the literature. The ninth column shows the level obtained by GHSSA (θ) and the tenth column shows the difference between the result obtained by GHSSA and the best level reported in the state of the art.

In 10 of the 19 evaluations carried out, the GHSSA algorithm was able to equal the best level reported in the state of the art. In six more cases, the solution found was close to the best solution (2, 3 or 4 additional rows) and in the three remaining cases a greater number of rows was required (CAs with ID 14, 18 and 19) which are the most difficult cases to solve for GHSSA, all of strength 3.

With experiments 2 and 3 it can be shown that GHSSA can be run on conventional desktop computers that can run in parallel between 4 and 8 run threads and thereby obtain similar or competitive results with those presented in the state of the art, several of which were generated on supercomputers.

Table 3. Comparison between GHSSA and the IPOG, IPOG-F, MiTS, SA-H and SA-VNS approaches

Id	MCA	IPOG	IPOG-F	MiTS	SA-H	SA-VNS	Best (β)	GHSSA (θ)	Δ
1	N15K11V5^1-3^8-2^2t2	17	20	15	15	15	15	15	0
2	N19K9V4^5-3^4t2	24	22	19	19	19	19	19	0
3	N20K75V4^1-3^39-2^35t2	26	28	22	20	20	20	22	+2
4	N21K21V5^1-4^4-3^11-2^5t2	25	27	22	21	21	21	23	+2
5	N30K61V4^15-3^17-2^29t2	34	35	30	30	30	30	30	0
6	N28K19V6^1-5^1-4^6-3^8-2^3t2	36	34	30	29	28	28	30	+2
7	N36K20V6^2-4^9-2^9t2	38	39	36	36	36	36	36	0
8	N37K16V6^4-4^5-2^7t2	44	42	38	38	37	37	37	0
9	N42K19V7^1-6^1-5^1-4^5-3^8-2^3t2	42	42	42	42	42	42	42	0
10	N46K15V6^6-5^5-3^4t2	55	53	50	–	46	46	50	+4
11	N45K18V6^7-4^8-2^3t2	55	55	47	47	45	45	49	+4
12	N50K19V6^9-4^3-2^7t2	64	60	51	51	50	50	53	+3
13	N64K8V8^2-7^2-6^2-5^2t2	68	64	64	64	64	64	64	0
14	N80K9V4^5-3^4t3	103	100	85	80	83	80	93	+13
15	N100K6V5^2-4^2-3^2t3	106	103	100	100	100	100	100	0
16	N360K7V10^1-6^2-4^3-3^1t3	372	361	360	360	360	360	360	0
17	N400K12V10^2-4^1-3^2-2^7t3	405	402	400	400	400	400	400	0
18	N376K15V6^6-5^5-3^4t3	452	426	–	–	376	376	426	+50
19	N500K8V8^2-7^2-6^2-5^2t3	593	554	540	535	500	500	578	+78

4 Conclusions and Future Work

In the construction of MCA/CAs for configurations where the number of variables and alphabet is high, Global-best Harmony Search algorithm (GHS) produces poor results.

In contrast, the proposed GHSSA algorithm was able to construct several configurations of MCA/CAs on relatively basic desktop or laptop computers. As a result, micro, small and medium software companies can use it to generate the MCA/CAs they use in testing software or hardware, without having to buy or rent supercomputers.

The performance of GHSSA improved considerably on the inclusion of Simulated Annealing as a local optimization technique along with two neighborhood construction schemes, thereby managing to construct configurations initially not possible with GHS alone. The memetic algorithm achieved with GHSSA (between the GHS and SA algorithms) made it possible to construct both uniform and mixed Covering Arrays, of simple and complex configuration, of different strengths ($2 \leq t \leq 6$) and whose results were like, or able to compete with those reported in the state of the art.

When building mixed CAs of complex configuration and very high strength, the proposed algorithm takes much more time. In some occasions, some tests executed for the construction of complex CAs were made using computers with very poor hardware and software configurations, exceeding the CPU and RAM load, which caused the operating system to interrupt the constructions that took several hours (even weeks) in execution.

A significant number of state-of-the-art algorithms for constructing CAs that have included in their implementations local searches with different neighborhood schemes and/or techniques to manage variable neighborhood, have obtained better results than those algorithms use no such searches. Hyper-heuristic [18] and Multiple Offspring Sampling [19] may thus be considered to represent the future in the construction of CAs and MCAs of variable strength and different variables and alphabets, work that the research group hopes to take on board.

References

1. Torres-Jimenez, J., Izquierdo-Marquez, I.: Survey of covering arrays. In: 2013 15th International Symposium on Symbolic and Numeric Algorithms for Scientific Computing (SYNASC), pp. 20–27 (2013)
2. George, H.A.: Constructing covering arrays using parallel computing and grid computing. Ph.D., Departamento de Sistemas Informáticos y Computación, Universitad Politécnica de Valencia, Valencia, Spain (2012)
3. Turban, R.C., Adviser-Colbourn, C.: Algorithms for Covering Arrays. Arizona State University, Tempe (2006)
4. Kacker, R.N., et al.: Combinatorial testing for software: an adaptation of design of experiments. Measurement **46**, 3745–3752 (2013)
5. Cohen, M.B., et al.: Constructing test suites for interaction testing. Presented at the Proceedings of the 25th International Conference on Software Engineering, Portland, Oregon (2003)
6. Nurmela, K.J.: Upper bounds for covering arrays by tabu search. Discret. Appl. Math. **138**, 143–152 (2004)

7. Stardom, J.: Metaheuristics and the search for covering and packing arrays [microform]. M. Sc. thesis, Simon Fraser University (2001)
8. Shiba, T., et al.: Using artificial life techniques to generate test cases for combinatorial testing. In: 2004 Proceedings of the 28th Annual International Computer Software and Applications Conference, COMPSAC 2004, vol. 1, pp. 72–77 (2004)
9. Omran, M.G.H., Mahdavi, M.: Global-best harmony search. Appl. Math. Comput. **198**, 643–656 (2008)
10. Bryce, R.C., Colbourn, C.J.: A density-based greedy algorithm for higher strength covering arrays. Softw. Test. Verif. Reliab. **19**, 37–53 (2009)
11. Walker Ii, R.A., Colbourn, C.J.: Tabu search for covering arrays using permutation vectors. J. Stat. Plan. Inference **139**, 69–80 (2009)
12. Forbes, M., et al.: Refining the in-parameter-order strategy for constructing covering arrays. J. Res. Natl. Inst. Stand. Technol. **113**, 287–297 (2008)
13. Torres-Jimenez, J., Rodriguez-Tello, E.: Simulated annealing for constructing binary covering arrays of variable strength. In: 2010 IEEE Congress on Evolutionary Computation (CEC), pp. 1–8 (2010)
14. Lei, Y., et al.: IPOG: a general strategy for t-way software testing. Presented at the Proceedings of the 14th Annual IEEE International Conference and Workshops on the Engineering of Computer-Based Systems (2007)
15. Rodriguez-Cristerna, A., Torres-Jimenez, J.: A simulated annealing with variable neighborhood search approach to construct mixed covering arrays. Electron. Notes Discret. Math. **39**, 249–256 (2012)
16. Hernández, A.L.G.: Un Algoritmo de Optimizacion Combinatoria para la Construccion de Covering Arrays Mixtos de Fuerza Variable. Ph.D., Laboratorio de Tecnologías de la Información, Centro de Investigación y de Estudios Avanzados del Instituto Politécnico Nacional (2013)
17. Avila-George, H., Torres-Jimenez, J., Hernández, V., Gonzalez-Hernandez, L.: Simulated annealing for constructing mixed covering arrays. In: Omatu, S., De Paz Santana, J.F., González, S.R., Molina, J.M., Bernardos, A.M., Rodríguez, J.M.C. (eds.) Distributed Computing and Artificial Intelligence. AISC, vol. 151, pp. 657–664. Springer, Heidelberg (2012). https://doi.org/10.1007/978-3-642-28765-7_79
18. Burke, E.K., et al.: Hyper-heuristics: a survey of the state of the art. J. Oper. Res. Soc. **64**, 1695–1724 (2013)
19. LaTorre, A., et al.: Multiple offspring sampling in large scale global optimization. In: 2012 IEEE Congress on Evolutionary Computation, pp. 1–8 (2012)

An Adaptive Hybrid Evolutionary Approach for a Project Scheduling Problem that Maximizes the Effectiveness of Human Resources

Virginia Yannibelli[1,2(✉)]

[1] ISISTAN Research Institute, UNCPBA University, Campus Universitario,
Paraje Arroyo Seco, 7000 Tandil, Argentina
virginia.yannibelli@isistan.unicen.edu.ar
[2] CONICET, National Council of Scientific and Technological Research,
Buenos Aires, Argentina

Abstract. In this paper, an adaptive hybrid evolutionary algorithm is proposed to solve a project scheduling problem. This problem considers a valuable optimization objective for project managers. This objective is maximizing the effectiveness of the sets of human resources assigned to the project activities. The adaptive hybrid evolutionary algorithm utilizes adaptive processes to develop the different stages of the evolutionary cycle (i.e., adaptive parent selection, survival selection, crossover, mutation and simulated annealing processes). These processes adapt their behavior according to the diversity of the algorithm's population. The utilization of these processes is meant to enhance the evolutionary search. The performance of the adaptive hybrid evolutionary algorithm is evaluated on six instance sets with different complexity levels, and then is compared with those of the algorithms previously reported in the literature for the addressed problem. The obtained results indicate that the adaptive hybrid evolutionary algorithm significantly outperforms the algorithms previously reported.

Keywords: Project scheduling · Human resource assignment
Multi-skilled resources · Evolutionary algorithms
Adaptive evolutionary algorithms · Hybrid evolutionary algorithms

1 Introduction

Project scheduling is a highly relevant and complex issue in most real-world organizations [1, 2]. Project scheduling generally involves defining feasible start times and human resource assignments for the project activities, such that a given optimization objective is achieved. Moreover, to define human resource assignments, it is necessary to consider the available knowledge about the effectiveness of human resources respecting the project activities. This is important since the development and the results of project activities mainly depend on the effectiveness of the human resources assigned to such activities [1, 2].

© Springer Nature Switzerland AG 2018
I. Batyrshin et al. (Eds.): MICAI 2018, LNAI 11288, pp. 33–49, 2018.
https://doi.org/10.1007/978-3-030-04491-6_3

In the past four decades, many kinds of project scheduling problems have been formally presented and addressed in the literature. However, to the best of the author's knowledge, only few project scheduling problems consider human resources with different effectiveness levels [3, 4], a very important aspect of real-world project scheduling. These project scheduling problems suppose very different assumptions in respect of the effectiveness of human resources.

The project scheduling problem presented in [5] supposes that the effectiveness level of a human resource depends on several factors inherent to its work context (i.e., the project activity to which the resource is assigned, the skill to which the resource is assigned within the project activity, the set of human resources assigned to the project activity, and the attributes of the resource). This assumption about the effectiveness of human resources is really valuable. This is because, in real-world project scheduling problems, human resources generally have different effectiveness levels with respect of different work contexts, and therefore, the effectiveness level of a human resource is considered with respect of the factors inherent to its work context [1, 2]. To the best of the author's knowledge, the influence of the work context on the effectiveness level of human resources is not considered in other project scheduling problems presented in the literature. The problem presented in [5] also considers a valuable optimization objective for project managers: maximizing the effectiveness of the sets of human resources assigned to the project activities.

The project scheduling problem presented in [5] is a variant of the known RCPSP (Resource Constrained Project Scheduling Problem) [6] and, therefore, is an NP-Hard optimization problem. Because of this, heuristic search and optimization algorithms are required to solve different problem instances in an acceptable amount of time. In this respect, to the best of the author's knowledge, four heuristic search and optimization algorithms have been presented so far in the literature to solve this problem. Specifically, a traditional evolutionary algorithm was presented in [5]. In [7], a traditional memetic algorithm was presented which includes a hill-climbing algorithm into the framework of an evolutionary algorithm. In [8], a hybrid evolutionary algorithm was presented which incorporates an adaptive simulated annealing algorithm within the framework of an evolutionary algorithm. In [9–11], a hybrid evolutionary algorithm was presented which utilizes semi-adaptive crossover and mutation processes, and an adaptive simulated annealing algorithm.

These four algorithms follow the stages of the traditional evolutionary cycle (i.e., parent selection, crossover, mutation, and survival selection stages) to develop the evolutionary search. Nevertheless, these algorithms use non-adaptive processes to carry out all or many of the stages of the evolutionary cycle. In this respect, the four algorithms use non-adaptive parent selection and survival selection processes, and the first three of these algorithms use non-adaptive crossover and mutation processes. The fourth of these algorithms uses semi-adaptive crossover and mutation processes; however, the adaptability of the behavior of these processes during the evolutionary search is very limited.

In this paper, the project scheduling problem presented in [5] is addressed with the aim of proposing a better heuristic search and optimization algorithm to solve it. In this respect, an adaptive hybrid evolutionary algorithm is proposed which utilizes adaptive processes to develop the different stages of the evolutionary cycle (i.e., adaptive parent selection, survival selection, crossover, mutation and simulated annealing processes). These processes adapt their behavior based on the diversity of the algorithm's population. The utilization of these adaptive processes is meant for improving the performance of the evolutionary search [12–14].

The above-mentioned adaptive hybrid evolutionary algorithm is proposed mainly because of the following reasons. Evolutionary algorithms with adaptive selection, crossover and mutation processes have been proven to be more effective than evolutionary algorithms with non-adaptive selection, crossover and mutation processes in the resolution of a wide variety of NP-Hard optimization problems [12–17]. Thus, the proposed adaptive hybrid evolutionary algorithm could outperform the heuristic search and optimization algorithms presented so far in the literature to solve the addressed problem.

The remainder of the paper is organized as follows. Section 2 presents a brief review of project scheduling problems reported in the literature which consider the effectiveness of human resources. Section 3 describes the project scheduling problem addressed. Section 4 presents the adaptive hybrid evolutionary algorithm proposed for the problem. Section 5 presents the computational experiments developed to evaluate the performance of the adaptive hybrid evolutionary algorithm and also an analysis of the results obtained. Finally, Sect. 6 presents the conclusions of the present work.

2 Related Works

Over the past four decades, different kinds of project scheduling problems which consider the effectiveness of human resources have been presented in the literature [3, 4]. However, these project scheduling problems suppose very different assumptions concerning the effectiveness of human resources. In this regards, to the best of the author's knowledge, only few project scheduling problems consider human resources with different effectiveness levels [3, 4], a very important aspect of real-world project scheduling problems. In this section, the focus is on reviewing the main assumptions about the effectiveness that have been considered in project scheduling problems presented in the literature.

In the multi-skill project scheduling problems presented in [19–23], project activities require a given number of skills for their development, and a given number of human resources for each skill required. The human resources available for the project activities master one or several skills. These problems suppose that the human resources that master a given skill have the same effectiveness level in respect of such skill.

In the skilled workforce project scheduling problems presented in [24–26], each project activity requires only one human resource with a given skill. Moreover, the human resources available for project activities master one or several skills. These problems suppose that the human resources which master a given skill have the same effectiveness level in relation to such skill.

The multi-skill project scheduling problem reported in [27] considers hierarchical levels of skills. In this regard, the problem supposes that the human resources which master a given skill have different effectiveness levels in respect of such skill. Moreover, the project activities of this problem require a given number of skills for their development, a given minimum level of effectiveness for each one of the skills required, and a given number of human resources for each pair skill-level. Then, this problem supposes that the human resource sets feasible for a given project activity have the same effectiveness level with respect to the development of such activity.

In the multi-skill project scheduling problems presented in [28–30], most project activities require only one human resource with a given skill. The human resources available for project activities master one or several skills. These problems suppose that the human resources which master a given skill have different effectiveness levels in respect of such skill. In addition, these problems suppose that the effectiveness level of a human resource in a given project activity only depends on the effectiveness level of the human resource with respect to the skill required for the activity.

In contrast with the project scheduling problems previously mentioned, the project scheduling problem presented in [5] supposes that the effectiveness level of a human resource depends on several factors inherent to its work context. Then, different effectiveness levels can be defined for each human resource regarding different work contexts. This assumption about the effectiveness of human resources is really important. This is because, in the context of real-world project scheduling problems, human resources have very different effectiveness levels in respect of different work contexts, and therefore, the effectiveness level of a human resource is considered in respect of the factors inherent to its work context [1, 2]. To the best of the author's knowledge, the influence of the work context on the effectiveness level of human resources is not considered in other project scheduling problems presented in the literature. Based on the mentioned above, the project scheduling problem presented in [5] supposes a valuable and novel assumption concerning the effectiveness level of human resources in the context of project scheduling problems.

3   Project Scheduling Problem Description

In this paper, the project scheduling problem introduced in [5] is addressed. A description of this problem is presented below.

Suppose that a project contains a set A of N activities, $A = \{1, ..., N\}$, that has to be scheduled. Specifically, a starting time and a human resource set have to be defined for each project activity of the set A. The duration, human resource requirements, and precedence relations of each project activity are known.

The duration of each project activity j is notated as d_j. Besides, it is considered that pre-emption of project activities is not allowed. This means that, when a project activity starts, it must be developed period by period until it is completed. Specifically, the d_j periods of time must be consecutive.

Among the project activities, there are precedence relations. This is because usually each project activity requires results generated by other project activities. Thus, the precedence relations establish that each project activity j cannot start until all its immediate predecessors, given by the set P_j, have completely finished.

To be developed, project activities require human resources skilled in different knowledge areas. Specifically, each project activity requires one or several skills and also a given number of human resources for each skill required.

It is considered that qualified workforce is available to develop the activities of the project. This workforce is made up of a number of human resources, and each human resource masters one or several skills.

Set SK contains the K skills required in order to develop the activities of the project, $SK = \{1, \ldots, K\}$, and set AR_k contains the available human resources with skill k. Then, the term $r_{j,k}$ represents the number of human resources with skill k required for activity j of the project. The values of the terms $r_{j,k}$ are known for each project activity.

It is considered that a human resource cannot take over more than one skill within a given activity, and also a human resource cannot be assigned more than one activity at the same time.

Based on the assumptions previously mentioned, a human resource can be assigned different project activities but not at the same time, can take over different skills required for a project activity but not simultaneously, and can belong to different possible sets of human resources for each activity.

Therefore, different work contexts can be defined for each available human resource. It is considered that the work context of a human resource r, denoted as $C_{r,j,k,g}$, is made up of four main components. In this respect, the first component refers to the project activity j which r is assigned (i.e., the complexity of j, the domain of j, etc.). The second component refers to the skill k which r is assigned within project activity j (i.e., the tasks associated to k within j). The third component is the set of human resources g that has been assigned j and that includes r (i.e., r must work in collaboration with the other human resources assigned to j). The fourth component refers to the attributes of r (i.e., his or her educational level regarding different knowledge areas, his or her level regarding different skills, his or her experience level regarding different tasks and domains, the kind of labor relation between r and the other human resources of g, etc.). In respect of the attributes of r, it is considered that these attributes could be quantified from available information about r (e.g., curriculum vitae of r, results obtained from evaluations made to r, information about the participation of r in already executed projects, etc.).

The four components previously mentioned are considered the main factors that determine the effectiveness level of a human resource. Because of this, it is assumed that the effectiveness level of a human resource depends on all the components of his or her work context. Then, different effectiveness levels can be considered for each human resource in respect of different work contexts.

The effectiveness level of a human resource r, in respect of a possible context $C_{r,j,k,g}$ for r, is notated as $e_{rCr,j,k,g}$. The term $e_{rCr,j,k,g}$ refers to how well r can take over, within activity j, the tasks associated to skill k, considering that r must work in collaboration with the other human resources of set g. The term $e_{rCr,j,k,g}$ takes a real value over the range $[0, 1]$. The values of the terms $e_{rCr,j,k,g}$ inherent to each human resource available for the project are known. It is considered that these values could be obtained from available information regarding the participation of the human resources in already carried out projects.

The problem of scheduling a project involves to determine feasible start times (i.e., the precedence relations among the project activities must not be violated) and feasible human resource assignments (i.e., the human resource requirements of project activities must be met) for project activities such that the optimization objective is achieved. In this respect, an optimization objective valuable for project managers is considered. This optimization objective implies maximizing the effectiveness of the sets of human resources assigned to the project activities. This objective is modeled by Formulas (1) and (2).

Formula (1) maximizes the effectiveness of the sets of human resources assigned to the N project activities. In this formula, set S contains all the feasible schedules for the project in question. The term $e(s)$ refers to the effectiveness level of the sets of human resources assigned to the project activities by schedule s. The term $R(j, s)$ refers to the set of human resources assigned to activity j by schedule s. The term $e_{R(j,s)}$ refers to the effectiveness level corresponding to $R(j, s)$.

Formula (2) estimates the effectiveness level of the set of human resources $R(j,s)$. This effectiveness level is estimated by calculating the mean effectiveness level of the human resources belonging to $R(j, s)$.

For a more detailed discussion of the project scheduling problem described here and, in particular, of Formulas (1) and (2), the readers are referred to the work [5] which has introduced this problem.

$$\max_{\forall s \in S} \left(e(s) = \sum_{j=1}^{N} e_{R(j,s)} \right) \tag{1}$$

$$e_{R(j,s)} = \frac{\sum_{r=1}^{|R(j,s)|} e_{r} C_{r,j,k(r,j,s),R(j,s)}}{|R(j,s)|} \tag{2}$$

4   Adaptive Hybrid Evolutionary Algorithm

To solve the addressed problem, an adaptive hybrid evolutionary algorithm is proposed. This algorithm utilizes adaptive processes to develop the different stages of the evolutionary cycle. These processes adapt their behavior according to the diversity of the evolutionary algorithm population, to promote either the exploration or exploitation of the search space. The utilization of these adaptive processes aims to enhance the performance of the evolutionary search [12–14].

The general behavior of the algorithm is described as follow. This algorithm is an iterative or generational process. This process starts from an initial population of solutions. Each solution encodes a feasible schedule for the project to be scheduled. Besides, each solution has a fitness value that represents the quality of the related schedule in respect of the optimization objective of the addressed problem. As mentioned in Sect. 3, such objective implies maximizing the effectiveness of the sets of human resources assigned to the project activities. The iterative process ends when a predefined number of iterations or generations is reached. Once this happens, the iterative process provides the best solution of the last population as a solution to the problem.

In each iteration, the algorithm develops the following stages. First, an adaptive parent selection process is applied in order to determine which solutions of the current population will compose the mating pool. Once the mating pool is composed, the solutions in the mating pool are paired, and a crossover process is applied to each pair of solutions with an adaptive crossover probability in order to generate new feasible ones. Then, a mutation process is applied to each solution generated by the crossover process, with an adaptive mutation probability. The mutation process is applied in order to introduce diversity in the new solutions generated by the crossover process. Then, an adaptive survival selection process is applied to create a new population from the solutions in the current population and the new solutions generated by crossover and mutation. Finally, an adaptive simulated annealing algorithm is applied to each solution of the new population, excepting the best solution of this population. The best solution remains in the population. Thus, the adaptive simulated annealing algorithm modifies the solutions of the new population.

4.1 Encoding of Solutions

In order to encode the solutions of the population, the encoding introduced in [5] for project schedules was used. By using this encoding, each solution is encoded by two lists with a length equal to N, considering that N is the number of activities in the project to be scheduled.

The first list is a traditional activity list. Each position on this list contains a different activity j of the project. Each activity j of the project can appear on this list in any position higher than the positions of all its predecessor activities. The activity list represents a feasible order in which the activities of the project can be added to the schedule.

The second list is an assigned resources list. This list contains information about the human resources of each skill k assigned to each activity of the project. Specifically, position j on this list contains a detail about the human resources of each skill k assigned to activity j of the project.

To decode or build the schedule related to the encoding previously described, the serial schedule generation method presented in [5] was used. By this method, each activity j is scheduled at the earliest possible time.

In order to generate the encoded solutions of the initial population according to the encoding previously described, the random generation process introduced in [5] was used. By using this process, a very diverse initial population is obtained. This is meant in order to avoid the premature convergence of the evolutionary search [12].

4.2 Fitness Function

To evaluate the encoded solutions, a fitness function specially designed was used. Considering a given encoded solution, the fitness function decodes the schedule s from the solution by using the serial schedule generation method mentioned in Sect. 4.1. Then, the fitness function calculates the value of the term $e(s)$ corresponding to s (Formulas (1) and (2)). This value defines the fitness value of the solution. Note that the term $e(s)$ takes a real value over the range $[0, \ldots, N]$.

To calculate the value of term $e(s)$, the fitness function uses the values of the terms $e_{rCr,j,k,g}$ inherent to s (Formula 2). As mentioned in Sect. 3, the values of the terms $e_{rCr,j,k,g}$ inherent to each available human resource r are known.

4.3 Adaptive Parent Selection Process

To develop the parent selection on the current population, an adaptive tournament selection process was defined. This process is an adaptive variant of the well-known tournament selection process with replacement [12].

In this process, the tournament size T is defined by Formula (3), where PD refers to the diversity of the current population, and PD_{MAX} refers to the maximum PD attainable. Then, T^H and T^L refer to the upper and lower bounds for the tournament size, respectively.

The term PD is defined by Formula (4), where f_{max} is the maximal fitness of the current population, f_{avg} is the average fitness of the current population, and $(f_{max} - f_{avg})$ is a measure of the diversity of the current population. This measure has been proposed by Srinivas and Patnaik [31], and is one of the population diversity measures most well-known in the literature [12].

The term PD_{MAX} is defined by Formula (5), where f_{MAX} and f_{MIN} represent to the maximum and minimum fitness values attainable, respectively. Note that f_{MAX} and f_{MIN} correspond to the upper and lower bounds of the fitness function described in Sect. 4.2.

By Formula (3), the tournament size T is adaptive according to the diversity of the current population. Specifically, when the population is very diverse, T is increased, promoting the selection of the solutions with high fitness values. This favors the exploitation of the search space. When the diversity of the population reduces, T is decreased, increasing the selection chances of the solutions with low fitness values. This is meant to preserve the diversity of the population and thus to favor the exploration of the search space, with the aim of avoiding the premature convergence of the evolutionary search.

$$T = \left\lceil \frac{PD}{PD_{MAX}} * \left(T^H - T^L\right) + T^L \right\rceil \tag{3}$$

$$PD = \left(f_{max} - f_{avg}\right) \tag{4}$$

$$PD_{MAX} = \left(f_{MAX} - f_{MIN}\right) \tag{5}$$

4.4 Adaptive Crossover and Adaptive Mutation Processes

In relation to the crossover process and the mutation process, processes feasible for the encoding of solutions were defined.

The crossover process is composed by a crossover operator feasible for activity lists and a crossover operator feasible for assigned resources lists. Regarding the crossover for activity lists, the one-point crossover operator for activity lists [18] was applied. Regarding the crossover for assigned resources lists, the uniform crossover operator was applied [12].

The mutation process is composed by a mutation operator feasible for activity lists and a mutation operator feasible for assigned resources lists. In relation to the mutation for activity lists, a variant of the simple shift operator for activity lists [18] was applied. In relation to the mutation for assigned resources lists, the random resetting operator [12] was applied.

These crossover and mutation processes are applied with adaptive crossover and mutation probabilities, respectively. In this regards, an adaptive crossover probability AP_c and an adaptive mutation probability AP_m were defined by Formulas (6)–(7). In these formulas, PD refers to the diversity of the current population, and PD_{MAX} refers to the maximum PD attainable, as was mentioned in Sect. 4.3. In Formula (6), the terms C^H and C^L represent to the upper and lower bounds for the crossover probability, respectively. In Formula (7), the terms M^H and M^L represent to the upper and lower bounds for the mutation probability, respectively. The term f_{max} is the maximal fitness of the population, f_{min} is the minimal fitness of the population, and f is the fitness of the solution to be mutated.

By Formula (6)–(7), AP_c and AP_m are adaptive according to the diversity of the current population. In this respect, when the diversity of the population reduces, AP_c and AP_m are increased, promoting the exploration of the search space. This is important for preventing the premature convergence of the evolutionary search. When the population is very diverse, AP_c and AP_m are decreased, promoting the exploitation of the search space. Therefore, probabilities AP_c and AP_m are adaptive according to the diversity of the population, to promote either the exploitation or exploration of the search space.

By Formula (7), AP_m is also adaptive according to the fitness of the solution to be mutated. In this respect, lower values of AP_m are defined for high-fitness solutions, and higher values of AP_m are defined for low-fitness solutions. This is meant in order to preserve high-fitness solutions, while disrupting low-fitness solutions to promote the exploration of the search space.

$$AP_c = \left(\frac{PD_{MAX} - PD}{PD_{MAX}}\right) * \left(C^H - C^L\right) + C^L \tag{6}$$

$$AP_m = \left(\frac{f_{max} - f}{f_{max} - f_{min}}\right) * \left(\frac{PD_{MAX} - PD}{PD_{MAX}}\right) * \left(M^H - M^L\right) + M^L \tag{7}$$

4.5   Adaptive Survival Selection Process

The survival selection process is applied to create a new population from the solutions in the current population (parent solutions) and the new solutions generated by crossover and mutation (offspring solutions).

To develop the survival selection, an adaptive deterministic crowding process was defined. This process is an adaptive variant of the well-known deterministic crowding process [12, 32].

In this process, offspring solutions compete with their respective parent solutions to be included in the new population. Specifically, each offspring competes with its closest parent, considering that the closeness between an offspring solution and a parent solution is defined based on the distance between their fitness values. When the fitness value of an offspring solution is better than that of its closest parent solution, the offspring solution is accepted for the new population. Otherwise, when the fitness value of an offspring solution is not better than that of its closest parent solution, the offspring solution is accepted for the new population with a probability $p_{offspring}$. If the offspring solution is not accepted, the parent solution is accepted for the new population directly.

The probability $p_{offspring}$ is defined as follows: $exp(-\Delta_{DC}/T)$, where Δ_{DC} is the difference between the fitness value of the parent solution and the fitness value of the offspring solution, and the term T is defined according to the diversity of the current population. Specifically, T is inversely proportional to the diversity of the current population, and is calculated as follows: $T = 1/PD$, where PD refers to the diversity of the current population, as was mentioned in Sect. 4.3.

By the definition of the probability $p_{offspring}$, this probability is adaptive based on the diversity of the current population. In this respect, when the population is very diverse, the value of T is very low, and therefore the probability $p_{offspring}$ of the process is also low. Thus, the process preserves the best solutions for the new population, favoring the exploitation of the search space. When the diversity of the population decreases, the value of T increases, and therefore the probability $p_{offspring}$ of the process also increases. Thus, the process introduces diversity into the new population, favoring the exploration of the search space. This is important to prevent the premature convergence of the evolutionary search.

4.6   Adaptive Simulated Annealing Algorithm

After obtaining a new population by the survival selection process, an adaptive simulated annealing algorithm was applied to each solution of this population, except to the best solution of this population which is maintained. This adaptive simulated annealing algorithm is a variant of the one presented in [9–11], and is described below.

The adaptive simulated annealing algorithm is an iterative process. This process starts from a given encoded solution s, and a given initial value T_0 for the temperature parameter. The iterative process ends when a given number of iterations I is reached, or the current value T_i of the temperature parameter is lower than or equal to 0. After this happens, the solution obtained by the process is provided.

In each iteration, the process generates a new encoded solution s' from the current encoded solution s by applying a move operator. Then, the process analyzes if the current solution s should be replaced by the new solution s'. When the fitness value of the current solution s is worse than that of the new solution s', the process replaces to the solution s by the new solution s'. Otherwise, when the fitness value of the current solution s is better than or equal to that of the new solution s', the process replaces to the solution s by the new solution s' with a probability $p_{new\_solution}$. This probability is defined as follows: $p_{new\_solution} = exp(-\Delta_{SA}/T_i)$, where Δ_{SA} is the difference between the fitness value of the current solution s and the fitness value of the new solution s', and T_i is the current value of the temperature parameter. The probability $p_{new\_solution}$ mainly depends on the current value T_i of the temperature parameter. If T_i is high, $p_{new\_solution}$ is also high, and if T_i is low, $p_{new\_solution}$ is also low. The value T_i of the temperature is reduced by a given cooling factor α at the end of each iteration.

The initial value T_0 of the temperature parameter is defined before applying the simulated annealing algorithm to the solutions of the population. In this case, the value T_0 is defined according to the diversity of the population. In particular, T_0 is inversely proportional to the diversity of the population, and is calculated as follows: $T_0 = 1/PD$, where PD refers to the diversity of the population, as mentioned in Sect. 4.3. By this definition of T_0, when the population is diverse, the value T_0 is low, and therefore the probability $p_{new\_solution}$ of the algorithm is also low. Thus, the algorithm fine-tunes the solutions of the population, promoting the exploitation of the search space. When the diversity of the population reduces, the value T_0 increases, and therefore the probability $p_{new\_solution}$ of the algorithm also increases. Thus, the algorithm introduces diversity into the population, promoting the exploration of the search space. This is important for avoiding the premature convergence of the evolutionary search. Based on the above-mentioned, the algorithm is adaptive according to the population diversity, in order to promote either the exploitation or exploration of the search space.

This simulated annealing algorithm utilizes a move operator in order to generate a new encoded solution from a given encoded solution. In this respect, a move operator feasible for the encoding of solutions was defined. This move operator is composed by a move operator feasible for activity lists and a move operator feasible for assigned resources lists. In respect of the move operator for activity lists, the adjacent pairwise interchange operator [18] was applied. For assigned resources lists, an operator which is a variant of the random resetting operator [12] was applied.

5 Computational Experiments

To develop the computational experiments, the six instance sets introduced in [7] were used. Table 1 presents the main characteristics of these six instance sets. Each instance set contains 40 instances. Each instance of these six instance sets contains information about a number of activities to be scheduled, and information about a number of available human resources for developing these activities. For a more detailed description of these instance sets, the readers are referred to [7].

Each instance of these six instance sets has a known optimal solution with a fitness value $e(s)$ equal to N (N refers to the number of activities in the instance). These know optimal solutions are considered here as references to evaluate the performance of the adaptive hybrid evolutionary algorithm.

The adaptive hybrid evolutionary algorithm was evaluated on the six instance sets. Specifically, the algorithm was run a number of 40 times on each instance of the six instance sets. In order to develop these runs, the parameter setting detailed in Table 2 was used. It is necessary to mention that such parameter setting was defined based on exhaustive preliminary experiments that showed that this setting led to the best and most stable results.

Table 1. Main characteristics of the instance sets introduced in [7].

Instance set	Activities per instance	Possible sets of human resources per activity	Instances
j30_5	30	1 to 5	40
j30_10	30	1 to 10	40
j60_5	60	1 to 5	40
j60_10	60	1 to 10	40
j120_5	120	1 to 5	40
j120_10	120	1 to 10	40

Table 2. Parameter setting of the adaptive hybrid evolutionary algorithm

Parameter	Value
Population size	90
Number of generations	300
Parent selection process	
T^H	10
T^L	4
Crossover process	
C^H	0.9
C^L	0.6
Mutation process	
M^H	0.3
M^L	0.01
Simulated annealing algorithm	
I (number of iterations)	25
α (cooling factor)	0.9

Table 3 presents the results obtained by the adaptive hybrid evolutionary algorithm for each of the six instance sets. The second column presents the average percentage deviation from the optimal value (Av. Dev. (%)) for each instance set. The third column presents the percentage of instances for which the optimal value was reached at least once among the 40 runs developed (Opt. (%)).

For the instance sets j30_5, j30_10, j60_5, j60_10 and j120_5 (i.e., the five less complex instance sets), the algorithm obtained Av. Dev. (%) values equal to 0% and Opt. (%) values equal to 100%. These results indicate that the algorithm has reached an optimal solution in each of the 40 runs developed on each instance of these sets.

For the instance set j120_10, the algorithm obtained an Av. Dev. (%) value equal to 0.01%. Considering that the instances of j120_10 have known optimal solutions with a fitness value $e(s)$ equal to 120, this result indicates that the average fitness value of the solutions obtained by the algorithm is 119.99. Therefore, the algorithm has reached very high-fitness solutions for the instances of j120_10. Moreover, the algorithm obtained an Opt. (%) value equal to 100% for j120_10. This result indicates that the algorithm has reached an optimal solution at least once among the 40 runs developed on each instance of j120_10.

Table 3. Results obtained by the adaptive hybrid evolutionary algorithm

Instance set	Av. Dev. (%)	Opt. (%)
j30_5	0	100
j30_10	0	100
j60_5	0	100
j60_10	0	100
j120_5	0	100
j120_10	0.01	100

5.1 Comparison with Competing Heuristic Algorithms

To the best of the author's knowledge, four heuristic search and optimization algorithms have been presented so far in the literature to solve the addressed problem: a traditional evolutionary algorithm [5], a traditional memetic algorithm [7] which incorporates a hill-climbing algorithm into the framework of an evolutionary algorithm, a hybrid evolutionary algorithm [8] which integrates an adaptive simulated annealing algorithm into the framework of an evolutionary algorithm, and a hybrid evolutionary algorithm [9–11] which utilizes semi-adaptive crossover and mutation processes as well as an adaptive simulated annealing algorithm.

The four algorithms above-mentioned utilize non-adaptive parent selection and survival selection processes to develop the evolutionary search. Besides, the first three of these algorithms use non-adaptive crossover and mutation processes. The fourth of these algorithms uses semi-adaptive crossover and mutation processes; however, the adaptability of the behavior of these processes during the evolutionary search is very limited.

In [7–9], the four algorithms above-mentioned have been evaluated on the six instance sets presented in Table 1. The results obtained by each of the four algorithms for these six instance sets are detailed in Table 4, as were reported in [7–9].

Table 4. Results obtained by the heuristic algorithms reported in the literature for the addressed problem

Instance set	Evolutionary algorithm [5]		Memetic algorithm [7]		Hybrid evolutionary algorithm [8]		Hybrid evolutionary algorithm [9–11]	
	Av. Dev. (%)	Opt. (%)	Av. Dev. (%)	Opt. (%)	Av. Dev. (%)	Opt. (%)	Av. Dev. (%)	Opt. (%)
j30_5	0	100	0	100	0	100	0	100
j30_10	0	100	0	100	0	100	0	100
j60_5	0.42	100	0	100	0	100	0	100
j60_10	0.59	100	0.1	100	0	100	0	100
j120_5	1.1	100	0.75	100	0.64	100	0.1	100
j120_10	1.29	100	0.91	100	0.8	100	0.36	100

According to the results in Table 4, the performance of the algorithm presented in [9–11] is better than those of the algorithms presented in [5, 7, 8]. Thus, the algorithm presented in [9–11] may be considered as the best algorithm presented so far in the literature for solving the addressed problem.

Below, the performance of the algorithm presented in [9–11] is compared with that of the adaptive hybrid evolutionary algorithm proposed here. For simplicity, the algorithm presented in [9–11] will be referred as algorithm HEA.

Comparing the results obtained by the algorithm HEA (as detailed in Table 4) with those obtained by the adaptive hybrid evolutionary algorithm (as detailed in Table 3), the following points may be mentioned. Both algorithms have obtained an optimal effectiveness level for j30_5, j30_10, j60_5 and j60_10 (i.e., the four less complex instance sets). However, the effectiveness level obtained by the adaptive hybrid evolutionary algorithm for j120_5 and j120_10 (i.e., the two more complex instance sets) is significantly higher than that obtained by the algorithm HEA. Therefore, the adaptive hybrid evolutionary algorithm outperforms the algorithm HEA on the more complex instance sets. This is mainly because of the following reasons.

The adaptive hybrid evolutionary algorithm uses adaptive processes (i.e., adaptive parent selection, crossover, mutation, and survival selection processes) to develop the evolutionary search. Such processes adapt their behavior according to the population diversity, to promote either the exploration or exploitation of the search space, and therefore, improve the performance of the evolutionary search. In contrast with the adaptive hybrid evolutionary algorithm, the algorithm HEA utilizes non-adaptive processes (i.e., non-adaptive parent selection and survival selection processes) and semi-adaptive processes (i.e., semi-adaptive crossover and mutation processes) to develop the evolutionary search. The non-adaptive processes disregard the population diversity, and are not able to adapt their behavior during the evolutionary search to improve the performance of the evolutionary search. The semi-adaptive processes consider the population diversity; however, the adaptability of the behavior of these processes during the evolutionary search is very limited. Based on the mentioned above, the adaptive hybrid evolutionary algorithm has significant advantages to develop the evolutionary search.

6 Conclusions and Future Work

In this paper, an adaptive hybrid evolutionary algorithm was proposed to solve the project scheduling problem introduced in [5]. This problem considers a valuable optimization objective for project managers. Such objective involves maximizing the effectiveness of the sets of human resources assigned to the project activities.

The proposed adaptive hybrid evolutionary algorithm uses adaptive processes to develop the different stages of the evolutionary cycle (i.e., adaptive parent selection, survival selection, crossover, mutation and simulated annealing processes). These processes adapt their behavior according to the diversity of the evolutionary algorithm population, to promote either the exploration or exploitation of the search space. The utilization of these adaptive processes is meant to improve the evolutionary search.

The performance of the adaptive hybrid evolutionary algorithm was evaluated on six instance sets with very different complexity levels. After that, the performance of this algorithm on these six instance sets was compared with those of the algorithms previously reported in the literature for solving the addressed problem. Based on the obtained results, it may be stated that the proposed adaptive hybrid evolutionary algorithm significantly outperforms the algorithms previously reported.

In future works, other adaptive processes will be evaluated into the framework of the hybrid evolutionary algorithm. In particular, other adaptive parent selection, adaptive survival selection, adaptive crossover and adaptive mutation processes will be evaluated. Besides, other population diversity measures will be evaluated to adapt the behavior of the adaptive processes during the evolutionary search.

References

1. Heerkens, G.R.: Project Management. McGraw-Hill, New York City (2002)
2. Wysocki, R.K.: Effective Project Management, 3rd edn. Wiley, Hoboken (2003)
3. De Bruecker, P., Van den Bergh, J., Beliën, J., Demeulemeester, E.: Workforce planning incorporating skills: state of the art. Eur. J. Oper. Res. **243**(1), 1–16 (2015)
4. Van den Bergh, J., Beliën, J., De Bruecker, P., Demeulemeester, E., De Boeck, L.: Personnel scheduling: a literature review. Eur. J. Oper. Res. **226**(3), 367–385 (2013)
5. Yannibelli, V., Amandi, A.: A knowledge-based evolutionary assistant to software development project scheduling. Expert Syst. Appl. **38**(7), 8403–8413 (2011)
6. Blazewicz, J., Lenstra, J., Rinnooy, K.A.: Scheduling subject to resource constraints: classification and complexity. Discrete Appl. Math. **5**, 11–24 (1983)

7. Yannibelli, V., Amandi, A.: A memetic approach to project scheduling that maximizes the effectiveness of the human resources assigned to project activities. In: Corchado, E., Snášel, V., Abraham, A., Woźniak, M., Graña, M., Cho, S.-B. (eds.) HAIS 2012. LNCS (LNAI), vol. 7208, pp. 159–173. Springer, Heidelberg (2012). https://doi.org/10.1007/978-3-642-28942-2_15

8. Yannibelli, V., Amandi, A.: A diversity-adaptive hybrid evolutionary algorithm to solve a project scheduling problem. In: Corchado, E., Lozano, José A., Quintián, H., Yin, H. (eds.) IDEAL 2014. LNCS, vol. 8669, pp. 412–423. Springer, Cham (2014). https://doi.org/10.1007/978-3-319-10840-7_50

9. Yannibelli, V., Amandi, A.: Hybrid evolutionary algorithm with adaptive crossover, mutation and simulated annealing processes to project scheduling. In: Jackowski, K., Burduk, R., Walkowiak, K., Woźniak, M., Yin, H. (eds.) IDEAL 2015. LNCS, vol. 9375, pp. 340–351. Springer, Cham (2015). https://doi.org/10.1007/978-3-319-24834-9_40

10. Yannibelli, V., Amandi, A.: Scheduling projects by a hybrid evolutionary algorithm with self-adaptive processes. In: Sidorov, G., Galicia-Haro, Sofía N. (eds.) MICAI 2015. LNCS (LNAI), vol. 9413, pp. 401–412. Springer, Cham (2015). https://doi.org/10.1007/978-3-319-27060-9_33

11. Yannibelli, V., Amandi, A.: Project Scheduling: a memetic algorithm with diversity-adaptive components that optimizes the effectiveness of human resources. Polibits **52**, 93–103 (2015)

12. Eiben, A.E., Smith, J.E.: Introduction to Evolutionary Computing. NCS. Springer, Heidelberg (2015). https://doi.org/10.1007/978-3-662-44874-8

13. Rodriguez, F.J., García-Martínez, C., Lozano, M.: Hybrid metaheuristics based on evolutionary algorithms and simulated annealing: taxonomy, comparison, and synergy test. IEEE Trans. Evol. Comput. **16**(6), 787–800 (2012)

14. Talbi, E. (ed.): Hybrid Metaheuristics. SCI, vol. 434. Springer, Heidelberg (2013)

15. Cheng, J., Zhang, G., Caraffini, F., Neri, F.: Multicriteria adaptive differential evolution for global numerical optimization. Integr. Comput.-Aided Eng. **22**(2), 103–117 (2015)

16. Wang, R., Zhang, Y., Zhang, L.: An adaptive neural network approach for operator functional state prediction using psychophysiological data. Integr. Comput.-Aided Eng. **23**, 81–97 (2016)

17. Zhu, Z., Xiao, J., Li, J.Q., Wang, F., Zhang, Q.: Global path planning of wheeled robots using multi-objective memetic algorithms. Integr. Comput.-Aided Eng. **22**(4), 387–404 (2015)

18. Kolisch, R., Hartmann, S.: Experimental investigation of heuristics for resource-constrained project scheduling: an update. Eur. J. Oper. Res. **174**, 23–37 (2006)

19. Li, H., Womer, N.K.: Solving stochastic resource-constrained project scheduling problems by closed-loop approximate dynamic programming. Eur. J. Oper. Res. **246**(1), 20–33 (2015)

20. Heimerl, C., Kolisch, R.: Scheduling and staffing multiple projects with a multi-skilled workforce. OR Spectrum **32**(4), 343–368 (2010)

21. Li, H., Womer, K.: Scheduling projects with multi-skilled personnel by a hybrid MILP/CP benders decomposition algorithm. J. Sched. **12**, 281–298 (2009)

22. Drezet, L.E., Billaut, J.C.: A project scheduling problem with labour constraints and time-dependent activities requirements. Int. J. Prod. Econ. **112**, 217–225 (2008)

23. Bellenguez, O., Néron, E.: A branch-and-bound method for solving multi-skill project scheduling problem. RAIRO – Oper. Res. **41**(2), 155–170 (2007)

24. Braekers, K., Hartl, R.F., Parragh, S.N., Tricoire, F.: A bi-objective home care scheduling problem: analyzing the trade-off between costs and client inconvenience. Eur. J. Oper. Res. **248**(2), 428–443 (2016)

25. Aickelin, U., Burke, E., Li, J.: An evolutionary squeaky wheel optimization approach to personnel scheduling. IEEE Trans. Evol. Comput. **13**(2), 433–443 (2009)

26. Valls, V., Pérez, A., Quintanilla, S.: Skilled workforce scheduling in service centers. Eur. J. Oper. Res. **193**(3), 791–804 (2009)
27. Bellenguez, O., Néron, E.: Lower bounds for the multi-skill project scheduling problem with hierarchical levels of skills. In: Burke, E., Trick, M. (eds.) PATAT 2004. LNCS, vol. 3616, pp. 229–243. Springer, Heidelberg (2005). https://doi.org/10.1007/11593577_14
28. Silva, T., De Souza, M., Saldanha, R., Burke, E.: Surgical scheduling with simultaneous employment of specialised human resources. Eur. J. Oper. Res. **245**(3), 719–730 (2015)
29. Gutjahr, W.J., Katzensteiner, S., Reiter, P., Stummer, Ch., Denk, M.: Competence-driven project portfolio selection, scheduling and staff assignment. CEJOR **16**(3), 281–306 (2008)
30. Hanne, T., Nickel, S.: A multiobjective evolutionary algorithm for scheduling and inspection planning in software development projects. Eur. J. Oper. Res. **167**, 663–678 (2005)
31. Srinivas, M., Patnaik, L.M.: Adaptive probabilities of crossover and mutation in genetic algorithms. IEEE Trans. Syst. Man Cybern. **24**(4), 656–667 (1994)
32. Mahfoud, S.W.: Crowding and preselection revised. Parallel Problem Solving from Nature **2**, 27–36 (1992)

Universal Swarm Optimizer
for Multi-objective Functions

Luis A. Márquez-Vega and Luis M. Torres-Treviño[✉]

Facultad de Ingeniería Mecánica y Eléctrica, Universidad Autónoma de Nuevo León,
Pedro de Alba s/n, 66455 San Nicolás de los Garza, Nuevo León, Mexico
luis.marquezvg@gmail.com, luis.torres.ciidit@gmail.com

Abstract. This paper presents the Universal Swarm Optimizer for
Multi-Objective Functions (USO), which is inspired in the zone-based
model proposed by Couzin that represents in a more realistic way the
behavior of biological species as fish schools and bird flocks. The algorithm is validated using 10 multi-objective benchmark problems and
a comparison with the Multi-Objective Particle Swarm Optimization
(MOPSO) is presented. The obtained results suggest that the proposed
algorithm is very competitive and presents interesting characteristics
which could be used to solve a wide range of optimization problems.

Keywords: Multi-objective optimization · Zone-based model
Swarm intelligence

1 Introduction

One of the objectives in engineering is to solve real-life problems, these problems
have the particularity of being multi-objective, this means that a solution for
the optimization of two or more objectives at the same time must be found, the
difficulty of this type of problems lies in the fact that the optimization of the
objectives is not closed to a single solution, as with single-objective problems, in
this case may be a set of solutions that optimize the objectives. Currently there
are many algorithms developed in order to give solution of this type of problems [5], among which the best known are Strenght-Pareto Evolutionary Algorithm (SPEA), Pareto Archieved Evolution Strategy (PAES), Pareto-frontier
Differential Evolution (PDE), Non-dominated Sorting Genetic Algorithm version
2 (NSGA-II), Multi-Objective Particle Swarm Optimization (MOPSO), Multi-
Objective Evolutionary Algorithm based on Decomposition (MOEA/D).

In the literature is observed that all of the algorithms previously mentioned
adequately optimize multi-objective problems, but the theorem called No Free
Lunch proposed by Wolpert and Macready [6] in 1997 proves that there is no
optimization algorithm which give solution to all problems efficiently. Starting
from this point is that the algorithm presented in this paper is proposed.

This algorithm is inspired in a model proposed by Couzin [2] which simulate
the behavior of individuals based on three main characteristics of biological

© Springer Nature Switzerland AG 2018
I. Batyrshin et al. (Eds.): MICAI 2018, LNAI 11288, pp. 50–61, 2018.
https://doi.org/10.1007/978-3-030-04491-6_4

species; repulsion, alignment and attractive tendencies towards other individuals based upon the position and orientation of individuals relative to one another, with this model is possible to configure the formation and the behavior of a swarm in order to solve a wide range of optimization problems.

The main idea of the proposed algorithm is to represent any swarm behavior in order to find, in a same algorithm, very different ways of giving an efficient solution to any optimization problem that may arise.

2 Methodology

In this section the basic concepts of Multi-Objective Optimization, the Multi-Objective Particle Swarm Optimization (MOPSO), the Universal Swarm Optimization for Multi-Objective functions (USO) and performance metrics are described.

2.1 Basic Concepts

As mentioned before, the optimal solution of a Multi-Objective problem is not closed to a single value, in this type of problems, the solutions can not be compared by relational operators. In this case, to decide if a solution is better than another is necessary determine if one solution dominates the other, this means that all the objective values $F(x)$ are better or equal than the objective values of other solution $F(y)$, this is called Pareto dominance. The following definitions mentioned in [5] are formulated as a minimization problem:

Definition 1. *Pareto Dominance:*
Suppose that $x, y \in \mathbb{R}^n$, is said that $x \succ y$ (x dominates y) iff:

$$\exists x, y \in \mathbb{R}^n \mid F(x) \leq F(y) \tag{1}$$

The definition of Pareto optimality is as follows:

Definition 2. *Pareto Optimality:*
A solution x_i^* is called a non-dominated solution iff:

$$\exists x_i^* \mid x_i^* \succ x_j \ \forall x_i, x_j \in \mathbb{R}^n \ i \neq j \tag{2}$$

A set including all the non-dominated solutions of a problem is called Solution Space (SS) and it is defined as follows:

Definition 3. *Solution space:*
The set of all Pareto-optimal solutions is called Solution Space as follows:

$$SS = \{x_1^*, x_2^*, ..., x_D^*\} \tag{3}$$

The Pareto front P_f is a set which contains the corresponding objectives values of Solution Space. The definition of the Pareto front is as follows:

Definition 4. *Pareto front:*
The set which contains the objective values of the solution space:

$$P_f := \{F(x_i^*) \mid x_i^* \in SS\} \tag{4}$$

2.2   Multiple-Objective Particle Swarm Optimization (MOPSO)

The Multi-Objective Particle Swarm Optimization (MOPSO) was presented by Coello as an extension of the Particle Swarm Optimization (PSO) in order to solve Multi-Objective problems. The algorithm presented in [1] is briefly explained as follows:

Main Algorithm. The algorithm of MOPSO is the following:

The position $X_i(t)$ and velocity $V_i(t)$ of each particle is initialized and evaluated, the non-dominated particles are saved in the repository Rep. Then the particles are located in the search space by generating hypercubes and the actual position is saved as the best position of the particle in Ib_i.

In every iteration the following must be done:

The speed of the particle is computed using the following expression:

$$V_i(t+1) = w * V_i(t) + \beta_1 * u() * (Ib_i - X_i(t)) + \beta_2 * u() * (Rep(h) - X_i(t)) \quad (5)$$

where w is the inertia weight, β_1 is the personal learning coefficient, β_2 is the global learning coefficient, $u()$ is a random number between $[0, 1]$ and $Rep(h)$ is the position of the leader taken from the repository, the index h is a randomly particle from a hypercube which was selected by Roulette Wheel Selection, hypercubes with more particles has a lower fitness value.

The new position of the particle is obtained by:

$$X_i(t+1) = X_i(t) + V_i(t+1) \quad (6)$$

Every new solution is evaluated in order to update the repository Rep and the best position of each particle.

External Repository. The external repository (or archive) has the objective of store the non-dominated particles along the iterations. The archive controller has the function of decide which particles maintain in the archive, maintaining only non-dominated particles. The adaptive grid is used to well distribute the non-dominated particles in the search space and is useful to decide which particle delete when the archive is full.

Use of a Mutation Operator. A mutation operator is necessary in the MOPSO due to the very high convergence speed of the PSO, if the algorithm has very high convergence speed may converge to a false Pareto front.

2.3   Universal Swarm Optimizer for Multi-objective Functions (USO)

As said before, the Universal Swarm Optimizer for Multi-Objective functions (USO) is inspired in the zone-based model, which is very useful because provide control on the formation and behavior of swarms. The mathematical formulation of the model is explained in [4] and is as follows:

Zone-Based Model. The mathematical formulation of the model is based on give to each particle the notion of repulsion, orientation and attraction towards other particles in the swarm.

The swarm is composed by N particles with positions $X_i(t)$ and unit directions $\hat{v}_i(t)$. Each particle travel through the search space at constant speed, and every iteration the particles determine new desired direction of travel by detecting particles within two zones. The first zone is dominated "zone of repulsion", is represented by a circle of radius ρ_r, in the center of this circle lies the particle i. The particle repel away from others within their zone of repulsion. The second zone is dominated "zone of orientation and attraction", is represented by an annulus of inner radius ρ_r and outer radius ρ_p, this zone has a blind area behind the particle, for which others particles are undetectable. The particles align with and are attracted towards particles within their zone of orientation and attraction.

For a given particle i, the set of particles in the zone of repulsion is represented by Z_i^R, and the set of particles in the zone of orientation and attraction is represented by Z_i^P.

If a particle i finds others particles within its zone of repulsion, its desired direction of travel is given by the following equation:

$$v_i(t+1) = -\sum_{j \in Z_i^R} \frac{X_j(t) - X_i(t)}{|X_j(t) - X_i(t)|} \tag{7}$$

If in the zone of repulsion the particle i does not find other particles, then it aligns with and feels an attraction towards others particles that are in its zone of orientation and attraction. In this case, its desired direction of travel is given by the following equation:

$$v_i(t+1) = \omega_a \frac{a_i(t)}{|a_i(t)|} + \omega_o \frac{o_i(t)}{|o_i(t)|} \tag{8}$$

where ω_a is the weight of attraction and ω_o is the weight of orientation, $a_i(t)$ and $o_i(t)$ are given in the Eqs. (9) and (10) respectively:

$$a_i(t) = \sum_{j \in Z_i^P} \frac{X_j(t) - X_i(t)}{|X_j(t) - X_i(t)|} \tag{9}$$

$$o_i(t) = \sum_{j \in Z_i^P} \hat{v}_j(t) \tag{10}$$

The desired direction of travel of a particle i must be normalized as the following equation:

$$\hat{v}_i(t+1) = \frac{v_i(t+1)}{|v_i(t+1)|} \tag{11}$$

If $v_i(t+1) = 0$, then the particle i maintains its previous direction of travel as its desired direction of travel.

Finally, each particle's position is updated as the following expression:

$$X_i(t+1) = X_i(t) + s * \hat{v}_i(t+1) \tag{12}$$

where s is a constant value of speed of travel.

With the previous formulations is clearly that, the formation and the behavior of the swarm can be easily manipulated by changing the radius of the zone of repulsion ρ_r and the outer radius of the zone of orientation and attraction ρ_p, as well, manipulating the weights of attraction w_a and orientation w_o different behaviors of the swarm are obtained, because the behavior of each particle is modified.

Main Algorithm. The algorithm of USO uses the previous mathematical formulations and also uses the archive controller and the adaptive grid proposed by Carlos C. Coello [1] in order to solve multi-objective problems. The main algorithm of USO is as follows:

Initialize the position $X_i(t)$ and direction $v_i(t)$ of each particle;
Evaluate $X(t)$ in order to obtain its objective values;
Add non-dominated particles in the repository Rep;
Use objective values of the particles as coordinates in the search space;
while $cycle \leq TotalCycles$ **do**
    for $i = 1, TotalParticles$ **do**
        if *Particle finds other particles in its zone of repulsion* **then**
          | Calculate desired direction as the equation (7);
        else if *Particle finds other particles in its zone of attraction* **then**
          | Calculate desired direction as the equation (8);
        else
          | Actual direction becomes its desired direction
        end
        The desired direction is normalized as equation (11);
        The new position the particle is obtained using the equation (13);
        Maintain the new position of the particle inside the search space;
        Evaluate the new position of the particle;
    end
    Verify non-dominated particles and update them in Rep;
end

Algorithm 1. Pseudocode of USO

$$X_i(t+1) = X_i(t) + s * \hat{v}_i(t+1) + \beta_2 * u() * (Rep(h) - X_i(t)) \tag{13}$$

where s is the constant speed of travel, β_2 is the global learning coefficient, $u()$ is a random number between $[0,1]$ and $Rep(h)$ is the position of the leader taken from the repository, the index h is selected as proposed in MOPSO.

Note that, another term was added in the Eq. (13) with respect to the Eq. (12), this is done with the purpose of giving the ability to the swarm to follow a leader and find better solutions in the search space.

2.4 Performance Metrics

For measure the performance of the algorithms, the following metrics has been selected:

In order to measure the convergence the Inverted Generational Distance (IGD) was selected and for measure the coverage the Spacing (SP) and Maximum Spread (MS) were selected.

The expression of IGD is in [5] as follows:

$$IGD = \frac{\sqrt{\sum_{i=1}^{n} d_i^2}}{n} \tag{14}$$

where n is the number of true Pareto optimal solutions and d_i indicates the Euclidean distance between the $i-th$ true Pareto optimal solution and its closest non-dominated solution obtained. The lower values of the IGD are preferable.

The corresponding expressions of SP and MS are in [3] as the Eqs. (15) and (17) respectively:

$$SP = \sqrt{\frac{1}{n-1} \sum_{i=1}^{n} (\bar{d} - d_i)^2} \tag{15}$$

$$d_i = \min_{j \in NDS \wedge j \neq i} \sum_{k=1}^{K} |f_k^i - f_k^j| \tag{16}$$

where d_i is shown in the Eq. (16), NDS is the set of non-dominated solutions obtained, $\bar{d}$ is the mean of all d_i, n is the size of the NDS and f_k^i is the function value of the $k-th$ objective function for solution i. The lower values of the SP are preferable.

$$MS = \sqrt{\frac{1}{K} \sum_{k=1}^{K} \left[\frac{\max_{i \in NDS} f_k^i - \min_{i \in NDS} f_k^i}{F_k^{max} - F_k^{min}} \right]^2} \tag{17}$$

where NDS is the set of non-dominated solutions obtained, K is the number of objectives, f_k^i is the function value of the $k-th$ objective function for solution i, F_k^{max} and F_k^{min} are the maximum and minimum value of the $k-th$ objective in the true Pareto front respectively. The values of the MS closer to 1 are preferable.

The best run was taken for the qualitative results and the computational time are also obtained in each unconstrained problem.

3 Experimental Setup

For the experiments is used as a benchmark 10 unconstrained problems proposed in CEC 2009 [7], because as said in [5] these test problems are considered as the most challenging test problems in the literature. Table 1 shows the mathematical formulation of each unconstrained problem:

The following considerations were taken in the test of each unconstrained problem:

Table 1. Mathematical formulation of test problems

Name	Mathematical formulation
$UP1$	$f_1 = x_1 + \frac{2}{|J_1|}\sum_{j \in J_1}[x_j - sin(6\pi x_1 + \frac{j\pi}{n})]^2$
	$f_2 = 1 - \sqrt{x_1} + \frac{2}{|J_2|}\sum_{j \in J_2}[x_j - sin(6\pi x_1 + \frac{j\pi}{n})]^2$
	$J_1 = \{j \mid j \ is \ odd \ and \ 2 \leq j \leq n\}$ and $J_2 = \{j \mid j \ is \ even \ and \ 2 \leq j \leq n\}$
$UP2$	$f_1 = x_1 + \frac{2}{|J_1|}\sum_{j \in J_1} y_j^2$
	$f_2 = 1 - \sqrt{x_1} + \frac{2}{|J_2|}\sum_{j \in J_2} z_j^2$
	$J_1 = \{j \mid j \ is \ odd \ and \ 2 \leq j \leq n\}$ and $J_2 = \{j \mid j \ is \ even \ and \ 2 \leq j \leq n\}$
	$y_j = x_j - [0.3x_1^2 cos(24\pi x_1 + \frac{4j\pi}{n}) + 0.6x_1]cos(6\pi x_1 + \frac{j\pi}{n})$
	$z_j = x_j - [0.3x_1^2 cos(24\pi x_1 + \frac{4j\pi}{n}) + 0.6x_1]sin(6\pi x_1 + \frac{j\pi}{n})$
$UP3$	$f_1 = x_1 + \frac{2}{|J_1|}(4\sum_{j \in J_1} y_j^2 - 2\prod_{j \in J_1} cos(\frac{20y_j\pi}{\sqrt{j}}) + 2)$
	$f_2 = 1 - \sqrt{x_1} + \frac{2}{|J_2|}(4\sum_{j \in J_2} y_j^2 - 2\prod_{j \in J_2} cos(\frac{20y_j\pi}{\sqrt{j}}) + 2)$
	$J_1 = \{j \mid j \ is \ odd \ and \ 2 \leq j \leq n\}$ and $J_2 = \{j \mid j \ is \ even \ and \ 2 \leq j \leq n\}$
	$y_j = x_j - x_1^{0.5(1+\frac{3(j-2)}{n-2})}$
$UP4$	$f_1 = x_1 + \frac{2}{|J_1|}\sum_{j \in J_1} h(y_j)$
	$f_2 = 1 - x_1^2 + \frac{2}{|J_2|}\sum_{j \in J_2} h(y_j)$
	$J_1 = \{j \mid j \ is \ odd \ and \ 2 \leq j \leq n\}$ and $J_2 = \{j \mid j \ is \ even \ and \ 2 \leq j \leq n\}$
	$y_j = x_j - sin(6\pi x_1 + \frac{j\pi}{n})$
	$h(t) = \frac{|t|}{1+e^{2|t|}}$
$UP5$	$f_1 = x_1 + (\frac{1}{2N} + \epsilon)\mid sin(2N\pi x_1)\mid + \frac{2}{|J_1|}\sum_{j \in J_1} h(y_j)$
	$f_2 = 1 - x_1 + (\frac{1}{2N} + \epsilon)\mid sin(2N\pi x_1)\mid + \frac{2}{|J_2|}\sum_{j \in J_2} h(y_j)$
	$J_1 = \{j \mid j \ is \ odd \ and \ 2 \leq j \leq n\}$ and $J_2 = \{j \mid j \ is \ even \ and \ 2 \leq j \leq n\}$, N is an integer, $\epsilon > 0$
	$y_j = x_j - sin(6\pi x_1 + \frac{j\pi}{n})$
	$h(t) = 2t^2 - cos(4\pi t) + 1$
$UP6$	$f_1 = x_1 + max\{0, 2(\frac{1}{2N} + \epsilon)\} + \frac{2}{|J_1|}(4\sum_{j \in J_1} y_j^2 - 2\prod_{j \in J_1} cos(\frac{20y_j\pi}{\sqrt{j}}) + 2)$
	$f_2 = 1 - x_1 + max\{0, 2(\frac{1}{2N} + \epsilon)\} + \frac{2}{|J_2|}(4\sum_{j \in J_2} y_j^2 - 2\prod_{j \in J_2} cos(\frac{20y_j\pi}{\sqrt{j}}) + 2)$
	$J_1 = \{j \mid j \ is \ odd \ and \ 2 \leq j \leq n\}$, N is an integer, $\epsilon > 0$ and $J_2 = \{j \mid j \ is \ even \ and \ 2 \leq j \leq n\}$
	$y_j = x_j - sin(6\pi x_1 + \frac{j\pi}{n})$
$UP7$	$f_1 = \sqrt[5]{x_1} + \frac{2}{|J_1|}\sum_{j \in J_1} y_j^2$
	$f_2 = 1 - \sqrt[5]{x_1} + \frac{2}{|J_2|}\sum_{j \in J_2} y_j^2$
	$J_1 = \{j \mid j \ is \ odd \ and \ 2 \leq j \leq n\}$ and $J_2 = \{j \mid j \ is \ even \ and \ 2 \leq j \leq n\}$
	$y_j = x_j - sin(6\pi x_1 + \frac{j\pi}{n})$
$UP8$	$f_1 = cos(0.5x_1\pi)cos(0.5x_2\pi) + \frac{2}{|J_1|}\sum_{j \in J_1}(x_j - 2x_2 sin(2\pi x_1 + \frac{j\pi}{n}))^2$
	$f_2 = cos(0.5x_1\pi)sin(0.5x_2\pi) + \frac{2}{|J_2|}\sum_{j \in J_2}(x_j - 2x_2 sin(2\pi x_1 + \frac{j\pi}{n}))^2$
	$f_3 = sin(0.5x_1\pi) + \frac{2}{|J_3|}\sum_{j \in J_3}(x_j - 2x_2 sin(2\pi x_1 + \frac{j\pi}{n}))^2$
	$J_1 = \{j \mid 3 \leq j \leq n, \ and \ j - 1 \ is \ a \ multiplication \ of \ 3\}$
	$J_2 = \{j \mid 3 \leq j \leq n, \ and \ j - 2 \ is \ a \ multiplication \ of \ 3\}$
	$J_3 = \{j \mid 3 \leq j \leq n, \ and \ j \ is \ a \ multiplication \ of \ 3\}$
$UP9$	$f_1 = 0.5[max\{0, (1+\epsilon)(1 - 4(2x_1 - 1)^2)\} + 2x_1]x_2 + \frac{2}{|J_1|}\sum_{j \in J_1}(x_j - 2x_2 sin(2\pi x_1 + \frac{j\pi}{n}))^2$
	$f_2 = 0.5[max\{0, (1+\epsilon)(1 - 4(2x_1 - 1)^2)\} - 2x_1 + 2]x_2 + \frac{2}{|J_2|}\sum_{j \in J_2}(x_j - 2x_2 sin(2\pi x_1 + \frac{j\pi}{n}))^2$
	$f_3 = 1 - x_2 + \frac{2}{|J_3|}\sum_{j \in J_3}(x_j - 2x_2 sin(2\pi x_1 + \frac{j\pi}{n}))^2$
	$J_1 = \{j \mid 3 \leq j \leq n, \ and \ j - 1 \ is \ a \ multiplication \ of \ 3\}$
	$J_2 = \{j \mid 3 \leq j \leq n, \ and \ j - 2 \ is \ a \ multiplication \ of \ 3\}$
	$J_3 = \{j \mid 3 \leq j \leq n, \ and \ j \ is \ a \ multiplication \ of \ 3\}$, $\epsilon > 0$
$UP10$	$f_1 = cos(0.5x_1\pi)cos(0.5x_2\pi) + \frac{2}{|J_1|}\sum_{j \in J_1}[4y_j^2 - cos(8\pi y_j) + 1]$
	$f_2 = cos(0.5x_1\pi)sin(0.5x_2\pi) + \frac{2}{|J_2|}\sum_{j \in J_2}[4y_j^2 - cos(8\pi y_j) + 1]$
	$f_3 = sin(0.5x_1\pi) + \frac{2}{|J_3|}\sum_{j \in J_3}[4y_j^2 - cos(8\pi y_j) + 1]$
	$J_1 = \{j \mid 3 \leq j \leq n, \ and \ j - 1 \ is \ a \ multiplication \ of \ 3\}$
	$J_2 = \{j \mid 3 \leq j \leq n, \ and \ j - 2 \ is \ a \ multiplication \ of \ 3\}$
	$J_3 = \{j \mid 3 \leq j \leq n, \ and \ j \ is \ a \ multiplication \ of \ 3\}$
	$y_j = x_j - 2x_2 sin(2\pi x_1 + \frac{j\pi}{n})$

- Total of particles: 100.
- Dimension of every particle: 10.
- Size of the repository: 100.
- Total of iterations: 250.
- Total of runs: 10.

The following parameters for MOPSO are chosen:

- $w = 0.5$: Inertial weight.
- $\beta_1 = 2$: Personal learning coefficient.
- $\beta_2 = 2$: Global learning coefficient.
- $nGrid = 10$: Number of grids per dimension.
- $\alpha = 0.1$: Inflation rate.
- $\beta = 2$: Leader selection pressure.
- $\gamma = 2$: Deletion selection pressure.
- $\mu = 0.1$: Mutation rate.

The following parameters for USO are chosen:

- $s = 0.5$: Speed.
- $\beta_2 = 2$: Global learning coefficient.
- $nGrid = 10$: Number of grids per dimension.
- $\alpha = 0.1$: Inflation rate.
- $\beta = 2$: Leader selection pressure.
- $\gamma = 2$: Deletion selection pressure.

Despite the previous parameters, the values for the radius of the zone of repulsion ρ_r, the outer radius of the zone of orientation and attraction ρ_p and the weights of attraction ω_a and orientation ω_o were separated in three different sets, in order to observe which formation of the swarm produces better performance in each unconstrained problem. Table 2 shows the different sets of these parameters that were proposed.

Table 2. Sets of parameters ρ_r, ρ_p, ω_a, ω_o

Parameter	Balanced swarm	Joined swarm	Dispersed swarm
ρ_r	0.2125	0.01	0.3000
ρ_p	0.3250	0.75	0.3005
ω_a	0.4750	0.85	0.5000
ω_o	0.5250	0.15	0.5000

These sets of parameters produces different behaviors on the swarms and report better results in different unconstrained problems, as follows:

- Balanced swarm: UP1, UP2 and UP3.
- Joined swarm: UP4, UP7, UP8, UP9 and UP10.
- Dispersed swarm: UP5 and UP6.

For example, in the joined swarm the size of the zone of repulsion is very small and the size of the zone of orientation and attraction is very big, and the weight of attraction has more relevance than the weight of orientation. In the dispersed swarm the zone of repulsion is much bigger than the zone of orientation and attraction, and the weights of attraction and orientation are the same.

4   Results

The qualitative results of the experiments are presented in the Fig. 1 and in the Tables 3 and 4 the statistical results of the performance metrics are presented.

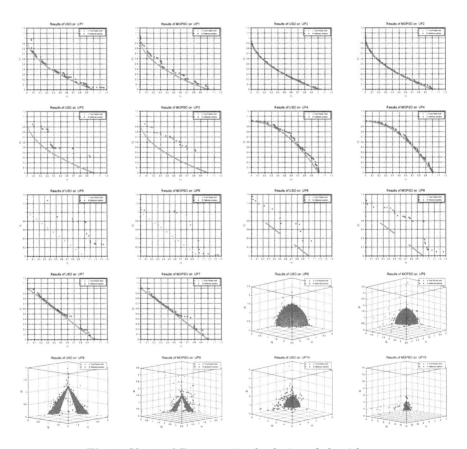

Fig. 1. Obtained Pareto optimal solution of algorithms.

Table 3. Results of performance metrics on UP1 to UP5

Problem	Unconstrained Problem 1							
Algorithm	USO				MOPSO			
Metric	*IGD*	*SP*	*MS*	*Time*	*IGD*	*SP*	*MS*	*Time*
Average	0.004952	0.014620	1.253199	0.640803	0.006120	0.029873	1.263222	0.230937
Median	0.004396	0.013403	1.170341	0.640685	0.006092	0.023860	1.155146	0.236246
Std.Dev.	0.001241	0.003425	0.221600	0.008723	0.000954	0.020206	0.237762	0.011448
Worst	0.006713	0.023612	1.787889	0.657886	0.007707	0.085604	1.650138	0.248765
Best	0.003595	0.011307	1.021803	0.621789	0.004750	0.012483	1.023743	0.212580
Problem	Unconstrained Problem 2							
Algorithm	USO				MOPSO			
Metric	*IGD*	*SP*	*MS*	*Time*	*IGD*	*SP*	*MS*	*Time*
Average	0.001581	0.012799	1.218252	0.668457	0.002287	0.032399	1.442005	0.224356
Median	0.001568	0.011556	1.129406	0.665624	0.002249	0.017188	1.374469	0.222012
Std.Dev.	0.000227	0.003084	0.229650	0.014282	0.000455	0.037100	0.261741	0.011936
Worst	0.002042	0.020730	1.754236	0.698916	0.002942	0.131925	2.049122	0.253709
Best	0.001229	0.010049	1.015813	0.648616	0.001587	0.010036	1.044396	0.210437
Problem	Unconstrained Problem 3							
Algorithm	USO				MOPSO			
Metric	*IGD*	*SP*	*MS*	*Time*	*IGD*	*SP*	*MS*	*Time*
Average	0.019740	0.040933	1.000000	0.613861	0.024021	0.031988	1.000000	0.223004
Median	0.019880	0.041765	1.000000	0.613861	0.023462	0.031158	1.000000	0.220272
Std.Dev.	0.001437	0.009838	0.000000	0.016121	0.001755	0.006923	0.000000	0.015317
Worst	0.021677	0.056233	1.000000	0.651872	0.028019	0.043922	1.000000	0.247346
Best	0.016658	0.022673	1.000000	0.591262	0.022237	0.023451	1.000000	0.194611
Problem	Unconstrained Problem 4							
Algorithm	USO				MOPSO			
Metric	*IGD*	*SP*	*MS*	*Time*	*IGD*	*SP*	*MS*	*Time*
Average	0.003438	0.010077	1.021429	0.841923	0.002668	0.010253	1.012140	0.263242
Median	0.003439	0.009768	1.020297	0.838589	0.002550	0.010411	1.013121	0.264408
Std.Dev.	0.000326	0.001078	0.011147	0.037250	0.000344	0.000938	0.006661	0.005156
Worst	0.004120	0.011830	1.047204	0.905150	0.003330	0.011492	1.024348	0.270564
Best	0.002993	0.008981	1.008772	0.789909	0.002269	0.008190	1.000000	0.250630
Problem	Unconstrained Problem 5							
Algorithm	USO				MOPSO			
Metric	*IGD*	*SP*	*MS*	*Time*	*IGD*	*SP*	*MS*	*Time*
Average	0.080650	0.076583	1.559152	0.610654	0.068987	0.079824	1.513806	0.197619
Median	0.082727	0.067430	1.521536	0.606329	0.065079	0.058064	1.481329	0.196785
Std.Dev.	0.016279	0.039176	0.272321	0.035243	0.015831	0.044339	0.162661	0.001683
Worst	0.107444	0.183906	2.048691	0.677859	0.114663	0.162488	1.960331	0.201184
Best	0.053259	0.040073	0.992411	0.568586	0.056124	0.036612	1.301013	0.196072

Table 4. Results of performance metrics on UP6 to UP10

Problem	Unconstrained Problem 6							
Algorithm	USO				MOPSO			
Metric	IGD	SP	MS	Time	IGD	SP	MS	Time
Average	0.043753	0.084577	1.397498	0.566111	0.042290	0.105398	1.684787	0.203235
Median	0.044915	0.062612	1.310586	0.565475	0.041177	0.076147	1.547352	0.203175
Std.Dev.	0.006680	0.072269	0.679772	0.022192	0.003311	0.083653	0.395895	0.001229
Worst	0.054447	0.290043	2.968650	0.600620	0.047360	0.323690	2.720809	0.205059
Best	0.031073	0.030365	1.018302	0.529400	0.037744	0.026471	1.262228	0.201032
Problem	Unconstrained Problem 7							
Algorithm	USO				MOPSO			
Metric	IGD	SP	MS	Time	IGD	SP	MS	Time
Average	0.003922	0.015096	1.184840	0.850778	0.004228	0.015810	1.130541	0.215190
Median	0.003781	0.014109	1.129395	0.834725	0.004038	0.013220	1.082481	0.215838
Std.Dev.	0.001045	0.004521	0.127901	0.045985	0.000786	0.007123	0.133488	0.004867
Worst	0.005895	0.023088	1.408987	0.924439	0.005553	0.034697	1.491917	0.223428
Best	0.002624	0.009003	1.052942	0.793543	0.002934	0.009834	1.013578	0.205314
Problem	Unconstrained Problem 8							
Algorithm	USO				MOPSO			
Metric	IGD	SP	MS	Time	IGD	SP	MS	Time
Average	0.001376	0.643688	2.857913	0.816707	0.001403	0.813903	3.730212	0.318885
Median	0.001327	0.653234	2.796426	0.821308	0.001355	0.761523	3.565615	0.319814
Std.Dev.	0.000213	0.272836	0.645440	0.016662	0.000170	0.253144	0.756444	0.009988
Worst	0.001719	1.276249	4.322127	0.831507	0.001648	1.328354	5.043005	0.336696
Best	0.001077	0.311949	1.889327	0.768573	0.001176	0.474463	2.897090	0.295071
Problem	Unconstrained Problem 9							
Algorithm	USO				MOPSO			
Metric	IGD	SP	MS	Time	IGD	SP	MS	Time
Average	0.000338	0.780253	3.327733	0.793085	0.000383	0.882015	3.976320	0.376058
Median	0.000356	0.910155	3.339535	0.792619	0.000360	0.838875	3.852437	0.385815
Std.Dev.	0.000098	0.306858	0.834101	0.005084	0.000157	0.416942	0.807900	0.015669
Worst	0.000499	1.247666	4.321281	0.801297	0.000620	1.570546	5.411682	0.390519
Best	0.000170	0.342260	1.761628	0.782678	0.000058	0.291160	2.702781	0.343512
Problem	Unconstrained Problem 10							
Algorithm	USO				MOPSO			
Metric	IGD	SP	MS	Time	IGD	SP	MS	Time
Average	0.005777	1.031892	7.479853	0.786711	0.005030	2.342577	11.66101	0.365809
Median	0.005794	1.007477	6.543000	0.785619	0.004894	2.295207	11.47160	0.372243
Std.Dev.	0.000825	0.454187	3.280149	0.006535	0.000806	1.269593	3.129219	0.033635
Worst	0.007343	2.102620	14.87943	0.796754	0.006526	5.035078	16.44938	0.404667
Best	0.004236	0.440281	3.878773	0.777164	0.003565	0.475477	7.460500	0.312650

5   Conclusion

As can be seen in the qualitative and statistical results, the proposed algorithm represents very good competence against the MOPSO algorithm, in the most problems the USO algorithm produces in general better statistical results than the MOPSO, only in UP4 and UP5, the proposed algorithm presents worse results in the average of IGD and MS at the same time, despite this USO presents the best found result of IGD and MS in UP5. One of the biggest challenges of this algorithm is the computational time required, which is much bigger than the computational time of MOPSO and this reduces relevance to the good results obtained in the others statistical performance metrics.

The strongest feature of this algorithm is the easy design of swarms with desired qualities, others characteristics must be studied more deeply, for example the optimal number of particles in the swarm because bigger swarms produce crowded places in which the vision of the particle is reduced.

More studies about this algorithm must be done in order to find more characteristics which help to improve the performance of the algorithm and reduce the computational time required. Experiments with more dimension per individual and more comparisons with others multi-objective algorithms must be done in order to have a wider view of the real scope of this algorithm.

References

1. Coello, C.A.C., Pulido, G.T., Lechuga, M.S.: Handling multiple objectives with particle swarm optimization. IEEE Trans. Evol. Comput. **8**(3), 256–279 (2004). https://doi.org/10.1109/TEVC.2004.826067
2. Couzin, I., Krause, J., James, R., Ruxton, G., Franks, N.: Collective memory and spatial sorting in animal groups. J. Theor. Biol. **218**(1), 1–11 (2002)
3. Samaei, F., Bashiri, M., Tavakkoli-Moghaddam, R.: A comparison of four multi-objective meta-heuristics for a capacitated location-routing problem. J. Ind Syst. Eng. **6**, 20–33 (2012)
4. Kolpas, A., Busch, M., Li, H., Couzin, I.D., Petzold, L., Moehlis, J.: How the spatial position of individuals affects their influence on swarms: a numerical comparison of two popular swarm dynamics models. PLoS ONE **8** (2013)
5. Mirjalili, S., Saremi, S., Mirjalili, S.M., dos Santos Coelho, L.: Multi-objective grey wolf optimizer: a novel algorithm for multi-criterion optimization. Expert Syst. Appl. **47**, 106–119 (2016)
6. Wolpert, D.H., Macready, W.G.: No free lunch theorems for optimization. Trans. Evol. Comput. **1**(1), 67–82 (1997). https://doi.org/10.1109/4235.585893
7. Zhang, Q., Zhou, A., Zhao, S., Suganthan, P.N., Liu, W., Tiwari, S.: Multiobjective optimization test instances for the CEC 2009 special session and competition (2009)

Broadcasting and Sharing of Parameters in an IoT Network by Means of a Fractal of Hilbert Using Swarm Intelligence

Jaime Moreno$^{(\boxtimes)}$, Oswaldo Morales, Ricardo Tejeida, and Juan Posadas

Instituto Politécnico Nacional, Escuela Superior de Ingeniería Mecánica y Eléctrica, Campus Zacatenco, Mexico city, Mexico
jemoreno@esimez.mx

Abstract. Nowadays, thousand and thousand of small devices, such as Microcontroller Units (MCU's), live around us. These MCU'not only interact with us turning on lights or identifying movement in a House but also they perform small and specific tasks such as sensing different parameters such as temperature, humidity, CO_2, adjustment of the environmental lights. In addition there is a huge kind of these MCU's like SmartPhones or small general purpose devices, ESP8266 or RaspberryPi3 or any kind of Internet of Things (IoT) devices. They are connected to internet to a central node and then they can share their information. The main goal of this article is to connect all the nodes in a fractal way without using a central one, just sharing some parameters with two adjacent nodes, but any member of these nodes knows the parameters of the rest of these devices even if they are not adjacent nodes. With a Hilbert fractal network we can access to the entire network in real time in a dynamic way since we can adapt and reconfigure the topology of the network when a new node is added using tools of Artificial Intelligence for its application in a Smart City.

Keywords: IoT · Adaptive algorithms · Swarm Intelligence
Hilbert fractal · ESP8266 · RaspberryPi3

1 Introduction

In recent years, the interconnection of Internet of Things (IoT) networks have received increasing interest from many academic and engineering fields. A hot topic is to Broadcasting and sharing of parameters in this kind of networks in a efficient way. So far, convenient and economical methods are very necessary. IoT networks has been considered as a new generation of networks that combine environmental sensing or temperature, humidity, CO_2, adjustment of the environmental lights with data transmission and processing through wireless communication techniques or Wireless Sensor Networks (WSN) [1,2], provide a better choice for broadcasting and sharing of parameters. Various sensors are

© Springer Nature Switzerland AG 2018
I. Batyrshin et al. (Eds.): MICAI 2018, LNAI 11288, pp. 62–75, 2018.
https://doi.org/10.1007/978-3-030-04491-6_5

attached in homes or wearable to acquire data containing physical or environmental states of these parameter. After sensor data collection, transmission, sharing processing and analyzing, structural problems such as adding new members in the network and communicate these parameters in a central way as a star configuration can be successfully predicted and adapted, so that the upcoming measurements caused by a conventional request may be answered [3, 4]. In addition, it is important to highlight that by 2020, The Internet of Things will have achieved *critical mass*. Linking enormous intelligence in the cloud to billions od mobile devices and having extremely inexpensive sensors and tag embedded in and on everything, will deliver enormous amount of new value to almost every human being. The full benefits, in terms of health, safety and convenience, will be enormous [4]. The perspective for this year, in this way will be 4 billion of connected people with a revenue of $4 Trillion USD because more than 25 millions of apps will be downloaded from the principal app-stores, thus we need approximately more than 25 billion MCU's which will generate 50 trillion of Gigabytes of Data inside the global IoT network.

Furthermore, The increasing amount of IoT devises in a common or a conventional House, Fig. 1 shows a usual Floor Plan. In a IoT network contains a WiFi Access Point (WAP) that connect all the IoT devises to a Internet or Intranet networking as the main node. If a MCU want to share some parameters with another MCU, these MCU's must communicate with the WAP. Even if they physically are very near each other if they do not have connection with the WAP they are not able to establish communication and sharing parameters.

Which is why in this paper, an effective method for broadcasting and sharing of parameters in WSN with Swarm Intelligence is proposed [5–7]. Here we use a large amount of networked sensors for realtime sensing and monitoring of different parameters in order to make efficient a IoT network and its application in a forthcoming Smart City. The further sections of this paper is organized as follows: General Description of the proposal is discussed in Sect. 2. In Sect. 3, a detailed description of Theoretical Framework is given describing Embedded Systems in the IoT network and Hilbert space-filling Curve. In Sect. 4, the description of the main scheme is given. In Sect. 5, simulation and performance comparison are presented to verify correctness and feasibility of the proposed scheme. Finally, some conclusions are drawn in Sect. 6.

2   General Description of the Proposal

This proposal is based in main work made by Beni and Wang [8] introduced the expression of Swarm Intelligence (SI) in their research of cellular robotic systems in 1993. The concept of SI is employed in work on Artifcial Intelligence (AI). SI is a computational intelligence technique which is based on the collective behavior of decentralized, self-organized systems. A typical SI system is made up of a group of simple agents which interact locally with each other and with the environment surrounding them. MCU's in an SI system follow simple rules and act without the control of any centralized entities. However, the social interactions between such MCU's may generate enormous benefits and often lead to a smart global behavior.

SI takes the full advantage of the swarm, therefore, it's able to provide optimized solutions, which ensure high robustness, flexibility and low cost, for large-scale sophisticated problems without a centralized control entity.

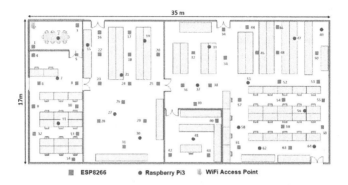

Fig. 1. Floor Plan for exemplifying the proposed algorithm.

Which is why, this proposal is divided in 3 main parts based on SI:

1. Selecting the main or initial MCU,
2. Generation of the topology in a given Floor Plant, and
3. Indexing all the MCU's inside the IoT network.

Henceforth, we use the Floor Plant depicted in Fig. 1, which consists of one WPA and several ESP8266 and RaspberryPi3, namely 64 embedded systems. In addition, in the first part the nearest MCU to the WPA is selected as the main MCU. Then, measuring the bandwidth in Mbps (mega bits per second) we can predict a general topology for a given Floor Plant, when a new MCU is added the algorithm recalculates a new topology. Ones the topology of the network is calculates every member of the IoT network is indexed by the Hilbert Fractal, then we measure the effectiveness of the methodology sharing parameter along the network and then Wi-Fi Peer-to-Peer (P2P) and SoftAP algorithms are configured.

3 Theoretical Framework

3.1 Embedded Systems in the IoT Network

NodeMCU ESP8266 embedded system depicted by Fig. 2(a) is a development card similar to Arduino, especially oriented to IoT. It is based on the System on Chip (SoC) ESP8266, a highly integrated chip, designed for the needs of a connected world. It integrates a powerful processor with 32 bit architecture (more powerful than the Arduino Due) and Wifi connectivity.

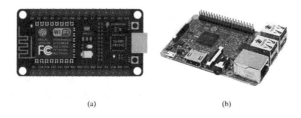

(a)   (b)

Fig. 2. Embedded systems in the IoT network (a) NodeMCU ESP8266 embedded system (MCU-ESP8266) and (b) Raspberry Pi 3 embedded system (MCU-RPi3)

For the development of applications you can choose between the Arduino and LUA languages. When working within the Arduino environment we can use a language we already know and make use of a simple IDE to use, in addition to making use of all the information about projects and libraries available on the Internet. The user community of Arduino is very active and supports platforms such as ESP8266. NodeMCU comes with a pre-installed firmware which allows us to work with the interpreted language LUA, sending commands through the serial port (CP2102). The NodeMCU and Wemos D1 mini cards are the most used platforms in IoT projects. It does not compete with Arduino, because they cover different objectives, it is even possible to program NodeMCU from the Arduino IDE. The NodeMCU card is specially designed to work in breadboard. It has an on-board voltage regulator that allows it to feed directly from the USB port. The input/output pins work at 3.3V. The CP2102 chip handles USB-Serial communication.

NodeMCU has the following technical specifications:

- Power Voltage (USB): 5 V DC
- Input/Output Voltage: 3.3 V DC
- SoC: ESP8266 (Module ESP-12)
- CPU: Tensilica Xtensa LX3 (32 bit)
- Clock Frequency: 80 MHz/160 MHz
- Instruction RAM: 32 KB
- Data RAM: 96 KB
- External Flash Memory: 4 MB
- GPIO Digital Pins: 17 (can be configured as PWM at 3.3 V)
- Analogue Pin ADC: 1 (0-1 V)
- UART: 2
- USB-Serial Chip: CP2102
- FCC certification
- Antenna in PCB
- 802.11 b/g/n
- Wi-Fi Direct (P2P), soft-AP
- Integrated TCP/IP Protocol Stack
- PLLs, regulators, DCXO and integrated power management
- Output power of + 19.5 dBm in 802.11b mode
- Leakage current less than 10 uA
- STBC, 1 × 1 MIMO, 2 × 1 MIMO
- A-MPDU & A-MSDU aggregation & 0.4 ms guard interval
- Wake up and transmit packets in < 2 ms
- Standby power consumption < 1.0 mW (DTIM3)

Raspberry Pi 3 embedded system is one the most advanced thin client solution of the brand so far, which is shown by Fig. 2(b). Since the first Raspberry, called Raspberry Pi Modelo B, was introduced in 2012, more than 8 million units have been sold. Then it had only 256 MB of RAM. The new versions have been adding improvements in their features, thus reaching the Raspberry 3 pi Model B (MCU-RPi3).

In this work we would like to talk to you about some of the features and advantages of the MCU-RPi3, in case you are looking for a thin client to work with. The Raspberry Pi 3 thin client includes a series of advantages and novelties compared to the previous model, the Raspberry Pi 2 Model B. Among them, it is worth mentioning some of its main features. First, it has a much faster and more powerful processor than the versions. It is an ARM Cortex A53 with 4 cores, 1.2 GHz and 64 bits. Its performance is at least 50% higher than version 2 of Raspberry. The new thin client of Raspberry fulfilled another of the wishes of its main clients: wireless Wifi connectivity and Bluetooth. The previous models could only be connected through an Ethernet cable or through wireless USB adapters, so this version makes things much easier. The MCU-RPi3 includes a Bluetooth 4.1 and an 802.11n wireless Wifi network card. For the rest, it should be said that the new model hardly changes the size and only adds some changes in the position of the LED lights. The MCU-RPi3 account, on the other hand, with the same features that already included version 2.1 GB of RAM, fourth USB ports, Ethernet and HDMI port, Micro SD, CSI camera, etc. It is also fully compatible with the Zero and 2 versions. The Raspberry brand recommends using it in schools or for general use. Other recommendations of the brand for integrated projects are the Pi Zero and the Model A +. One of the great advantages of the MCU-RPi3 is precisely its price. Many companies, schools and centers use thin clients to create a more secure structure, but it can be given all kinds of uses: from creating a mini mail server to a low-end FM station.

The full specs for the Raspberry Pi 3 include:

- CPU: Quad-core 64-bit ARM Cortex A53 clocked at 1.2 GHz
- GPU: 400 MHz VideoCore IV multimedia
- Memory: 1 GB LPDDR2-900 SDRAM (i.e. 900 MHz)
- USB ports: 4
- Video outputs: HDMI, composite video (PAL and NTSC) via 3.5 mm jack
- Network: 10/100 Mbps Ethernet and 802.11n Wireless LAN
- Peripherals: 17 GPIO plus specific functions, and HAT ID bus
- Bluetooth: 4.1
- Power source: 5 V via MicroUSB or GPIO header
- Size: 85.60 mm × 56.5 mm
- Weight: 45 g (1.6 oz)

Wi-Fi Peer-to-Peer (P2P) and SoftAP allow easy, direct connection among Wi-Fi devices, anywhere. There is no need for an Access Point. Through negotiation, one device becomes the Group Owner and the other the client. Initiation of the connection is designed to be quick and easy. Implementation does not require addition of new hardware, allowing existing smxWiFi users to add them to their products with just a software update. Wi-Fi Peer-to-Peer is a protocol designed to replace the legacy ad-hoc protocol for interconnecting Wi-Fi devices. Improvements over ad-hoc include easier connection and the latest security, 802.11i. It was designed to satisfy the growing need for dynamic connections among groups of electronics. It allows OEMs to create products that can interoperate with each other and with common consumer devices, such as phones, tablets, cameras, and printers. As with any Wi-Fi device, the range is up to 200m, making connection convenient even when devices are not in immediate proximity to one another.

SoftAP gives a device limited Access Point capabilities. When combined with Wi-Fi P2P, it allows the device to be the Group Owner. Without it, a device can participate only as a client.

3.2 Hilbert Space-Filling Curve

The Hilbert curve is an iterated function that is represented by a parallel rewriting system, concretely a L-system L-system. In general, a L-system structure is a tuple of four elements:

1. *Alphabet*: the variables or symbols to be replaced.
2. *Constants*: set of symbols that remain fixed.
3. *Axiom* or *initiator*: the initial state of the system.
4. *Production rules*: how variables are replaced.

In order to describe the Hilbert curve alphabet let us denote the upper left, lower left, lower right, and upper right quadrants as $\mathcal{W}$, $\mathcal{X}$, $\mathcal{Y}$ and $\mathcal{Z}$, respectively, and the variables as $\mathcal{U}$ (*up*, $\mathcal{W} \to \mathcal{X} \to \mathcal{Y} \to \mathcal{Z}$), $\mathcal{L}$ (*left*, $\mathcal{W} \to \mathcal{Z} \to \mathcal{Y} \to \mathcal{X}$), $\mathcal{R}$ (*right*, $\mathcal{Z} \to \mathcal{W} \to \mathcal{X} \to \mathcal{Y}$), and $\mathcal{D}$ (*down*, $\mathcal{X} \to \mathcal{W} \to \mathcal{Z} \to \mathcal{Y}$). Where $\to$ indicates a movement from a certain quadrant to another. Each variable represents not only a trajectory followed through the quadrants, but also a set of 4^m transformed pixels in m level. The structure of our Hilbert Curve representation does not need fixed symbols, since it is just a linear indexing of pixels.

Fig. 3. First three levels of a Hilbert fractal curve. Axiom $= \mathcal{U}$ employed for this work.

The original work by Hilbert [9] proposes an axiom with a $\mathcal{D}$ trajectory, while we propose to start with an $\mathcal{U}$ trajectory (Fig. 3). Our proposal is based on the most of the energy is concentrated in the nearest MCU, namely at the right or left. The first three levels of a Hilbert Curve are portrayed in left-to-right order by Fig. 3. In this way the production rules of the Hilbert Curve are defined by:

- $\mathcal{U}$ is changed by the string $\mathcal{L}\mathcal{U}\mathcal{U}\mathcal{R}$
- $\mathcal{L}$ by $\mathcal{U}\mathcal{L}\mathcal{L}\mathcal{D}$
- $\mathcal{R}$ by $\mathcal{D}\mathcal{R}\mathcal{R}\mathcal{U}$
- $\mathcal{D}$ by $\mathcal{R}\mathcal{D}\mathcal{D}\mathcal{L}$.

In this way high order curves are recursively generated replacing each former level curve with the four later level curves. The Hilbert Curve has the property of remaining in an area as long as possible before moving to a neighboring spatial region. Hence, correlation between neighbor pixels is maximized, which is an important property in image compression processes. The higher the correlation at the preprocessing, the more efficient the data transmission.

4 Broadcasting and Sharing of Parameters in an IoT Network

4.1 Selecting the Main or Initial MCU

Our algorithm considers that the WPA is on random position inside the room, but with connection with at least one MCU in the IoT network. At the beginning all MCU's try to be connected to the WPA, if they do not establish the connection, the WiFi Direct option is enabled. MCU's that connect to the WPA disable the Direct WIFI mode until instructed otherwise. So, the WPA sets up with the broadcasting mode sending packages in order to measure the bandwidth of every member in the IoT network. Henceforth, we use a MCU Network Identifier (MNI), which contains the type of MCU, either MCU-ESP8266 or MCU-RPi3, and byte tag for identification inside the IoT network. For example a MNI=MCU-ESP8266-i refers a ESP8266, which was identified as the *ith* embedded system inside the room, no matter the type or characteristics of the MCU. Due to the material with which the room walls are made, connection is made to only with some MCU's, in quantity, these must be less or equal than 4^n. Hence, The IoT network is made up of 4^n elements, as much, that must be interconnected and it is defined by Eq. 1.

$$IoTe = \sum_{i=1}^{4^n} MCU_i \tag{1}$$

where $IoTe$ is the number of embedded systems in the network, i is index of the MCU Network Identifier, and n is the level of the Hilbert fractal, which is defined by the Eq. 2.

$$n = \lceil \log_4 (IoTe) \rceil \tag{2}$$

Which is why we can define the $MCU_{i=1}$ as the nearest MCU connected with the WPA because it get the highest link speed of the $IoTe$ links. So, MCU_1 is indexed as the first node in the network.

4.2 Generation of the Topology in a Given Floor Plant

Ones $MCU_{i=1}$ is identified as the main node in the IoT network. Hence, the $IoTe$ embedded systems, $MCU_{i=1}$ included, enable WiFi Direct mode and full a table with the bandwidth of the nodes are next to them, in order to generate a List of Reachable Nodes (LRN). Every MCU generates a particular LNR, then a the proposed IoT network can generate 4^n LNR's. In addition, every LRN$_i$, contains the bandwidth of all MCU_i, which it establish connection. In order to belong to the IoT network every MCU_i must connect to at least one link with other MCU_i LRN's are shared and $\sum_{i=1}^{4^n} MCU_i$ know the way to any node in the network, namely everyone knows the topology of the network. This second step has a paradigm neither optimized nor intelligent. So, it is important to employ some tools of SI to improve these initial results.

4.3   Indexing All the MCU's Inside the IoT Network

Thus, each node has an initial index in the network which, for the time being, is not optimized. To optimize it, all the elements must be numbered according to the Hilbert fractal and the position where each node passes.

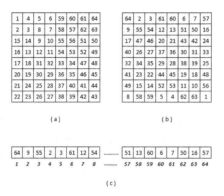

Fig. 4. (a) Matrix θ, (b) Matrix $\mathcal{H}$, and (c) Vector $\overrightarrow{\mathcal{H}}$

A linear indexing is developed in order to identify the MCU_i matrix array into a vector. Let us define the Microcontroller Units matrix array as $\mathcal{H}$ and the interleaved resultant vector as $\overrightarrow{\mathcal{H}}$, being $2^n \times 2^n$ be the size of $\mathcal{H}$ and 4^n the size of $\overrightarrow{\mathcal{H}}$, where n is the Hilbert curve level. Algorithm 1 generates a Hilbert mapping matrix θ with level n, expressing each curve as four consecutive indexes. The level n of θ is acquired concatenating four different θ transformations in the previous level $n-1$. Algorithm 1 generates the Hilbert mapping matrix θ, where $\overrightarrow{\beta}$ refers a $180°$ rotation of β and β^T is the linear algebraic transpose of β. Figure 4 shows an example of the mapping matrix θ at level $n = 3$. Thus, each MCU_i at $\mathcal{H}_{(i,j)}$ is stored and ordered at $\overrightarrow{\mathcal{H}}_{\theta_{(i,j)}}$, being $\theta_{(i,j)}$ the location index of it into $\overrightarrow{\mathcal{H}}$. In addition, Fig. 4(b) shows the position inside a room and the MNI

Algorithm 1. Function to generate Hilbert mapping matrix θ of size $2^n \times 2^n$.

Input: n
Output: θ
1 if $n = 1$ then
2    $\theta = \begin{matrix} 1 & 4 \\ 2 & 3 \end{matrix}$
3 else
4    $\beta = $ **Algorithm 1** $(n-1)$
5    $\theta = \begin{matrix} \beta^T & (\widetilde{\beta})^T + (3 \times 4^{n-1}) \\ \beta + 4^{n-1} & \beta + (2 \times 4^{n-1}) \end{matrix}$

---

of each MCU_i, namely, the best way to communicate for MCU_i is the MCU_{i+1}, giving as a result the increment of the bandwidth.

That is why the nodes i are consecutive and communicate with the node $i+1$. In the case that in the current topology there is no MCU_i in a certain position i, the index that is closest to $i-1$ is searched and this MCU_i is considered as a No-Significant Node (NSN) otherwise it is considered as Significant Node or (SN). It is important to mention that the node MCU_1 is always SN and it is the only one that establishes a connection with the WPA. In the case that this node MCU_{i+1} is NSN, MCU_i searches a SN incrementing i until finding it. In order to know the significance of the IoT, we propose a String of Network Significance (SNS) with 4^n bits or 2^{2n-3}bytes; 0 for NSN and 1 for SN. The SNS is broadcasted to $\sum_{i=1}^{4^n} MCU_i$ and it takes few nanoseconds. Every MCU_i measures or evaluates certain parameters such as temperature, humidity, CO_2, adjustment of the environmental lights. To differentiate them, we propose a 1-byte marker that specifically indicates which parameter that a sensor measures or MS and its value in Celsius degrees or voltage, for instance. Table 1 shows some examples of parameters that a MCU_i can measures, also a 1-byte MS can indicate up to 256 parameters.

Table 1. Some examples of marker of sensor or MS.

Label	Parameter
00H	Temperature
01H	Humidity
02H	Passive infrared sensors
03H	RFID
04H	Door control
05H	Temperature control
06H	Biorhythm
07H	Biosensors

Each parameter value that measures an MCU_i is represented by a two-byte long Marker, which is divided in three parts: Sign, Exponent ε and Mantissa μ (Fig. 5).

The most significant bit of the marker is the sign of parameter, whether 0 for positive or 1 for negative. The ten least significant bits are employed for the allocation of μ, which is defined by [10] as:

$$\mu = \left\lfloor 2^{10} \left(\frac{Parameter}{2^{R-\varepsilon}} - 1 \right) + \frac{1}{2} \right\rfloor \tag{3}$$

Equation (4) expresses how ε is obtained, which is stored at the 5 remaining bits of the marker

$$\varepsilon = R - \lceil \log_2 |Parameter| \rceil \tag{4}$$

where R is the number of bits used to represent the highest value of a certain parameter, defined as

$$R = \lceil \log_2 \left[\max \left\{ Parameter \right\} \right] \rceil . \tag{5}$$

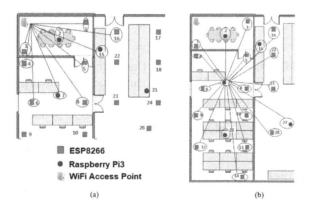

Fig. 5. Structure of the sensor marker or SM.

5   Experimental Results

In this section we extend the exposed in previous section. Also, we use the Floor Plant of the Fig. 1 and the following amount of MCU's:

– 46 MCU-ESP8266
– 18 MCU-RPi3

It is important to note that we use a WPA to connect the IoT network to Internet as initial node with following features: Tenda N300 Wireless WIFI Router WI-FI Repeater Booster Extender 802.11 b/g/n RJ45 with 4 Ports at 300 Mbps.

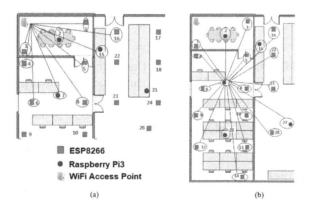

(a)   (b)

Fig. 6. Experimental results: (a) estimation of link speed every member in the IoT network and (b) reachable Nodes for MCU-RPi3-7

From Fig. 1, Fig. 6(a) shows that the WPA is on the upper right corner. At the beginning all the MCU's try to be connected to the WPA, if they do not establish the connection, the WiFi Direct option is enabled. MCU's that connect to the WPA disable the Direct WIFI mode until instructed otherwise. So, the WPA sets up with the broadcasting mode sending packages in order to measure the bandwidth of every member in the IoT network. Henceforth, we use a MCU Network Identifier (MNI), which contains the type of MCU, either MCU-ESP8266 or MCU-RPi3, and byte tag for identification inside the IoT intranet. For example a MNI = MCU-ESP8266-8 refers a ESP8266, which was identified as the eighth embedded system inside the room, no matter the type or characteristics of the MCU. Due to the material with which the walls are made, connection is made to only seventeen MCU's the speeds depicted in Table 2, we only show the first ten MCU connected to the WPA. Which is why we can define the MCU-ESP8266-1 as the nearest MCU connected with the WPA because it get a link speed of 199.37 Mbps and the furthest MCU is MCU-ESP8266-16 with 53.14 Mbps. MCU-ESP8266-1 is indexed as the first node in the network.

Table 2. Bandwidth (Mbps) of the first ten MCU's inside the IoT network

MCU network identifier	Linear distance (m)	Actual distance (m)	Bandwidth (Mbps)
MCU-ESP8266-1	3	3	199.37
MCU-ESP8266-4	4	12	117.12
MCU-ESP8266-6	7.5	13	107.98
MCU-RPi3-7	7	11	126.26
MCU-ESP8266-8	9	11.5	121.69
MCU-ESP8266-5	6.5	6.5	167.38
MCU-RPi3-2	3.5	3.5	194.80
MCU-RPi3-15	8	16	80.56
MCU-ESP8266-16	8.5	19	53.14
MCU-ESP8266-3	6	6	171.95

Ones MCU-ESP8266-1 is identified as the main node or Node 1 in the IoT network, all the MCU's, MCU-ESP8266-1 included, enable WiFi Direct mode and full a table with the bandwidth of the nodes are next to them, in order to generate a List of Reachable Nodes (LRN). Every MCU generates a particular LNR, for instance, Fig. 6(b) shows a visual representation of the MNI's whom the MCU-RPi3-7 can connect. According to Fig. 1 our example generates 64 LNR's.

In addition, Table 3 shows the LRN-7, i.e. the bandwidth of the nineteen MCU's, which it establish connection the MCU-RPi3-7. As we can see MCU-ESP8266-6 and MCU-ESP8266-8 obtain the two highest bandwidth for Node 7, with 203.94 Mbps and 201.20 Mbps, respectively.

In addition, Fig. 4(b) shows the position inside a room and the MNI of each MCU_i, namely, the best way to communicate for MCU_{64} is the MCU_9, increasing the bandwidth and so on. In addition, Fig. 7 shows a visual interpretation of the linear indexing making by means of a $n = 3$ Hilbert Fractal.

Table 3. Seventh list of reachable nodes or LRN$_7$ generated by the MCU-RPi3-7.

MNI	Linear distance (m)	Actual distance (m)	Bandwidth (Mbps)
MCU-ESP8266-1	5.50	10.40	131.74
MCU-ESP8266-4	4.50	4.50	185.66
MCU-ESP8266-6	2.50	2.50	203.94
MCU-ESP8266-9	5.00	5.00	181.09
MCU-ESP8266-12	8.00	8.00	153.67
MCU-RPi3-11	6.00	6.00	171.95
MCU-ESP8266-14	10.50	10.50	130.83
MCU-ESP8266-13	7.50	7.50	158.24
MCU-ESP8266-10	4.50	4.50	185.66
MCU-ESP8266-28	8.20	10.40	131.74
MCU-RPi3-27	9.30	9.70	138.14
MCU-ESP8266-8	2.80	2.80	201.20
MCU-ESP8266-23	6.00	6.00	171.95
MCU-ESP8266-22	7.20	9.80	137.22
MCU-ESP8266-16	8.40	13.50	103.41
MCU-RPi3-15	6.30	11.25	123.97
MCU-ESP8266-5	4.45	4.45	186.12
MCU-ESP8266-3	7.25	7.25	160.53
MCU-RPi3-2	5.30	8.75	146.82

Fig. 7. Floor Plan for conventional house.

We perform one thousand requests from the farthest nodes, i.e. from MCU_{64} to MCU_1 and from MCU_8 to MCU_{57}. On the average the link $MCU_{64} \leftrightarrow MCU_1$ obtains a bandwidth of 194.69 Mbps and for the one $MCU_8 \leftrightarrow MCU_{57}$ 192.35 Mbps. In both in a central topology, we cannot connect neither $MCU_{64} \leftrightarrow MCU_1$ nor $MCU_8 \leftrightarrow MCU_{57}$.

6   Conclusions

Reducing energy consumption is a compulsory objective in the design of any communication protocol for Wireless Sensor Networks. Most of this energy can be saved through member aggregation, given that most of the sensed information is redundant due to geographically collocated sensors. Therefore, efficient broadcasting scheme to consider sharing parameters have been proposed. However, they still suffer from the high communication cost and have not fully resolved the sharing parameters. We proposed a lightweight broadcasting algorithm in wireless sensor networks. From the performance analysis, we can confirm that the IoT network reconfiguration scheme improves both the network lifetime and efficient parameter information than the traditional schemes. Also, we obtain some parameters connected in a certain MCU_i of all the IoT network in microseconds, what meets the definition of real time definition. In addition, this proposal make use of Swarm Intelligence, since every MCU_i learns of the rest of the members of the IoT network and knows what happen even it is not physically connected to other member.

Acknowledgment. This article is supported by National Polytechnic Institute (Instituto Poliécnico Nacional) of Mexico by means of Project No. 20180514 granted by Secretariat of Graduate and Research, National Council of Science and Technology of Mexico (CONACyT). The research described in this work was carried out at the Superior School of Mechanical and Electrical Engineering (Escuela Superior de Ingeniería Mecánica y Eléctrica), Campus Zacatenco.

References

1. Hassan, T., Aslam, S., Jang, J.W.: Fully automated multi-resolution channels and multithreaded spectrum allocation protocol for IoT based sensor nets. IEEE Access **6**, 22545–22556 (2018)
2. Jo, O., Kim, Y.K., Kim, J.: Internet of things for smart railway: feasibility and applications. IEEE Internet Things J. **5**(2), 482–490 (2018)
3. Sandoval, R.M., Garcia-Sanchez, A.J., Garcia-Haro, J.: Improving RSSI-based path-loss models accuracy for critical infrastructures: a smart grid substation case-study. IEEE Trans. Ind. Inform. **14**(5), 2230–2240 (2018)
4. Zhang, C., Ge, J., Pan, M., Gong, F., Men, J.: One stone two birds: a joint thing and relay selection for diverse IoT networks. IEEE Trans. Veh. Technol. **67**(6), 5424–5434 (2018)
5. Li, T., Yuan, J., Torlak, M.: Network throughput optimization for random access narrowband cognitive radio internet of things (NB-CR-IoT). IEEE Internet Things J. **5**(3), 1436–1448 (2018)
6. Sharma, P.K., Chen, M.Y., Park, J.H.: A software defined fog node based distributed blockchain cloud architecture for IoT. IEEE Access **6**, 115–124 (2018)
7. Taghizadeh, S., Bobarshad, H., Elbiaze, H.: CLRPL: context-aware and load balancing RPL for IoT networks under heavy and highly dynamic load. IEEE Access **6**, 23277–23291 (2018)

8. Beni, G., Wang, J.: Swarm intelligence in cellular robotic systems. In: Dario, P., Sandini, G., Aebischer, P. (eds.) Robots Biological Systems: Towards a New Bionics?. NATO ASI Series (Series F: Computer and Systems Sciences), vol. 102, pp. 703–712. Springer, Berlin (1993). https://doi.org/10.1007/978-3-642-58069-7_38
9. Hilbert, D.: Über die stetige Abbildung einer Linie auf ein Flächenstück. Math. Ann. **38**(3), 459–460 (1891)
10. Boliek, M., Christopoulos, C., Majani, E.: Information Technology: JPEG2000 Image Coding System, JPEG 2000 Part I final committee draft version 1.0 ed., ISO/IEC JTC1/SC29 WG1, JPEG 2000, April 2000

Solid Waste Collection in Ciudad Universitaria-UNAM Using a VRP Approach and Max-Min Ant System Algorithm

Katya Rodriguez-Vazquez[⊠], Beatriz Aurora Garro[⊠], and Elizabeth Mancera[⊠]

IIMAS, UNAM, Ciudad Universitaria, Mexico City, Mexico
{katya.rodriguez,beatriz.garro}@iimas.unam.mx,
elizabethal_20@hotmail.com

Abstract. The collection of solid waste is a very important problem for most of the modern cities of the world. The solution to this problem requires to apply optimization techniques capable of design the best path routes that guarantee to collect all the waste minimizing the cost. Several computation techniques could be applied to solve this problem and one of the most suitable could be swarm optimization such as ant colony optimization. In this paper, we propose a methodology for searching a set of collection paths of solid waste that optimize the distance of a tour in Ciudad Universitaria (UNAM). This methodology uses a vehicle routing problem (VRP) approach combined with Max-Min Ant System algorithm. To assess the accuracy of the proposal, we select the scholar circuit in the area of Ciudad Universitaria. The results shown a shortest distance travelled and better distribution than the empiric route used actually for the cleaning service.

Keywords: Vehicle routing problem (VRP)
Max-Min Ant System (Max-Min AS) algorithm · Collection of solid waste
Optimization path routes

1 Introduction

The application of mathematical models to solve real routing problems has gained attention in the last years due to the positive impact reflected in costs, time and quality of service. Within these real problems, we could mention the optimization of waste collection systems in urban areas [10, 14, 15]. The main objectives of a waste collection system are the reduction of the use of vehicles for transport, minimize the distance traveled and cover a service to a client, where profits are maximized and the time spent for collecting the waste is minimal.

The collection of waste can be solved considering containers or sidewalks represented in graphs as nodes and edges. For the first case, it could be applied a vehicle routing problem (VRP) approach. For the second case, can be applied a capacity arc routing problem (CARP). In the case of VRP with only one vehicle, the problem is reduced to the classic traveling salesman problem (TSP) [2, 8] and in the case of CARP, this can be posed as the Chinese postman problem [6]. The latter considers a

© Springer Nature Switzerland AG 2018
I. Batyrshin et al. (Eds.): MICAI 2018, LNAI 11288, pp. 76–85, 2018.
https://doi.org/10.1007/978-3-030-04491-6_6

mailman whose task is to deliver correspondence in a neighborhood. To achieve this, it is necessary to leave the post office, go through all the streets (arcs and return to said office), finding the shortest possible distance without repeating arcs and with the possibility of going through the same node several times (intersections between arcs).

Several exact methods have been used to solve the problem of waste collection [7, 12, 13], however the application of these methods is limited because complex problems such as VRP and TSP become NP-hard. In this case, the exact methods are not capable of finding an optimal solution in a reasonable time. For that reason, it is necessary to apply other type of techniques that allow to find solutions near to the optimum value in a reasonable time. For example, more efficient techniques such as heuristics, which, although they do not always find the optimal solutions, are able to solve the problem in a polynomial time and the solutions found are good approximations to it.

In the literature, the problem of waste collection has been studied and addressed from the two approaches mentioned above (VRP and CARP). For example in [12], the authors apply a CARP approach for minimizing the costs of transporting waste in Brussels, by using a branching and boundary method. In [1] the authors present a TSP model to design collection routes in southern Buenos Aires.

On the other hand, some heuristics such as swarm intelligence has been apply to the waste collection problem. Ant Colony Optimization (ACO) is a population-based approach which has been successfully applied to several NP-hard combinatorial optimization problems. For example, Hornig and Fuentealba apply an Ant Colony Optimization (ACO) algorithm to the collection of waste by containers, in a municipality of Chile [8]. Karadimas et al., also apply an ACO algorithm to establish collection routes in an area of Greece [11].

Our problem is focused on the routing of vehicles for collecting solid waste in Ciudad Universitaria (UNAM). Actually, this problem was solved empirically, but it is not efficient in costs and time [9]. It should be noted that a poorly planned waste collection problem generates traffic problems, unhealthiness, high monetary and time costs. For this reason, it was proposed to investigate the issue and generate a solution to the planning of such routing in the UNAM-CU. This work is important due to the application to a real problem, where optimization techniques have not been used to solve an important routing problem in UNAM-CU.

In this paper, we propose a methodology for searching a set of collection paths of solid waste that optimize the distance of a tour in Ciudad Universitaria (UNAM) using a vehicle routing problem (VRP) approach combined with Max-Min Ant System algorithm. To assess the accuracy of the proposal, we select the scholar circuit in the area of Ciudad Universitaria. The results shown a shortest distance travelled and better distribution than the empiric route used actually for the cleaning service.

2 Basic Concepts

2.1 Max-Min Ant System

Ant Colony Optimization (ACO) Algorithm was proposed by Marco Dorigo [2–5]. ACO algorithm is inspired in the foraging behavior of real ant colonies. Its functioning is based on the indirect communication among the individuals of a colony of agents that emerge from the trails of a chemical substance (called pheromone) which real ants use for communication. This information is modified by the ants to reflect their experience accumulated while solving a particular problem.

To solve our collection problem we use the Max-Min Ant System which is a variant of ACO that have given the best results [16]. In general, the steps in these algorithms are the following:

1. Initialize the pheromone traces and parameters using an initial solution obtained with the heuristic of the nearest neighbor.
2. As long as the unemployment criterion is not met:
   a. Generate a solution for each ant.
   b. Improve the best solution with local search heuristics.
   c. Update pheromone levels.
3. Return the best solution so far.

In the Max-Min Ant System (MMAS) algorithm, the heuristic of the nearest neighbor is used to obtain an initial solution by iteratively adding collection points to the path of a vehicle. The route starts at the auxiliary deposit and the nearest collection point is chosen as long as the capacity restriction is not violated. In case of any collection point is not allow to visit, the route is closed when an auxiliary deposit is added and then a new route starts. It ends up generating routes once all the collection points have been visited and the total distance traveled called C_{nn} is calculated. Once the above is done, the initial pheromone trace is calculated as $\tau_0 = 1/nC_{nn}$ or as $\tau_0 = 1/C_{nn}$.

Initially artificial ants (k is equal to the number of collection points) are randomly located at a collection point, which would be the first node to visit after leaving the auxiliary deposit. The above is done with the purpose of having more exploration at the beginning of the algorithm.

Each of the artificial ants builds a feasible solution to the problem by successively adding collection points until they have visited all of them. Each ant has a memory of the points visited, so it is not possible for it to visit a point more than once. Likewise, the ant moves to a collection point, as long as the capacity restriction is not violated. In this way, in conjunction with the previous feature, any set of routes they create is feasible. Particularly, the ant returns to the deposit (ends a route) when the capacity restriction is met or when all collection points have been visited.

To decide which pickup point to move to, the MMAS algorithms use a transition rules that combine the pheromone trail (τ) and the heuristic information (η) to choose with some probability p_{ij} the next point j to visit from the point i. In the transition rule, the parameters α and β are used, which determine the importance of pheromone and heuristic information, respectively [17].

On the other hand, η_{ij} is the reciprocal of the minimum distance between i and j. The previous function defines the heuristic information in a simple way, also, the closer two points are the greater than the value of the heuristic information.

The transition rule for the MMAS algorithm is defined as Eq. (1)

$$p_{ij}^k = \begin{cases} \dfrac{\tau_{ij}^\alpha \eta_{ij}^\beta}{\sum_{u \in N_i^k} \tau_{iu}^\alpha \eta_{iu}^\beta} & if \quad j \in N_i^k \\ 0 & otherwise \end{cases} \tag{1}$$

On the other hand, N_i^k denotes the feasible set of collection points neighboring point i. In case the ant k has already visited all the points in the list or cannot visit them because it violates the capacity constraint, then a point that is not in the list N_i^k chosen such that it maximizes the function $\tau_{ij}^\alpha \eta_{ij}^\beta$ but neither does it violate the capacity restriction. If it was possible to select a point i, it is added to the route currently being built by the ant. If it cannot go to another collection point without violating the capacity restriction, then the ant closes that route by going to the auxiliary deposit and creating a new route. For this new route, select a random collection point that will be the immediate point to visit after the deposit.

It is important to mention that the pheromone trace is bounded inferiorly and superiorly; the initial value of pheromone is defined as the upper limit of the same and secondly, pheromone levels are reinitialized if during a certain number of iterations, the solution does not improve. The pheromone update is defined by Eq. (2).

$$\tau_{ij} = \left[(1 - \rho)\tau_{ij} + \Delta\tau_{ij}^S \right]_{\tau_{min}}^{\tau_{max}} \tag{2}$$

The term $(1 - \rho)\tau_{ij}$ corresponds to an evaporation of the pheromone at a constant rate ρ, while $\Delta\tau_{ij}^S$ refers to the pheromone deposit. This deposit is a function of the quality of the best global solution (s_{bs}) or best iteration solution (s_{is}). When it is decided to use s_{bs}, the exploration is limited because the solutions tend to be concentrated near that value. On the other hand, if only (s_{is}) is used, the probability of exploring the search space increases, since the solutions can change considerably from iteration to iteration. Assuming that L_s is the total distance of s_{bs} or s_{is}, $\Delta\tau_{ij}^S$ is calculated by Eq. (3)

$$\Delta\tau_{ij}^S = \begin{cases} \frac{1}{Ls} & if \quad (i,j) \, belongs \, to \, the \, best \, path \\ 0 & otherwise \end{cases} \tag{3}$$

Additionally, τ_{max} and τ_{min}, are the upper and lower limit of the pheromone levels respectively. The lower limit allows all connections to have at least a minimum amount (τ_{min}) of pheromone. This allow to assign a small probability greater to zero to the connections producing greater exploration in the space of search. On the other hand, the difference of pheromone values between connections decreases and with this, the probability of choosing components with different pheromone quantities is also reduced and therefore, there will be more exploration in the search space. The upper limit prevents the amount of pheromone from growing without measure, which prevents premature convergence to suboptimal [16].

Finally, to carry out the global update, a single ant is taken into account, which generated the best overall solution. With this update we seek to reinforce the amount of pheromone in those shorter tours. Equation (4) is used to add the pheromone to the arcs, after each iteration.

$$\tau_{ij} = \begin{cases} (1 - \rho)\tau_{ij} + \rho\Delta\tau_{ij}^{best} & \text{if } (i,j) \text{ belongs to the best path} \\ \tau_{ij} & \text{otherwise} \end{cases} \quad (4)$$

3 Proposed Methodology

In this work we try to find an optimal collection route design in Ciudad Universitaria. The objective is to minimize the total distance traveled, which in turn would result in a decrease in fuel costs, deterioration of vehicles and time savings.

Currently, the collection is carried out manually and mechanically. The manual collection only takes place in an area of the university called "Casco central", while in the rest of the university the collection is mechanical. To carry out the mechanical collection, there are 8 collection vehicles and a truck, whose capacity is 15.3 m³ and 6.0 m³, respectively. Regardless of the number of vehicles available, only five are used, which are assigned to six inorganic waste collection routes; the truck is used for the collection of organic waste. It should be noted that this work only focuses on the collection of inorganic waste, which we will refer to as solid waste.

To collect solid waste, two trips are usually made on each of the six routes: one in the morning around 6:00 a.m and another in the afternoon, around 5:00 p.m from Monday to Friday. These sixroutes are: CCU1 Circuit, CCU2 Circuit, Scholar Circuit, Scientific Circuit, Exterior Circuit and Sports Circuit. The first two circuits cover an area called Cultural Zone, while the remaining 4 routes cover an area known as the Scholar Zone. In the case of the Cultural Zone, there is only one vehicle that makes a tour in the morning and another in the afternoon, also begins and ends its route in a deposit called "Taller de Mantenimiento del Centro Cultural". While, the other four vehicles attend the Scholar Zone and start and finish each route in the deposit called "Talleres Centrales de Conservación". It is worth mentioning that, in both cases, the truck must leave the waste to a deposit of the government of Mexico City, known as "Estación de Transferencia" before returning. This station is located far from the CU limits, but the vehicles generally go in the morning and in the afternoon to discharge the collected waste and then return to CU. If the solid waste has not been collected in its entirety, the routes are returned to CU and the missing items are collected, however, they are no longer taken back to the "Estación de Transferencia", as the quantities are small.

Table 1 shows the general features (total distance traveled, number of collection points and total volume of waste generated) of each of the four routes.

Table 1 shows the collection points and the capacity of the vehicles. A single vehicle is not able to collect in a single trip the amount of waste generated in the scholar zone. Given the above, it is necessary to make an allocation of the collection points in several routes. In that sense, our problem consists of finding a set of routes that minimize the total distance traveled, under the following restrictions:

- There is a maximum of 4 vehicles available, which translates into a maximum of one partition in four circuits.

Table 1. Features of collecting routes

| | Scholar zone | | | | |
	Sport circuit	Exterior circuit	Scientist circuit	Scholar circuit	Total
Collection points	16	22	23	24	**85**
% collection points	18.8	25.9	27.1	28.2	**100**
Distance traveled (m)	43,407	38,698	32,081	41,808	**155,994**
% Distance traveled	27.8	24.8	20.6	26.8	**100**
Volume collected (m³)	11.3	16.3	14.3	18.4	**60**
% Volume collected	18.7	27.0	23.7	30.5	**100**

- At all collection points, the available volumes have a maximum capacity of 16.5 m^3, that is, in each route, the maximum waste collected would be 16.5 m^3.
- Each vehicle or route started in the corresponding deposit and must return to it after having discharged the waste in the "Estación de transferencia".
- The direction of the roads in Ciudad Universitaria must be taken into account, since in general the physical route between two points and the distance between them is the same when going from A to B than from B to A.

Thus, the objective of the problem is to find a set of routes of minimum cost, so that each collection point is visited only once and the capacity of the vehicle is not exceeded by the total of the collected waste, furthermore each route must start and end in the corresponding deposit. The foregoing points out that the problem is solved as asymmetric capacity vehicle routing problem (ACVRP) [18].

The mathematical model for our routing problem, which is based on the classic VRP, is described with Eqs. (5–12)

$$Min \sum_{i \in V} \sum_{j \in V} \sum_{k \in V} c_{ij} x_{ij}^k \tag{5}$$

Subject to

$$\sum_{k \in K} y_i^k = 1 \qquad \forall i \in V \tag{6}$$

$$\sum_{k \in K} y_0^k = K \tag{7}$$

$$\sum_{\substack{j \in V \\ j \neq i}} x_{ij}^k = \sum_{\substack{j \in V \\ j \neq i}} x_{ji}^k = y_i^k \quad \forall i \in V, k \in K \tag{8}$$

$$u_{ik} - u_{jk} + Q x_{ik}^k \leq Q - q_j \qquad \forall i,j \in V', i \neq j, k \in K \tag{9}$$

$$q_i \leq u_i^k \leq Q \quad \forall i \in V', k \in K \tag{10}$$

$$x_{ij}^k \in \{0,1\} \qquad \forall i,j \in V', k \in K \tag{11}$$

$$y_i^k \in \{0,1\} \qquad \forall i \in V', k \in K \tag{12}$$

Where x_{ij} is a binary variable that takes the value of one if the arc (i, j) is in the solution path of the vehicle k and y_{ij} takes the value of one if the collection point i is visited by the vehicle k. Each collection point i has a demand associated $q_i^k \in R^+$ with $q_i^k = 0 \forall i \epsilon V'$ nonnegative, which will be satisfied by a single vehicle k.

4   Experimental Results

In this section, we present the results obtained by applying the Max-Min Ant System (MMAS) algorithm to the collection problem in the Scholar Zone of Ciudad Universitaria (CU), when seen as an ACVRP. First the results obtained with the MMAS algorithms are shown and then a comparison is made with the routes designed by the "Dirección General de Obras y Conservación-CU" (DGOyC), unit in charge of coordinating the collection of solid waste of CU.

The parameters for the MMAS algorithm were set as: $m = n - 1$, $\alpha = 1$, $\beta = 3$, $\tau_o = 1/nC_{nn}$, $\rho = 0.1$, LC = 15. It is important to mention that the global solution was used to update the pheromone and the restart mechanism was activate after 250 iterations if the solution is not improved. The number of iteration was set to 5000 and 30 experiments were performed to validate statistically the results.

Table 2 shows the average results obtained with MMAS algorithm.

Table 2. Average results obtained with MMAS in the scholar zone

Algorithm	μ	σ	mín
MMAS	123,768.90	1,232.12	120,746

In Table 3, several characteristics related to each of the four routes generated by the proposed approach are observed. The first characteristic represents the number of points grouped in the routes. The above information is used to analyze the distribution of collection points in each generated route. It can be said that it is better to have routes with a similar number of collection points between them, since the difference between visits made by each vehicle does not vary greatly from one route to another. The percentages tell us that 20% of the total collection points (85) are in Route 2, 24.7% in Route 4, 27.1% in Route 3 and 28.2% in Route 1.

Table 3. Features of the routes obtained with the MMAS algorithm.

	MMAS				
	Route 1	Route 2	Route 3	Route 4	Total
Collection points	24	17	23	21	**85**
% collection points	28.2	20.0	27.1	24.7	**100**
Distance traveled (m)	33,265	28,566	27,711	31,204	**120,746**
% Distance traveled	27.6	23.7	22.9	25.8	**100**
Volume collected (m^3)	15	15	15	15	**60**
% Volume Collected	24.9	25.2	25.2	24.7	**100**
# of points Sport C.	16	–	–	–	**16**
# of points exterior C.	6	5	7	4	**22**
# of points scientific C.	–	2	15	6	**23**
# of points scholar C.	2	10	1	11	**24**

Not only the distribution of collection points is important, but also the distance traveled in each route. The percentages of distance traveled, which are observed in Table 3, are calculated with respect to the total distance traveled between the four routes. Given the above, for the MMAS algorithm, the distance covered in Route 3 corresponds to 22.9% of the total distance, for Route 2 it is 23.7%, for Route 4 25.8% and for Route 1 27.6%. As for the ACS algorithm, Route 2 covers 24.2% of the total distance, Route 3 covers 24.3%, Route 1 covers 25.4% and Route 4 covers 26.1%. Although the differences between the percentages of the MMAS and ACS algorithm are very small, the previous figures suggest that the trucks on the routes generated with ACS travel more similar distances, in comparison with the MMAS algorithm. In fact, in the case of the ACS, the difference between the route that travels the shortest distance and the one that travels a greater distance is 1.9%; while for the MMAS algorithm the difference is 4.7%.

It is also important to analyze what percentage of waste is collected in each route, since it is important that there is an equitable distribution of the volume collected. The above is because it is better that there are no vehicles that are very empty or that are filled to their maximum capacity. In this area, analyzing the MMAS algorithm, Route 4 collects 24.7% of the total waste generated in the Scholar Zone, Route 1 collects 24.9% and Routes 2 and 3 collect 25.2%, each one.

On the other hand, it is worth mentioning that in all the routes the deposit is the node with name Conservation Central Workshops.

Once the best solutions obtained with the MMAS algorithm was generated, it is also important to make a comparison with the design made by the DGOyC. First, in the routes of the DGOyC, the percentages of collection points grouped in the routes are: 18.8% for the Sport Circuit, 25.9% for the Exterior Circuit, 27.1% for the Scientific

Circuit and 28.2% for the Scholar Circuit (Table 1). These results suggest that with MMAS algorithms the distribution of collection points is better compared to the distribution made by the DGOyC. It is also observed that the difference between the smallest route and the largest one is 8.2% for MMAS, while the difference in the design of the DGOyC is 9.4%.

As for the volume collected, Table 1 shows the percentages of the DGOyC: 18.7% of the waste is collected in the Sport Circuit, 23.7% in the Scientific Circuit, 27.1% in the Exterior Circuit and 30.5% in the Scholar Circuit. Again, the differences between these percentages are greater than those obtained with the new route design generated with the MMAS algorithm (Table 3). Furthermore, with the new route design obtained with the MMAS, the waste collection between the four routes is distributed more equitably.

Finally, it is necessary to compare the total distance covered with the route design of the DGOyC and MMAS algorithm. The total distance traveled from the routes of the Scholar Zone, associated with the design of the DGOyC is 155,994 meters and for MMAS is 120,743 meters. Thus, the reduction obtained with MMAS is 22.6%.

5 Conclusions

In this article, a real problem of routing was solved for waste collection in UNAM-CU. This problem was posed as an asymmetric CVRP. To solve it, we used the MMAS Algorithm.

Particularly, within the Scholar Zone, the improvement obtained with respect to the design of the DGOyC is 22.6%.

The results show that improvements can be made in the solid waste collection system that currently operates in CU. Well, it was mainly possible to minimize the total distance traveled and consequently the cost derived from the collection. It can be said that using the ACO algorithms to solve this type of problem is feasible, since in addition to improving the collection system, the implementation is simple and robust; since you can make modifications to the algorithms in such a way that there is a mapping between them and the real problem that you are looking to solve.

Acknowledgments. The authors would like to thank Dirección General de Obras y Conservación, UNAM. This work was supported by the SECITI under Project SECITI/064/2016.

References

1. Bonomo, F., Durán, G., Larumbe, F., Marenco, J.: A method for optimiz-ing waste collection using mathematical programming: a Buenos Aires case study. Waste Manag. Res. **30**(3), 311–324 (2012)
2. Dorigo, M., Birattari, M., Stützle, T.: Ant colony optimization-artificial ants as a computational intelligence technique. IEEE Comput. Intell. Mag. **1**, 28–39 (2006)
3. Dorigo, M., Gambardella, L.M.: Ant colony system: a cooperative learning approach to the traveling salesman problem. IEEE Trans. Evol. Comput. **1**(1), 53–66 (1997)

4. Dorigo, M., Maniezzo, V., Colorni, A.: Ant system: an autocatalytic optimizing process. Université Libre de Bruxelles (1991)
5. Dorigo, M.: Optimization, learning and natural algorithms. Ph.D. thesis, Dip. Electtronica e Informazion, Politecnico di Milano Italy (1992)
6. Filipiak, K.A., Abdel-Malek, L., Hsieh, H.-N., Meegoda, J.N.: Optimization of municipal solid waste collection system: case study. Pract. Period. Hazard. Toxic Radioact. Waste Manag. **13**(3), 210–216 (2009)
7. Hemmelmayr, V., et al.: A heuristic solution method for node routing based solid waste collection problems. J. Heuristics **19**(2), 129–156 (2013)
8. Hornig, E.S., Fuentealba, N.R.: Modelo ACO para la recolección de residuos por contenedores. Rev. Chil. Ing. **17**(2), 236–243 (2009)
9. http://www.obras.unam.mx/Pagina/index.php
10. Ismail, Z., Loh, S.: Ant colony optimization for solving solid waste collection scheduling problems. J. Math. Stat. **5**(3), 199 (2009)
11. Karadimas, N.V., et al.: Optimal solid waste collection routes identified by the ant colony system algorithm. Waste Manag. Res. **25**(2), 139–147 (2007)
12. Kulcar, T.: Optimizing solid waste collection in Brussels. Eur. J. Oper. Res. **90**(1), 71–77 (1996)
13. Mansini, R., et al.: A linear programming model for the separate refuse collection service. Comput. Oper. Res. **25**(7), 659–673 (1998)
14. Mourao, M., Almeida, M.T.: Lower-bounding and heuristic methods for a refuse collection vehicle routing problem. Eur. J. Oper. Res. **121**(2), 420–434 (2000)
15. Ogwueleka, T.C.: Route optimization for solid waste collection: Onitsha (Nigeria) case study. J. Appl. Sci. Environ. Manag. **13**(2), 37–40 (2009)
16. Stützle, T., Hoos, H.H.: Improving the ant system: a detailed report on the MAX– MIN ant system. Technical report AIDA–96–12, FG Intellektik, FB Informatik, TU Darmstadt, Germany, August 1996 (1996)
17. Dorigo, M., Di Caro, G., Gambardella, L.M.: Ant algorithms for discrete optimization. Artif. Life **5**(2), 137–172 (1999)
18. Dantzig, G.B., Ramser, J.H.: The truck dispatching problem. Manag. Sci. **6**(1), 80–91 (1959)

Selection of Characteristics and Classification of DNA Microarrays Using Bioinspired Algorithms and the Generalized Neuron

Flor Alejandra Romero-Montiel$^{(\boxtimes)}$ and Katya Rodríguez-Vázquez

Instituto de Investigaciones en Matemticas Aplicadas y Sistemas, IIMAS, UNAM,
Mexico City, Mexico
f.alejandra.r.m@gmail.com, katya.rodriguez@iimas.unam.mx

Abstract. DNA microarrays are used for the massive quantification of gene expression. This analysis allows to diagnose, identify and classify different diseases. This is a computationally challenging task due to the large number of genes and a relatively small number of samples.

Some papers applied the generalized neuron (GN) to solve approximation functions, to calculate density estimates, prediction and classification problems [1,2].

In this work we show how a GN can be used in the task of microarray classification. The proposed methodology is as follows: first reducing the dimensionality of the genes using a genetic algorithm, then the generalized neuron is trained using one bioinspired algorithms: Particle Swarm Optimization, Genetic Algorithm and Differential Evolution. Finally the precision of the methodology it is tested by classifying three databases of DNA microarrays: *Leukemia benchmarck ALL − AML*, *Colon Tumor* and *Prostate cancer*.

Keywords: Microarrays · Genetic algorithms · PSO
Differential evolution · Neural networks · Pattern recognition

1 Introduction

DNA microarrays allow the quantification of thousands of gene expressions, providing a complete picture of the genetic alterations related to different diseases. An early and accurate prognosis makes it much easier to select the best treatment for a specific patient to resist a disease. The DNA microarray usually has thousands of gene expressions for each biological sample, in contrast to the small number of samples, this poses a disadvantage when performing a learning and recognition task, making it a challenge for the current classification methods. It has been shown that the presence of redundant and irrelevant information affects the classification accuracy of machine learning algorithms, therefore it is necessary to select the relevant genes. Several techniques based on supervised learning

I. Batyrshin et al. (Eds.): MICAI 2018, LNAI 11288, pp. 86–97, 2018.
https://doi.org/10.1007/978-3-030-04491-6_7

for the classification of microarrays has been used; in [3] the results obtained by comparing an Artificial Neural Network (ANN), a Support Vector Machines, Naive Bayes, Logistic Regression, k-Nearest Neighbors and Classification Trees, reporting a significant classification accuracy of 98% using ANN. A two-phase hybrid model for cancer classification is proposed in [4], which integrates the selection of characteristics based on the correlation with improved binary PSO (iBPSO). This model selects a set of low dimensionality genes using a Naive Bayes classifier with cross validation. The subsets of genes are very small (up to 1.5%) In this work the problem of selection of characteristics and classification of DNA microarrays is analyzed. The methodology used consists of first reducing the dimensionality of the genes using a genetic algorithm, then classifying the data with a generalized neuron (GN) where some parameters are determined by three different bioinspired algorithms: Particle Swarm Optimization, Genetic Algorithm and Differential Evolution. The data sets used will be described in Sect. 2. Following with the ins and outs of GN in Sect. 3. Genetic algorithms, PSO and differential evolution are described in Sect. 4, the proposed methodology and the results are detailed in Sects. 5 and 6 respectively, to finalize the conclusions in Sect. 7.

2   Data Sets

Leukemia benchmark ALL − AML contains the measurements corresponding to *ALL* and *AML* samples of bone marrow and peripheral blood. Originally composed of 38 samples for training (27 *ALL and* 11 *AML*) and 34 samples for test (20 *ALL and* 14 *AML*) where each sample contains information of 7129 gene expressions. It was performed a random partition with the 70% of samples for training and the remaining 30% for the test, there are 50 samples in the training set and 22 in the test set.

The database *colon Tumor* contains 62 samples, of which 22 are positive for Tumor in colon and 40 negative. Performing a partition as in the previous data set, there are 43 samples for training and 19 for testing. Each sample contains 2000 genes.

Prostate cancer consists of 102 samples for training (52 with prostate tumor and 50 with "normal" prostate tumor) The test set contains 25 samples with prostate tumor and 9 normal, each sample contains 12,600 genes. Using the previous partition, there are 95 samples for training and 41 for testing.

3   Generalized Neuron (GN)

The generalized neuron (GN) was proposed in [5], has been applied in problems of classification and approximation of functions [6].

The general structure of the model are two functions of aggregation (sum and product) and two transfer functions (sigmoidal and gaussian). In the Fig. 1 the structure of the generalized neuron can be observed.

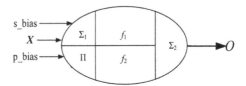

Fig. 1. Generalized neuron

Unlike the common neuron model that has either aggregation functions $\prod$ (product) or $\sum$ (sum), the generalized neuron model has both aggregation functions.

The characteristic sigmoidal function (f_1) is used with the addition function $\sum_1$, while the Gaussian function (f_2) is used with the product function $\prod$. The output of the $\sum_1$ part with the sigmoidal activation function is represented by Eq. 1.

$$O_\Sigma = f_1(S_{GNN}) = \frac{1}{1 + exp(-\lambda_s * S_{GNN})} \tag{1}$$

In this case, $S_{GNN} = \sum W_{\Sigma i} X_i + X_{0\Sigma}$, where λ_s is the gain factor, W_Σ represent the input weights and $X_{0\Sigma}$ is the sum-bias. The output neural using *product* aggregation function is represented by Eq. 2.

$$O_\Pi = f_2(P_{GNN}) = exp(-\lambda_p * (P_{GNN})^2) \tag{2}$$

Also, for this case, $P_{GNN} = \prod W_{\Pi i} X_i * X_{0\Pi}$, where λ_p also is the gain factor, W_Π represent the input weights and $X_{0\Pi}$ is the product-bias. Finally, the total output of the GN is represented by Eq. 3.

$$O_{GNN} = W * O_\Sigma + (1 - W) * O_\Pi \tag{3}$$

4   Bioinspired Algorithms

Bio-inspired algorithms are a field of study that combines engineering, social behavior and emergency. They are effective optimization algorithms, since they are inspired by the behavior of intelligent animals. This type of algorithm finds a set of possible solutions distributed in a search space, these solutions change in terms of different operators that help improve the solution of the problem [7].

The solutions are evaluated using a fitness function until the best solution is found.

In this section, we describe the algorithms used to select the set of genes, as well as those used for the training of a generalized neuron.

4.1   Genetic Algorithms

Genetic algorithms (GA) work on a set of potential solutions, called *population*. This population is composed of a series of solutions called *individuals* and an individual is made up of a series of positions that represent each of the variables involved in the optimization processes and which are called *chromosomes*. These chromosomes are composed of a chain of symbols that in many cases is presented in binary numbers, although it is also possible to use hexadecimal, octal, real coding, etc. In a genetic algorithm, each individual is defined as a data structure that represents a possible solution to the search space of the problem. Evolution strategies work on individuals, which represent the solutions to the problem, so these *evolve* through *generations*. Within the population, each individual is differentiated according to their value of *aptitude* or *fitness*, which is obtained using some measures according to the problem to be solved. To obtain the next generations, new individuals are created, called *children*, using two basic evolution strategies such as the crossover operator and the mutation operator. In Algorithm 1 the pseudocode of the simple genetic algorithm is shown.

Algorithm 1. Simple Genetic Algorithm

```
S ← Generate an initial population S
while the stop condition is not reached do
    evaluate each individual of the population S
    for i = 1 until size (S)/2 do
        Select two individuals from the previous generation.
        Cross with certain probability the two individuals obtaining two
        descendants.
        Mutation of individuals.
        Insert the two mutated descendants in the new generation.
    end for
end while
return The best solution found.
```

4.2 Particle Swarm Optimization

The particle swarm optimization (PSO) algorithm is a method for the optimization of continuous non-linear functions proposed by Kennedy and Eberhart [8]. Is a metaphor of the social behavior, for the case of a bird flocking, is inspired on the movement of the flock in the search of food.

PSO is easy to implement and less time consuming process as there are no crossover and mutation operators like Genetic Algorithm. The potential solutions in PSO are called as particles, which moves in the problem space, finds the optimum solution and stores it in the memory. Each particle moves towards personal and global best position with a certain velocity and modifies it after each iteration. The velocity and position of ith particle in space are updated as follows

$$v_i^{k+1} = w * v_i^k + \varphi_1 * rand_1 * (pBest_i - x_i^k) + \varphi_2 * rand_2 * (g_i - x_i^k) \quad (4)$$

$$x_i^{k+1} = x_i^k + v_i^{k+1} \quad (5)$$

The Eq. 4 reflects the update of the velocity vector of each particle i in each iteration k. The *cognitive* component is modeled by the factor $\varphi_1 * rand_1 * (pBest_i - x_i^k)$ and represents the distance between the current position and the best known position for that particle, that is, the decision that the particle influenced by his own experience throughout his life. The *social* component is modeled by $\varphi_2 * rand_2 * (g_i - x_i^k)$ and represents the distance between the current position and the best neighborhood position, that is, the decision that the particle will take according to the influence that the rest of the cluster exerts on her. A more detailed description of each factor is made below:

- v_i^k: velocity of the particle i in the k iteration
- w: inertia factor
- φ_1,φ_2: they are learning rates (weights) that control the *cognitive* and *social* components
- $rand_1,rand_2$: random numbers between 0 and 1.
- x_i^k: current position of the particle i in the k iteration
- $pBest_i$: best position (solution) found by the i particle so far
- g_i: represents the position of the particle with the best $pBest\_fitness$ of the environment of p_i (*lBest* or *localbest*) or of the whole cluster (*gBest* or *globalbest*).

The Eq. 5 models the movement of each particle i in each iteration k.

In Algorithm 2 the pseudocode of the global version is shown

Algorithm 2. `Global PSO`

   $S \leftarrow$ `Initialize Cumulus`
  while `The stop condition is not reached` **do**
    for $i = 1$ *until* $size(S)$ **do**
      `evaluate each particle` x_i `of the cluster` S
      if `fitness`(x_i) `it is better than fitness`$(pBest_i)$ **then**
        $pBest_i \leftarrow x_i$; `fitness`$(pBest_i) \leftarrow$ `fitness`(x_i)
      end if
      if `fitness`$(pBest_i)$ `it is better than fitness`$(gBest)$ **then**
        $gBest \leftarrow pBest_i$; `fitness`$(gBest) \leftarrow$ `fitness`$(pBest_i)$
      end if
    end for
    for $i = 1$ *until* $size(S)$ **do**
      $v_i \leftarrow w * v_i + \varphi_1 * rand_1 * (pBest_i - x_i) + \varphi_2 * rand_2 * (gBest - x_i)$
      $x_i \leftarrow x_i + v_i$
    end for
  end while
  return `The best solution found`

4.3   Differential Evolution

The differential evolution (DE) algorithm is a optimization technique proposed in [9]. It has few parameters -CR (crossover rate) and F (mutation rate)- and it converges to the optimum faster than others evolutionary techniques. The pseudo code of this algorithm is shown in Algorithm 3.

Algorithm 3. Differential evolution

```
G= 0
```
Initialize the population $x_{i,G}$ $\forall i, i = 1, ..., M$
Evaluate the population $f(x_{i,G} \forall i, i = 1, ...M)$
for $G = 1$ to MAX NUM ITERACIONES do
   for $i = 1$ to M do
      Select $r_1 \neq r_2 \neq r_3$ randomly.
      $j_{rand} = randint(1, D)$
      for $i = 1$ to M do
         if $rand_j[0,1) < CR$ o $j = j_{rand}$ then
            Calculate the value $u_{i,j,G+1} = X_{r3,j,G} + F * (x_{r1,j,G} - x_{r2,j,G})$
         else
            $u_{i,j,G+1} = x_{i,j,G}$
         end if
      end for
      if $f(u_{i,G+1}) \leq f(x_{i,G})$ then
         Select $x_{i,G+1} = u_{i,G+1}$
      else
         Select $x_{i,G+1} = x_{i,G}$
      end if
   end for
   G=G+1
end for
return   The best solution found

---

5   Methodology to Classify DNA Microarrays

The methodology used in this work to perform a task of binary classification of DNA microarrays is divided into two stages:

- The first one dedicated to selecting the set of genes that best describe the DNA microarray.
- The second focused on training a generalized neuron to improve the accuracy of the classification task.

and the schematic representation can be seen in the Fig. 2.

The first step proposes a reduction in the dimensionality of the DNA microarray, reducing the number of characteristics, with which the generalized neuron

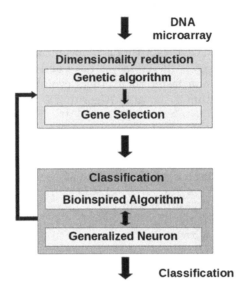

Fig. 2. Scheme of the proposed methodology

will be trained. The second step is to use this information to train a generalized neuron and perform the classification. These two steps are repeated until the maximum number of iterations is reached. A more detailed description of the stages are described below:

5.1 Dimensionality Reduction

According to [10], the selection of the set with the best genes can be defined in terms of an optimization problem. In this work a genetic algorithm was used. To explore the solution space, individuals are subsets of genes of the fixed-length DNA microarray. The fitness function used, is the minimum number of elements wrongly labeled when using those genes to train a GN, that is, with the proposed base different GN models are trained, each of them with a different classification result, taking as fitness value the classification with fewer errors. An example of coding is:

$$[19, 58, 7, 325, 205] \tag{6}$$

Vector entries vary between 0 and the number of features that have the original base minus one, which indicates the indexes of the columns of the original data set that will be taken to form a new database. The cross operator used is, given the individuals $Ind_1 = [I_{11}, I_{12}, \ldots, I_{1n}]$ and $Ind_2 = [I_{21}, I_{22}, \ldots, I_{2n}]$, two descendants are generated, $H_1 = Ind_1 + \alpha * (Ind_2 - Ind_1)$ and $H_2 = Ind_2 + \alpha * (Ind_1 - Ind_2)$, with $\alpha \in [-0.25, 1.25]$.

The mutation is a unary operator, given an individual $Ind_1 = [I_{11}, I_{12}, \ldots, I_{1n}]$ this has a 10% chance of mutating (in most cases it is low), if the individual mutates, each entry I_{1k} will have the $\frac{1}{n}$ probability of mutating, that is to say: $Im_{1k} = I_{1k} + (\beta - 0.5) * (p_2 - p_1) * 0.1$ where $\beta \in (0, 1)$, $p_2 = 4$ and $p_1 = -4$. After applying these operators, the floor function is applied to all the individuals, checking in each individual that the entries are all different from each other, otherwise the case will be replaced with a random one.

5.2 Classification Using a GN

Once proposed the set of characteristics, these genes will form a new database, which is partitioned into two sets: training and testing. The training set has the 70% and the test set the remaining 30%. This partition is done randomly to ensure that the sets contain elements with both labels. Then the GN is trained with a bioinspired algorithm, with real coding.

The solutions (individuals), generated with the bioinspirated algorithm, encode the structure of GN in terms of synaptic weights (W_Σ, W_Π, W), bias and parameters of the activation function (λ) for each type of neuron $(\Sigma$ and $\prod)$.

An example of coding is:

$$[0.98, 2.12, -3.45, -0.34, 1.52,$$
$$0.872, -3.369, 2.548, -0.125, 1.417, \qquad (7)$$
$$-2.834, 0.723, -1.396, 2.196, -3.592]$$

The entries are in the range $(-4, 4)$, if the number of characteristics in the base is 5, the length of the individuals is $(2 * 5) + 5$. The cross and mutation are those mentioned above.

The step function is applied to the output of the GN to determine the class to which it belongs. In the Fig. 3 the graph of that function is shown. The fitness function is based on the number of mislabeled items, the confusion matrix is obtained by comparing the actual labels with the proposed labels, the number of false negatives and false positives is the value of the fitness function that is sought to be minimized.

$$U(t) = \begin{cases} 0, & t \le 0 \\ 1, & t > 0 \end{cases}$$

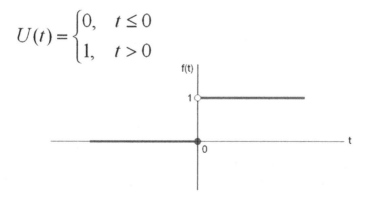

Fig. 3. Graph of the step function of Heavisidee

This proposal can be applied only in cases of binary classification, if it were necessary to choose between three or more labels, it would be necessary to add more generalized neurons.

6 Experimental Results

In this section, the results obtained will be analyzed with the proposed methodology to determine the accuracy, using three bases of different microarrays.

To select the best set of genes using the genetic algorithm, with whole coding, the following parameters were defined: population size $= 30$ and maximum number of cycles $= 30$, selection per tournament, crossing probability $= 100\%$, probability of mutation of the individual $= 10\%$, probability of mutation of each gene $= 1/(number\ of\ genes)$ and elitism.

– GA and GA. The best results obtained are shown in the Table 1.

Table 1. # It.= Number of iterations, Tra.= Training

Data set	Genes	# It.	Precision		Recall		f1-score	
			Tra.	Test	Tra.	Test	Tra.	Test
ALL-AML (7129)	15	30	0.91	0.96	0.90	0.95	0.90	0.95
ALL-AML (7129)	25	50	0.85	1.0	0.84	1.0	0.83	1.0
Colon Tumor (2000)	10	30	0.89	0.95	0.88	0.95	0.88	0.95
Colon Tumor (2000)	15	50	0.82	1.0	0.81	1.0	0.81	1.0
Prostate cancer (12600)	40	30	0.95	0.90	0.95	0.90	0.95	0.90
Prostate cancer (12600)	20	50	0.87	0.93	0.87	0.93	0.87	0.93

– GA and PSO. The best results obtained are shown in the Table 2.

Table 2. # It.= Number of iterations, Tra.= Training

Data set	Genes	# It.	Precision		Recall		f1-score	
			Tra.	Test	Tra.	Test	Tra.	Test
ALL-AML (7129)	3	30	0.90	1.0	0.88	1.0	0.87	1.0
ALL-AML (7129)	15	50	0.88	1.0	0.88	1.0	0.88	1.0
Colon Tumor (2000)	20	30	0.81	1.0	0.81	1.0	0.81	1.0
Colon Tumor (2000)	5	50	0.84	1.0	0.84	1.0	0.83	1.0
Prostate cancer (12600)	30	30	0.88	0.95	0.86	0.95	0.86	0.95
Prostate cancer (12600)	10	50	0.92	1.0	0.92	1.0	0.92	1.0

– GA and ED. The best results obtained are shown in the Table 3.

Table 3. # It.= Number of iterations, Tra.= Training

Data set	Genes	# It.	Precision		Recall		f1-score	
			Tra.	Test	Tra.	Test	Tra.	Test
ALL-AML (7129)	15	30	1.0	0.96	1.0	0.95	1.0	0.96
ALL-AML (7129)	3	50	0.87	1.0	0.86	1.0	0.85	1.0
Colon Tumor (2000)	5	30	0.89	1.0	0.86	1.0	0.85	1.0
Colon Tumor (2000)	10	50	0.91	1.0	0.91	1.0	0.91	1.0
Prostate cancer (12600)	10	30	0.95	0.96	0.95	0.95	0.95	0.95
Prostate cancer (12600)	20	50	0.95	0.96	0.95	0.95	0.95	0.95

7   Conclusions and Future Work

The different experiments allow to determine the behavior of the proposed methodology in the classification of DNA microarrays.

During the first stage, a dimensionality reduction was successfully applied to the data sets *Leukemia benchmarck ALL − AML*, *Colon Tumor* and *Prostate cancer* to select the set of genes that best describes a disease in particular, using a bioinspired algorithm. The problem of dimensionality reduction can be treated as an optimization problem, because the dimensional decrease of a DNA microarray can be seen as a combinatorial problem that tries to find among thousands of genes the most relevant.

The results obtained, in most of the cases, used less than one percent of the genes to perform a detection or classification task, coinciding with the results reported in the literature.

The set of genes of interest in each case is found in a very wide search space, for example for the data set *colon Tumor* where each sample contains 2000 genes, when looking for subsets of 5 genes, the search space consist of 265 335 665 000 400 solutions, when looking for subsets with 10 elements the search space $2.7589878594600562 \times 10^{26}$ which is greater than the age of the universe in seconds. Unable to evaluate all possible solutions, making use of bioinspired algorithms is a viable option.

In the second stage, the performance of the GN was evaluated, the results obtained showed that the whole data set was solved with a good precision, so the bioinspired algorithms are adequate techniques to train a GN. These GN were trained using the set of genes proposed in the first stage.

It can be observed that the GN training can be done using different algorithms, obtaining similar results.

The number of weights used to train the model is reduced thanks to the decrease in the dimensionality of the microarrays and the compact structure of the GN, which allows limiting the numerical complexity of the problem while maintaining a good performance.

The GN trained with the proposed methodology is able to detect, predict and classify a disease with an acceptable precision.

Comparing the results obtained with those described in [3], we can see that they are as good as and the GN has a lower number of weights to train.

Comparing the results obtained when training with the three bioinspired algorithms and subsets of the base *Leukemia ALL − AML* it can be concluded that an adequate subset should contain between 15 and 20 genes. A suitable subset for the base *colon Tumor*, should contain between 30 and 40 genes and for the base *Prostate cancer* the subset should contain between 10 and 15 genes.

The subsets of indices obtained are all different from each other, even though the percentage of precision is very similar, which could suggest that there are multiple optima.

The experiments confirm that the bioinspired algorithms depend to a large extent on the initialization of the population and on the number of possible solutions that it evaluates.

As future work, we intend to use this proposal to find the subsets of genes that best describe a disease.

Acknowledgments. The authors thank the IIMAS headquarters Mrida and Dr. Ernesto Perez Rueda for their valuable comments. Alejandra Romero thanks CONA-CYT for the scholarship received.

References

1. Rizwan, M., Jamil, M., Kothari, D.P.: Generalized neural network approach for global solar energy estimation in India. IEEE Trans. Sustain. Energy **3**, 576–584 (2012)
2. Kiran, R., Jetti, S.R., Venayagamoorthy, G.K.: Online training of a generalized neuron with particle swarm optimization. In: International Joint Conference on Neural Networks, IJCNN 2006, pp. 5088–5095. IEEE (2006)
3. Dwivedi, A.K.: Artificial neural network model for effective cancer classification using microarray gene expression data. Neural Comput. Appl. **29**(12), 1545–1554 (2018)
4. Jain, I., Jain, V.K., Jain, R.: Correlation feature selection based improved-binary particle swarm optimization for gene selection and cancer classification. Appl. Soft Comput. **62**, 203–215 (2018)
5. Kulkarni, R.V., Venayagamoorthy, G.K.: Feedforward and recurrent architectures: generalized neuron. Neural Netw. **22**, 1011–1017 (2009)
6. Chaturvedi, D.K., Man, M., Singh, R.K., Kalra, P.K.: Improved generalized neuron model for short-term load forecasting. Soft Comput. **8**(1), 10–18 (2003)
7. Medina, C.H.D.: Algoritmos bioinspirados. Ph.D. thesis, Instituto Politécnico Nacional. Centro de Investigación en Computación (2011)

8. Eberhart, R.C., Shi, Y., Kennedy, J.: Swarm Intelligence. Elsevier, Amsterdam (2001)
9. Storn, R., Price, K.: Differential evolution-a simple and efficient adaptive scheme for global optimization over continuous spaces. ICSI, Berkeley (1995)
10. Garro, B.A., Rodríguez, K., Vázquez, R.A.: Classification of DNA microarrays using artificial neural networks and ABC algorithm. Appl. Soft Comput. **38**, 548–560 (2016)

Supervised and Unsupervised Neural Networks: Experimental Study for Anomaly Detection in Electrical Consumption

Joel García[1](✉), Erik Zamora[1](✉), and Humberto Sossa[1,2](✉)

[1] Instituto Politécnico Nacional - Centro de Investigación en Computación,
Av. Juan de Dios Batiz S/N, GAM, 07738 Mexico City, Mexico
`joel.gaid.we@gmail.com`, {`ezamorag,hsossa`}`@cic.ipn.mx`
[2] Tecnológico de Monterrey, Campus Guadalajara, Av. Gral. Ramón Corona 2514,
45138 Zapopan, Jalisco, Mexico

Abstract. Households are responsible for more than 40% of the global electricity consumption [7]. The analysis of this consumption to find unexpected behaviours could have a great impact on saving electricity. This research presents an experimental study of supervised and unsupervised neural networks for anomaly detection in electrical consumption. Multilayer perceptrons and autoencoders are used for each approach, respectively. In order to select the most suitable neural model in each case, there is a comparison of various architectures. The proposed methods are evaluated using real-world data from an individual home electric power usage dataset. The performance is compared with a traditional statistical procedure. Experiments show that the supervised approach has a significant improvement in anomaly detection rate. We evaluate different possible feature sets. The results demonstrate that temporal data and measures of consumption patterns such as mean, standard deviation and percentiles are necessary to achieve higher accuracy.

Keywords: Anomaly detection · Neural networks · Supervised Unsupervised · Statistic

1 Introduction

Energy production has had a considerable impact on the environment. Moreover, different international agencies, such as OECD [11] and UNEP [7], estimated that the global energy consumption will increase by 53% in the future. Saving energy and reducing losses caused by fraud are today's challenges. In China, for instance, numerous companies have adopted strategies and technologies to be able to detect anomalous behaviours of a user's power consumption [14]. With the development of internet of things and artificial intelligence, it is possible to take measures of household consumption and analyze them for anomalous behaviour patterns [3].

© Springer Nature Switzerland AG 2018
I. Batyrshin et al. (Eds.): MICAI 2018, LNAI 11288, pp. 98–109, 2018.
https://doi.org/10.1007/978-3-030-04491-6_8

Anomaly detection is a problem that has been widely studied with many techniques. To be able to apply the correct one, it is necessary to identify the sort of anomaly. In [5] Chandola, Banerjee and Kumar presented a research about anomaly types found in real applications and the most suitable method of attacking them. Anomalies are value groups which do not match with the whole dataset. They can be classified within three categories: point, contextual and collective anomalies. Point anomalies refer to individual data values that can be considered as anomalous with respect to the remainder. Meanwhile, contextual anomalies take advantage of complementary information related to the background of the problem in order to discover irregularities. Collective anomalies are similar to the contextual type. Collective refers to the fact that a collection of instances can be considered anomalous with respect to the rest of the elements. Our research considers the contextual anomalies in electrical consumption, because they need extra knowledge besides the usage measurements, to be detected. According to Chandola *et al.* [5], for anomaly detection in power consumption the most suitable methods are: Parametric Statistic Modelling, Neural Networks, Bayesian Networks, Rule-based Systems and Nearest Neighbor based techniques. The focus of this paper is on two neural network approaches: supervised and unsupervised learning. Furthermore, we present a feature selection study. The results are compared with a traditional statistical method based on the two-sigma rule of normal distributions.

The remaining sections of the paper are organized as follows: Sect. 2 presents a brief description of related work to the detection of electrical consumption anomalies. Section 3 describes the neural network methods proposed and the statistical procedure. Section 4 outlines the experiments and Sect. 5 gives the results obtained. Finally, Sect. 6 states the conclusion and future work.

2   Related Work

Diverse techniques have been used for identifying anomalies; statistics procedures and machine learning are some of them. In [2] and [12], two research works of 2016 and 2017 respectively, an unsupervised learning technique is presented. Both cases use autoencoders as the pattern learner engine. The difference is the composition of the feature sets. In [12] an 8-feature set is used while in [2] there is an increase in the number of temporal and generated data that results in a 15-feature set. So, in this research, we analyze feature sets by evaluating three possible groups, two based on the ones used in [12] and [2], and a third new feature set with 10 variables.

In [13] Lyu *et al.* propose a hyperellipsoidal clustering algorithm for anomaly detection focused on fog computing architecture. Jui-Sheng Chou, Abdi Suryadinata Telaga implemented a neural network called ARIMA (Auto-Regressive Integrated Moving Average) to identify contextual anomalies [6]. The network learns the normal patterns of energy usage, then it uses the two-sigma rule to detect anomalies. An interesting fact is that, the ARIMA architecture has only one hidden layer. Thus, it is convenient to explore more complex models. In this study, we test various deep neural architectures and compare their accuracy. In [18] Yijia and Hang developed a waveform feature extraction model for anomaly detection. It uses line loss and power analysis. Ouyang, Sun and Yue proposed a time series feature extraction algorithm and a classification model to detect anomalous consumption [15]. They treat the problem as with point anomalies and exclude other information besides the consumer's usage. On the contrary, we consider temporal information and generated data such as mean, standard deviation or percentiles since a point value could be abnormal in one context but not in another.

3   Methodology

In this section, we introduce the components and techniques used in the supervised and unsupervised learning method, and the statistical procedure. Autoencoder and ROC curve are used by the unsupervised approach, while multilayer perceptrons correspond to the supervised approach.

3.1   Autoencoder

The unsupervised method for anomaly detection is based on the neural model known as autoencoder or autoassociator [4]. The structure of an autoencoder has two parts; an encoder and a decoder. The goal of this network is to encode an input vector x so as to obtain a representation $c(x)$. Then it should be possible to reconstruct the input data patterns from its representation. The cost function for this model is the Mean Squared Error MSE given in (1):

$$MSE = \frac{1}{D} \sum_{i=1}^{D} (X_i - \hat{X}_i)^2 \tag{1}$$

where $\hat{X}_i$, predicted value; X_i, observed value; and D, sample size.

We expect the autoencoder to capture the statistical structure of the input signals. The Reconstruction Error RE, defined in (2), is the difference between the current input and the corresponding output of the network. It is a metric that shows how similar those elements are.

$$RE = \sqrt{\sum_{j=1}^{D} (X_j - \hat{X}_j)^2} \tag{2}$$

where X, input vector of D different variables; and $\hat{X}$, vector of D different variables of the constructed output by the autoencoder.

Then, it is necessary to establish a threshold for the Reconstruction Error. A small threshold might result in large False Positives (FP), which are normal patterns identified as anomalies. On the other hand, big thresholds achieve higher False Negative (FN) or anomalous values unidentified.

3.2   Receiver Operating Characteristics

The Receiver Operating Characteristics (ROC) curve has been widely used since 1950 in fields such as electric signal detection theory and medical diagnostic [10]. In a ROC graph, x-axis represents the False Positive Rate FPR (or 1-Specificity) and the y-axis the True Positive Rate TPR (or sensitivity). All possible thresholds are then evaluated by plotting the FPR and TPR values obtained with them. Hence, we can select a threshold that maximizes the TPR and at the same time minimizes the FPR. This policy would correspond to point $(0, 1)$ in the ROC curve, as shown in Fig. 1. In this work, we focus on that policy.

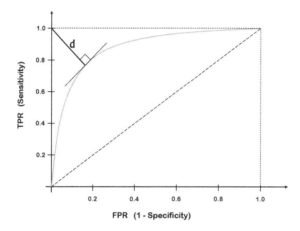

Fig. 1. ROC curve analysis for determination of optimal threshold.

Once the desired point in the plot has been established, the most suitable threshold can be determined by calculating the shortest distance, from that point to the curve, with the Pythagoras theorem given in (3):

$$d^2 = (1 - \text{sensitivity})^2 + (1 - \text{specificity})^2 \tag{3}$$

where d is the shortest distance from a desired point to the ROC curve.

In this research, we evaluate 7 autoencoder models for anomaly detection. The threshold that optimises TPR and FPR is calculated in each case.

3.3   Multilayer Perceptron

A Multilayer Perceptron (MLP) is a class of artificial neural network formed by layers of perceptrons. MLPs usually have an input layer, some hidden layers and an output layer. Every perceptron is a linear classifier. It produces a unique output y by executing a linear combination of an input vector X, multiplied for a weights vector W^T, and adding a bias b, as shown in (4):

$$y = \varphi(\sum_{i=1}^{n} w_i x_i + b) = \varphi(W^T X + b) \tag{4}$$

Where φ is an activation function which is typically the hyperbolic tangent or the logistic sigmoid. In this work, we use the hyperbolic tangent function since it can output positive and negative values, which is relevant for the application.

MLPs frequently perform classification and regression tasks. This network learns the dependencies between inputs and outputs in supervised learning through the backpropagation training algorithm [1]. During the training process, weights and bias are modified in order to minimized a certain loss function. The loss or objective function can be defined in different ways; this implementation uses the Mean Squared Error shown in Eq. (1).

3.4   Statistical Procedure

The proposed models, based on neural networks, were compared with a statistical technique. This traditional procedure employs the properties of gaussian distributions. This probability distribution is extensively used when modelling nature, social and psychological problems [9]. The properties that the statistical method exploits are related to the standard deviation and coverage of the normal distribution.

The 3-sigma rule states three facts; the first establishes that, in the range $[\mu - \sigma, \mu + \sigma]$ is approximately 68.26% of the distribution. About 95.44% of the data is within two standard deviations of the median $[\mu - 2\sigma, \mu + \sigma]$. This characteristic is commonly named as two-sigma rule. Finally, 99.74% of the distribution data falls in the range $[\mu - 3\sigma, \mu + 3\sigma]$. Considering the interval which contains about 95% of the instances, the remaining 5% can be considered abnormal values as they are out of the expected range. In this context, an anomalous element is a value that does not match with the normal pattern of the tenant electricity consumption. We used the two-sigma rule to label data of the consumption dataset for the supervised neural network experiment (Fig. 2).

3.5   Metrics

The approaches seen in this paper utilize the F1-score as the metric to assess the model's performance. The F1-score, given by (5), relates the precision P (6) and recall R (7) of every model. Precision is also known as the positive predicted value while recall is also referred to as sensitivity:

$$F1_{score} = 2\frac{PR}{P + R} \tag{5}$$

$$P = \frac{TP}{TP + FP} \tag{6}$$

$$R = \frac{TP}{TP + FN} \tag{7}$$

where P, precision; R, recall; TP, number of true positives; FP, number of false positives; and FN, number of false negatives.

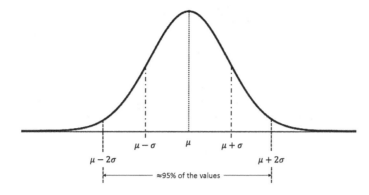

Fig. 2. Normal graph with distribution data properties.

4   Experiments

In this section, four experiments are presented: feature sets evaluation, supervised method, unsupervised method evaluation and the statistical procedure implementation.

4.1   Dataset

The information correspond to an individual household electric power consumption database generated by UC Irvine Machine Learning Repository [8]. It contains measurements of electric power consumption in a home with one-minute sampling rate over a period of almost 4 years. Data was split into two groups, measurements of 2007 are used for training and validation while values of 2008 are used for testing.

4.2   Features Evaluation

Feature selection is the most important part in contextual anomaly detection. Choosing the right variables can lead to high anomaly detection rates. In the neural networks approach evaluation, we built three distinct feature sets: A, B and C. Feature sets A and C are based on the sets used in [12] and [2], respectively. The third set, B, was proposed with an average number of features. The three sets are shown in Tables 1, 2 and 3. Every feature set is built with data of three categories: **temporal information** such as season, day of the week, month and so on; a 12 hourly values **consumption window**; and **generated data** derived from the window consumption such as mean, standard deviation, subtraction between elements of the window, interquartile range and percentiles. Although all three sets have the 12 hourly consumption window, they contain a different number of temporal values and generated data.

The unit circle projection used for temporal features consists of a representation to encode periodic variables [17]. The mapping is calculated by a sine and cosine component, as shown in formula (8):

$$\hat{t}_{n,k} = \{cos(\frac{2\pi t_{n,k}}{p_k}), sin(\frac{2\pi t_{n,k}}{p_k})\} \tag{8}$$

where $k = \{month, \, day \, of \, week, \, week \, of \, year, \, ...\}$; $\hat{t}_{n,k}$, periodic value; p_k, known period. For $k = month \rightarrow p_k = 12$, for $k = day \, of \, week \rightarrow p_k = 7$ and so on. In contrast to other encodings, such as one-hot [16], which increases the input dimensionality, the unit circle representation requires only two dimensions per periodic value. One-hot encoding maps categorical variables to integer values and then represents each one with a binary vector. In each vector, only the element whose index equals the integer value is marked as one, the remainders are marked as zero.

Table 1. Feature set A.

Feature	Description
Month of year	Unit circle projection with $p_k = 12$
Day of week	Unit circle projection with $p_k = 7$
Week of year	Unit circle projection with $p_k = 53$
Hour of day	Unit circle projection with $p_k = 24$
Electricity consumption window	Window with 12 samples of energy active in one hour $S_j, \{j = 1, ..., 12\}$
First difference	Difference between last and first elements of a consumption window $S_{12} - S_1$

Table 4 shows the highest F1-scores derived from the supervised and unsupervised method. The biggest difference between the sets can be observed in the unsupervised approach. In this experiment, the third feature group outperforms the first two sets by having a 0.899 F1-score. However, with the supervised implementation the difference is smaller. Although the third set continues being the best, the remaining sets almost overlap in their performance.

4.3   Unsupervised Method

In this approach, seven autoencoder models were evaluated. As can be seen in Table 5, their structure varies from one layer for the encoder and the decoder part, up to three layers in each. The number of neurons in each layer is in terms of n. It depends on the characteristic vector length, which is different for every proposed set. That is, n is 21 for feature set A, 28 for set B and 33 for set C. In the model M4, the number of neurons is 33 for set A and B, and it is 40 for set C. In order to calculate the F1-score for anomaly detection, we have used the ROC curve analysis to obtain a threshold for the reconstruction error in each autoencoder model. Table 6 shows the scores achieved for the seven models with the three feature sets. For each feature set there is a model that accomplishes the best performance. For set A the advisable model is M7 while for set B it is M6 and M5 for set C.

Table 2. Feature set B.

Feature	Description
Day of year	Unit circle projection with $p_k = 365$
Season	1–4 Winter, spring, summer, autumn
Month of year	Unit circle projection with $p_k = 12$
Day of week	Unit circle projection with $p_k = 7$
Hour of day	Unit circle projection with $p_k = 24$
Electricity consumption window	Window with 12 samples of energy active in one hour $S\{j = 1, ..., 12\}$
$\bar{x}$	Mean of sensor data values in each consumption window
s	Standard deviation of sensor data values in each consumption window
$S_{12} - S_1$	Difference between last and first elements of a consumption window
$\bar{x}_i - \bar{x}_{(i-1)}$	Difference between the means of ith and $(i-1)^{st}$ consumption windows

4.4   Supervised Method

In this section, four models of MLP were designed following the pattern shown in Table 7. As in the unsupervised experiment, the numbers of neuron instances depend on the characteristic vector length (n is 21 for set A, 28 for set B and 33 for set C). The models were trained for classification with labeled data of 2007. We used the two-sigma rule of normal distribution for labeling. Then the models were tested with data of 2008. Table 8 reports F1-scores accomplished with the supervised approach. In this case, all MLPs have accuracy better than that obtained by the unsupervised model based on autoencoders. All models have a score above 0.93. Moreover, the difference between the highest values with each set is just one percentage point. That is, set A with M4 had 0.947 while set B with M1 obtained 0.957 and set C with M2 gave 0.967.

5   Results and Discussion

The supervised approach outperforms the results of the unsupervised method and the statistical procedure. Besides, this method can be implemented on embedded systems due to the light structure of MLPs. In the supervised methods, we combine a statistical part and a neural network technique. This produces what can be named an hybrid model; at first, data rows are labeled with the two sigma rule and then, a multilayer perceptron is trained to execute clasiffication. The feature set C performs better in both, supervised and unsupervised learning, as demonstrated by having the greatest F1-score values. This improvement is achieved by the increase of temporal and generated data. So we conclude that

Table 3. Feature set C.

Feature	Description
Day of year	Unit circle projection with $p_k = 365$
Season	1–4 Winter, spring, summer, autumn
Month of year	Unit circle projection with $p_k = 12$
Day of week	Unit circle projection with $p_k = 7$
Hour of day	Unit circle projection with $p_k = 24$
Electricity consumption window	Window with 12 samples of energy active in one hour $S\{j = 1, ..., 12\}$
$\bar{x}$	Mean of sensor data values in each consumption window
s	Standard deviation of sensor data values in each consumption window
$S_{12} - S_1$	Difference between last and first elements of a consumption window
$\bar{x}_i - \bar{x}_{(i-1)}$	Difference between the means of ith and $(i-1)^{st}$ consumption windows
$\bar{x}_{(i+1)} - \bar{x}_i$	Difference between the means of $(i+1)^{st}$ and i^{th} consumption windows
Q1	First quartile of the sensor data values in each window
Q2	Median of the sensor data values in each window
Q3	Third quartile of the sensor data values in each window
IQR	Interquartile range of the sensor data values in each window

Table 4. Performance of feature sets in neural networks approach.

Approach	F1-score		
	Feature set A	Feature set B	Feature set C
Unsupervised	0.540	0.791	**0.899**
Supervised	0.948	0.957	**0.967**

in order to have a high anomaly detection rate, additional information like the previously mentioned is needed. Contextual or temporal data represent relevant information concerning the individual consumption. This pattern depends on the moment it was measured, for example the season and the hour of the day, but also, generated data such as the mean and standard deviation of an hourly consumption reading can vary from one user to another. Analysing consumption in buildings to find unexpected behaviours is especially useful; it can lead users

Table 5. Structure of autoencoder models.

Model	Autoencoder					
	Encoder			Decoder		
	Number of neurons in layer					
	1	2	3	1	2	3
M1	$n/2$	-	-	n	-	-
M2	$2*n/3$	-	-	n	-	-
M3	n	-	-	n	-	-
M4	33 or 40	-	-	33 or 40	-	-
M5	$n/2$	$n/3$	-	$n/2$	n	-
M6	$2*n/3$	$n/2$	-	$2*n/3$	n	-
M7	$2*n/3$	$n/2$	$n/3$	$n/2$	$2*n/3$	n

Table 6. F1-scores of unsupervised approach.

Model	F1-score		
	Feature set A	Feature set B	Feature set C
M1	0.457	0.762	0.807
M2	0.473	0.648	0.654
M3	0.518	0.588	0.863
M4	0.463	0.656	0.869
M5	0.530	0.789	**0.899**
M6	0.458	**0.791**	0.786
M7	**0.540**	0.782	0.793

Table 7. Structure of MLPs proposed.

Model	MLP			
	Number of neurons in layer			
	1	2	3	4
M1	$n/2$	2	-	-
M2	$n/2$	$n/3$	2	-
M3	$2*n/3$	$n/2$	2	-
M4	$2*n/3$	$n/2$	$n/3$	2

Table 8. F1-scores of supervised approach MLPs.

Model	F1-score		
	Feature set A	Feature set B	Feature set C
M1	0.942	**0.957**	0.966
M2	0.948	0.955	**0.967**
M3	0.939	0.932	0.951
M4	**0.947**	0.936	0.947

to save energy and for companies to avoid frauds. These are some ways to reduce the environmental impact of the electricity generation and usage.

6   Conclusion and Future Work

In this research, supervised and unsupervised approaches of neural networks for anomaly detection were researched. We compare them with a statistical procedure. Furthermore, feature selection has been studied. The results show that supervised learning outperforms the accuracy of unsupervised and statistical methods. Moreover, the feature set C performs best. The fact that the supervised approach takes advantage of statistical methods to label data before training is what leads to the improvement showed in the results regarding the anomaly detection rate. Because the usage behaviour can differ between buildings, the proposed methods for detecting anomalies in electricity consumption works individually for a certain home. In this sense, a topic for further research is the generation of models that can generalize the consumption in a group of houses. That way, we might decrease the resources occupied in training the neural network models. Future works could test the proposed neural network techniques with other real datasets. Additionally, other methods such as PCA or Support Vector Machines could be researched.

Acknowledgments. E. Zamora and H. Sossa would like to acknowledge CIC-IPN for the support to carry out this research. This work was economically supported by SIP-IPN (grant numbers 20180180 and 20180730) and CONACYT (grant number 65) as well as the Red Temática de CONACYT de Inteligencia Computacional Aplicada. J. García acknowledges CONACYT for the scholarship granted towards pursuing his MSc studies.

References

1. A beginner's guide to multilayer perceptrons. deeplearning4j.org/multilayer perceptron. Accessed 25 May 2018
2. Araya, D.B., Grolinger, K., ElYamany, H.F., Capretz, M.A., Bitsuamlak, G.: Collective contextual anomaly detection framework for smart buildings, pp. 511–518. IEEE (2016)

3. Ashton, K.: That "Internet of Things" thing. RFiD J. **22**, 97–114 (2009)
4. Bengio, Y., et al.: Learning deep architectures for AI. Found. Trends® Mach. Learn. **2**(1), 1–127 (2009)
5. Chandola, V., Banerjee, A., Kumar, V.: Anomaly detection: a survey. ACM Comput. Surv. (CSUR) **41**(3), 15 (2009)
6. Chou, J.-S., Telaga, A.S.: Real-time detection of anomalous power consumption. Renew. Sustain. Energy Rev. **33**, 400–411 (2014)
7. Costa, A., Keane, M.M., Raftery, P., O'Donnell, J.: Key factors methodology: a novel support to the decision making process of the building energy manager in defining optimal operation strategies. Energy Build. **49**, 158–163 (2012)
8. Dheeru, D., Taniskidou, E.K.: UCI machine learning repository (2017)
9. Gómez Chacón, I.M., et al.: Educación matemática y ciudadanía (2010)
10. Hajian-Tilaki, K.: Receiver operating characteristic (ROC) curve analysis for medical diagnostic test evaluation. Casp. J. Intern. Med. **4**(2), 627 (2013)
11. IEA: World energy outlook 2011 executive summary (2011)
12. Tasfi, N.L., Higashino, W.A., Grolinger, K., Capretz, M.A.: Deep neural networks with confidence sampling for electrical anomaly detection, June 2017
13. Lyu, L., Jin, J., Rajasegarar, S., He, X., Palaniswami, M.: Fog-empowered anomaly detection in iot using hyperellipsoidal clustering. IEEE Internet Things J. **4**(5), 1174–1184 (2017)
14. Ouyang, Z., Sun, X., Chen, J., Yue, D., Zhang, T.: Multi-view stacking ensemble for power consumption anomaly detection in the context of industrial internet of things. IEEE Access **6**, 9623–9631 (2018)
15. Ouyang, Z., Sun, X., Yue, D.: Hierarchical time series feature extraction for power consumption anomaly detection. In: Li, K., Xue, Y., Cui, S., Niu, Q., Yang, Z., Luk, P. (eds.) LSMS/ICSEE-2017. CCIS, vol. 763, pp. 267–275. Springer, Singapore (2017). https://doi.org/10.1007/978-981-10-6364-0_27
16. Pedregosa, F., et al.: Scikit-learn: machine learning in Python. J. Mach. Learn. Res. **12**, 2825–2830 (2011)
17. Stover, C.: Unit circle
18. Yijia, T., Hang, G.: Anomaly detection of power consumption based on waveform feature recognition, pp. 587–591. IEEE (2016)

Artificial Neural Networks and Common Spatial Patterns for the Recognition of Motor Information from EEG Signals

Carlos Daniel Virgilio Gonzalez[1](✉), Juan Humberto Sossa Azuela[1,2], and Javier M. Antelis[2]

[1] Centro de Investigación en Computación – Instituto Politécnico Nacional, Av. Juan de Dios Bátiz and M. Othón de Mendizabal, 07738 Mexico City, Mexico
danielvg92@gmail.com, hsossa@cic.ipn.mx
[2] Tecnológico de Monterrey, Escuela de Ingeniería y Ciencias, Av. General Ramón Corona 2514, Zapopan, Jalisco, Mexico
mauricio.antelis@itesm.mx

Abstract. This paper proposes the use of two models of neural networks (Multi Layer Perceptron and Dendrite Morphological Neural Network) for the recognition of voluntary movements from electroencephalographic (EEG) signals. The proposal consisted of three main stages: organization of EEG signals, feature extraction and execution of classification algorithms. The EEG signals were recorded from eighteen healthy subjects performing self-paced reaching movements. Three classification scenarios were evaluated in each participant: Relax versus Intention, Relax versus Execution and Intention versus Execution. The feature extraction stage was carried out by applying an algorithm known as Common Spatial Pattern, in addition to the statistical methods called Root Mean Square, Variance, Standard Deviation and Mean. The results showed that the models of neural networks provided decoding accuracies above chance level, whereby, it is able to detect a movement prior its execution. On the basis of these results, the neural networks are a powerful promising classification technique that can be used to enhance performance in the recognition of motor tasks for BCI systems based on electroencephalographic signals.

Keywords: Brain computer interface · EEG signals · Motor task
Common Spatial Pattern · Dendrite Morphological Neural Network
Multilayer Perceptron

1 Introduction

Brain-Computer Interfaces (BCI) is a promising research field which intends to use non-invasive techniques, allowing communication between humans and computers by brain activity recording, recorded at the surface of the scalp with electroencephalography (EEG). The main purpose for BCI is to enable communication for people with severe disabilities. It is a new alternative to provide

© Springer Nature Switzerland AG 2018
I. Batyrshin et al. (Eds.): MICAI 2018, LNAI 11288, pp. 110–122, 2018.
https://doi.org/10.1007/978-3-030-04491-6_9

people suffering partial or complete motor impairments. In a BCI system, electrical activity within the cerebral cortex is detected through the use of an electrode cap, followed by a EEG signals conversion process in order to obtain instructions to control external devices. In recent years, BCI has been a major focus in the field of brain science and biomedical engineering and it has been developed into a new multidisciplinary cross technology. The key component of a BCI system is how it extracts EEG features to improve the recognition accuracy. At present, the most commonly used pattern recognition methods include the Linear Discriminant Analysis (LDA), K-Nearest Neighbor (KNN) classification algorithm, Naive Bayes (NB) and Support Vector Machine (SVM).

The EEG-based BCI typically operates in cue-based synchronous protocols [1]. For example, in a motor imagery mental (MI) task the user imagines the movement of a limb during a well-establish period of time while the EEG activity is recorded. After the MI is finished, a set of attributes is computed from the recorded EEG signals which are provided to a classification algorithm to recognize the moved limb. Finally, the classifier output is used as a command in an application. The synchronous BCIs requires a few seconds to collect brain signals while the user is carrying out the mental task. After this, the system recognizes the mental task, which takes a few milliseconds. Therefore, the user first performs the mental task and after it is finished the output command is generated, as a consequence, there is a noticeable time interval since the initiation of the mental task and the response produced in the application, which makes that output movements are not seen natural by the user.

To solve these problems, we propose to approach the problem in a different way: Instead of detecting the moment in which the mental task is performed, the moment when the subject is prepared to perform the task itself is detected by using two models of neural networks artificial (ANN).

This can be done because there is a specific area in the human brain that is responsible for this: Premotor cortex [17]. This area is adjacent to the well-known area of the motor cortex. The functions of the premotor cortex are generally divided into two regions: lateral and medial. The neurons in the premotor cortex have responses that are linked to the occurrence of movements as in the primary motor area. However, in contrast to the neurons in the primary motor area, instead of directly ordering the initiation of a movement, these neurons seem to encode the intention to perform a particular movement; therefore, they seem to be particularly involved in the selection/preparation of movements (Fig. 1).

Therefore, here we investigated the recognition of motor execution and motor intention using EEG brain signals recorded in BCI settings. The study applies the ANN models for the recognition of motor tasks from EEG signals recorded in an asynchronous BCI experiment. To do so, EEG signals were recorded from eighteen healthy subjects performing self-paced reaching movements and the ANN classification algorithm was applied to recognize between movement states in the following two-class classification scenarios, relax versus intention, relax versus execution and intention versus execution.

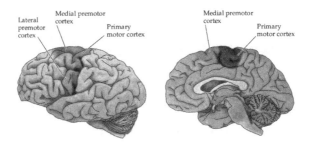

Fig. 1. The primary motor cortex and the premotor area seen in lateral and medial views [17].

BCI are commonly used in different applications, such as: spelling system [7], control of a humanoid robot (NAO) [14], control of a cursor on a screen [15], lighting control [13], control of an object in a virtual environment [4], including the control of a wheelchair [12]. In the topic of *signal processing* there are also different ways of approaching the problem, for example in [21] they use frequency analysis of signals acquired from electrodes placed in the central sulcus of the brain, this method depended on the amplitude of the brain signal. For large amplitudes the cursor moves upwards and for the smallest amplitudes it moves downwards. Another method widely used in the processing of brain signals is the Wavelet Transform, in addition to the Common Spatial Pattern filtering method to classify patterns of 2 kinds of Motor Imagery signals [10]. Other extraction methods are used, such as the Autoregressive Model [11], Power Band [11], Independent Component Analysis (ICA), Principal Component Analysis (PCA), among others. The CSP (Common Spatial Pattern) filtering has already been used in the classification of brain signals and it is currently very popular as it improves the performance of them [3,5,16,22].

2   Acquisition of the Brain Signals

The database used in this work was obtained from eighteen healthy subjects: six males and twelve females (age range 18–23 years). All participants were right-handed students from Tecnológico de Monterrey, Escuela de Ingeniería y Ciencias, they were new in EEG recording or BCI experiments, and none of them presented known neurological or motor deficit. EEG signals were recorded from 21 electrodes according to the 10-10 international system at scalp locations Fp1, Fp2, F7, F3, Fz, F4, F8, T3, C3, Cz, C4, T4, T5, P3, pz, P4, T6, O1, O2, A1, A2. The ground was placed on Fpz while the reference was placed on the left earlobe. The impedance of these electrodes was kept below $5\,k\Omega$. In addition, electromyographic (EMG) signals from both arms were also recorded. The EMG signals were recorded from the biceps using a bipolar montage. The impedance of these electrodes was kept below $20\,k\Omega$. EEG and EMG signals were recorded at a sampling frequency of $2048\,Hz$ with a notch filter at $60\,Hz$ to remove the power-line interference. The motor task consisted of the execution of self-paced reaching movements performed individually with the left or the right arm-hand [9]. For the execution of this experiment, the participants were also seated in front of a computer screen with the forearms relaxed on the chair's arms without causing any muscle tension. The experiment consisted of trials and was controlled by visual cues presented on the screen. A trial consisted of the time sequence depicted in Fig. 2.

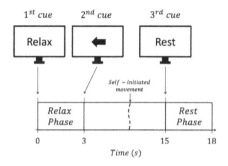

Fig. 2. Illustration of the temporal sequence of a trial during the execution of the experiment [2].

The first cue showed the text "Relax" and indicated to maintain the forearms in the initial position and to relax without moving any body part. This lasted three seconds. The second cue showed the image of an arrow pointing to the left or right and instructed to move the corresponding arm-hand towards the center of the screen. However, the participants were requested not to initiate the movement immediately but first to stay relaxed for about five seconds and then to self-initiate the movement. This second cue was showed by twelve seconds. Because the participants decided when to start the reaching movement, then the actual movement onset is different across trials. The third cue showed the text "Rest" and instructed to rest, blink and move if necessary. This also lasted three seconds. For each participant, the experiment was executed in four blocks of 24 trials each. To avoid fatigue, they were encouraged to rest between blocks as necessary.

The preprocessing aimed to obtain a set of trials trimmed from the first cue up to one second after the movement initiation (estimated with the EMG activity) as well as to clean the resulted EEG traces. First, the sampling rate of the recorded signals was decreased to 256 Hz by simply keeping every 8th sample starting from the first sample. The resulted sampling frequency of 256 Hz is commonly used for EEG signal analysis as it captures all the motor-related brain rhythms. Then, a zero-phase shift four-order Butterworth bandpass-filter from 0.5 to 60 Hz and a common average reference filter were applied to the EEG signals. Then, the last three seconds of each trial was discharged, which resulted in 15 s-long segments that initiated at the first visual cue and finished at the occurrence the third visual cue. Subsequently, the time instant of the reaching movement onset was estimated using the EMG signals. This was carried out for each trial independently by applying the Hilbert transform to the EMG signal of the moving arm. The EMG-based movement onset was then used to discharge trials with fast or slow movement initiation. In specific, we excluded trials that started 3 s before or 11 s after the presentation of the second visual cue (the one that instructed to perform the reaching movement). In addition, the EMG-based movement onset was used as the reference to construct the time axis of each trial. In consequence, $t = 0$ represents the initiation of the reaching movement (same across all trials) while t_{ini} represents the trial's initiation time (different across all trials). Finally, all trials were trimmed from t_{ini} up to 1 s relative to the EMG-based movement onset. As a result, all trials have the same time for movement initiation and trial's end but different trial's initiation time and trial's length. Figure 3 illustrates the temporal axis across trials.

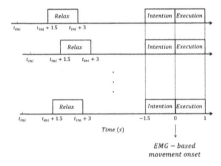

Fig. 3. Illustration of the time axis across trials after the processing step. Note that all trials are referenced ($t = 0$) to the EMG-based movement onset. All trials have different initiation time (t_{ini}) [2].

The database had already been made a selection of the best channels through the r-squared analysis, the process is reported in [2]. It was used to examine significant differences in the spectral power-based features between different phases (e.g. relax versus intention) and subsequently to select the channels with the higher discriminative power. The final number of channels selected was nine: F3, Fz, F4, C3, Cz, C4, P3, Pz and P4.

3   Feature Extraction

In this work, the spatial filtering method was applied to brain signals sampled on the time. As mentioned above, each trial is composed of 9 channels, which were filtered to obtain 9 channels different from the original ones. Later, the calculation of 4 statistical methods of each channel was carried out: Root Mean Square (RMS), Variance, Standard Deviation (SD) and Mean. Finally, a set of

36 features (9 channels × 4 Statistical Method = 36) for each trial was obtained and a stage of feature reduction was carried out using the Principal Component Analysis (PCA) algorithm which allowed us to perform the subsequent classification only using ten features. This process was made with the signals of each subject. In this work, the spatial filter is calculated using the Common Spatial Pattern method, which will be explained below.

3.1   Common Spatial Pattern (CSP)

The spatial filtering is an important part of detecting neuromodulatory changes over the motor cortex. The most used and successful method of discriminating such changes is Common Spatial Pattern (CSP). It has been mainly used in face recognition, object recognition, and EEG anomalies detection. Also, it has been successfully applied to BCI. The CSP algorithm is a kind of multidimensional statistics which has often been applied to EEG signals feature extraction and analysis in two-class multichannel methods. CSP finds optimal spatial filters that maximize the ratio of average variances between two different classes. Computationally, CSP is solved by simultaneous diagonalization of the covariance matrices of the two classes [5]. Figure 4 shows an example of the effect of applying spatial filtering on brain signals. On the left side (Channel 19 and 23), there are two EEG channels that refer to the brain activity of a test subject when performing mental tasks one (Channel 19 and 23 - Up) and two (Channel 19 and 23 - Down). As shown in the picture, there is no clear difference between the signals of both mental tasks. However, new signals are obtained when the CSP filtering is applied (CSP1 and CSP2) which show a greater difference: in CSP1 the mental task one shows a greater variance while the mental task two shows a smaller variation. The opposite occurs in CSP2.

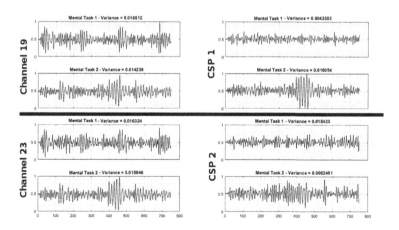

Fig. 4. Effect of CSP filter on patterns.

4   Classification

Four different classification algorithms were employed in this work, K-Nearest Neighbors (KNN), Support Vector Machines (SVM), Multilayer Perceptron (MLP) and Dendrite Morphological Neural Network (DMNN). Below is explained how the DMNN works, it is a method that is not commonly used with brain signals. For each classification scenario, the performance was assessed by a 10-fold cross-validation process, which was applied independently for the set of trials of each participant. In each case, the set of trials was randomly partitioned into ten subsets which were used to build mutually exclusive training and test sets. Nine of the subsets were used to train the classifier while the remaining set was used to measure performance. This process was repeated until the ten combinations of train and test sets were exhausted. To measure performance, the metrics decoding accuracy or DA, F1-score or f1, true positive rate or TPR and true negative rate or TNR were computed. DA was computed as:

$$DA = \frac{TP + TN}{TP + TN + FP + FN}$$

where TP, TN, FP and FN are true positives, true negatives, false positives, and false negatives, respectively. f1 is the weighted average of the precision and recall and was computed as [8]:

$$f1 = 2 \times \frac{precision \times recall}{precision + recall}$$

where $precision = TP/(TP+FP)$ and $recall = TP/(TP+FN)$. Finally, TPR and TNR were computed as:

$$TPR = \frac{TP}{TP + FN} \quad TNR = \frac{TN}{TN + FP}$$

To assess the significance of the classification results, we computed the analytic statistical significance chance level and we performed statistical tests to measure differences between this significance chance level and the distribution of classification accuracies. The significant chance level of the decoding accuracy or DA_{sig} was computed using the cumulative binomial distribution [6] at the confidence level of $\alpha = 0.05$.

4.1   Multilayer Perceptron (MLP)

The multilayer perceptron is an artificial neural network formed by multiple layers, this allows it to solve problems that are not linearly separable, which is the main limitation of the perceptron, it can be totally or locally connected. It is considered a neural network consisting of L layers, without counting the input layer, by taking a random layer called ℓ which has N_ℓ neurons $X_1^{(\ell)}$, $X_2^{(\ell)}$, $\ldots$, $X_{N_\ell}^{(\ell)}$, each with a transfer function called $f^{(\ell)}$. This transfer function can be different in each layer and when using the delta rule as a base, this function must be differentiable without having to be linear. This model of ANN was trained using the backpropagation method.

4.2 Dendrite Morphological Neural Networks (DMNN)

In order to produce closed separation surfaces to discriminate data from different classes, artificial neural networks (ANN) such as Dendrite Morphological Neural Networks (DMNN) have been proposed [18]. The difference in these models of neural networks is the incorporation of a computational structure in the dendrites of the neurons. The use of dendrites has several advantages [19]; (i) increases the computational power of the neural model thus no hidden layers are required; (ii) allows the capability of multiclass discrimination; (iii) produces closed separation surfaces between classes.

The architecture of a DMNN is illustrated in Fig. 5. The model consists of n input neurons (number of attributes), m class neurons (number of classes) and a selection unit (final output). All input neurons are connected to each class neuron through d dendrites $D_1, ..., D_d$. Note that each input neuron has at most two weighted connections on a given dendrite, one excitatory and other inhibitory, which are represented as black and white dots, respectively. The weights between neuron N_i and dendrite D_k in the class neuron M_j are denoted by ω_{ijk}^l, where $l = 1$ represents excitatory input while $l = 0$ represents inhibitory input. The value for these weights are unknown and have to be learned from a training dataset. The output of dendrite D_k in the class neuron M_j is $\tau_k^j(x)$, which depends on the vector of attributes and the weights. Each class neuron provides an output value $j(x)$ which is computed from the values of all dendrites. Finally, a selection unit determines the class label y from all the values provided by the class neurons.

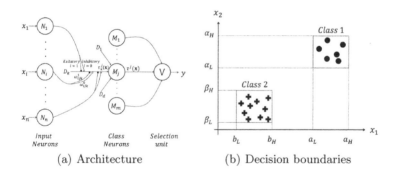

(a) Architecture   (b) Decision boundaries

Fig. 5. Dendrite Morphological Neural Networks.

The training algorithm for this network was reported in [19], which uses an approximation of type divide and conquer to generate hypercubes.

5 Results: Recognition of Movement Intention from EEG Signals

As a result of the preprocessing step, the total number of trials across all subjects was on average 94 ± 2 (minimum of 89 and maximum of 96). Therefore, the significant chance level was computed with the number of classes $N_C = 2$ and the minimum number of trials across all participants $N_T = 89$ resulting in $DA_{sig} = 62\%$. The time of the trial initiation across all subjects was on average -10.12 ± 1.57 s (minimum of -13.94 s and maximum of -6.02 s).

Table 1 shows the distributions of DA computed for each participant in the three classification scenarios with the four proposed classification methods. In scenario relax versus intention, fifteen subjects (1–3, 5–14 and 17–18) showed that the median of their distributions of DA (with the best classification method) are significantly different and higher than the DA_{sig}. In scenario relax versus execution, the median of the distributions of DA (with the best classification method) were significantly different and above chance level DA_{sig} in all of the participants. For the last scenario, intention versus execution, the median of the distributions of DA (with the best classification method) were significantly different and greater than the DA_{sig} for all the participants.

Table 1. DA computed in all classification scenarios in all subjects.

Subject	Relax vs intention				Relax vs execution				Intention vs execution			
	KNN	SVM	DMNN	MLP	KNN	SVM	DMNN	MLP	KNN	SVM	DMNN	MLP
S01	70.00%	56.11%	67.22%	**74.44%**	78.33%	60.00%	74.44%	**84.44%**	69.44%	65.00%	53.89%	**80.56%**
S02	59.94%	64.11%	54.67%	**67.39%**	75.44%	57.28%	64.78%	**77.61%**	65.11%	56.39%	70.00%	**77.78%**
S03	63.06%	52.22%	61.88%	**65.76%**	74.24%	54.51%	74.17%	**79.17%**	77.78%	59.58%	69.03%	76.60%
S04	55.33%	54.94%	50.94%	**56.61%**	75.44%	56.06%	64.39%	**75.72%**	61.06%	56.89%	60.11%	**71.17%**
S05	63.67%	53.83%	59.78%	**63.72%**	80.11%	70.50%	71.61%	**82.11%**	69.78%	68.17%	60.11%	**75.61%**
S06	61.56%	54.33%	59.67%	**66.94%**	78.89%	65.00%	74.72%	**82.61%**	75.89%	62.28%	66.78%	74.83%
S07	72.67%	60.56%	63.44%	**72.78%**	82.33%	63.22%	73.61%	**85.89%**	69.61%	53.72%	66.72%	**78.17%**
S08	67.33%	55.39%	64.56%	**75.89%**	76.72%	57.00%	71.56%	**82.28%**	61.50%	54.33%	64.11%	**74.67%**
S09	55.39%	54.00%	58.89%	**62.06%**	73.94%	56.67%	63.56%	**85.56%**	84.44%	64.17%	81.39%	**92.17%**
S10	61.28%	59.83%	59.94%	**65.94%**	76.06%	65.22%	66.94%	**84.39%**	64.56%	60.33%	56.11%	**69.89%**
S11	**65.17%**	58.56%	59.06%	63.56%	79.00%	63.44%	69.89%	**84.39%**	71.06%	57.56%	63.94%	**73.67%**
S12	67.00%	53.78%	65.89%	**76.67%**	85.22%	63.94%	78.22%	**86.72%**	78.89%	57.44%	68.28%	78.33%
S13	66.83%	61.67%	60.94%	**68.50%**	78.17%	66.56%	73.33%	**85.89%**	68.56%	56.39%	70.94%	**82.33%**
S14	62.50%	53.17%	62.94%	**66.67%**	70.06%	58.78%	66.33%	**83.17%**	69.17%	58.17%	59.56%	**84.50%**
S15	**52.44%**	51.06%	51.44%	47.28%	67.83%	59.56%	61.50%	**74.94%**	69.89%	57.50%	63.22%	**70.89%**
S16	**57.50%**	52.28%	57.00%	50.06%	64.06%	56.50%	58.50%	**71.17%**	62.83%	52.17%	57.83%	**67.00%**
S17	63.89%	55.11%	59.00%	**67.17%**	74.44%	61.94%	64.06%	**77.61%**	63.67%	**68.28%**	51.17%	59.28%
S18	67.61%	57.28%	63.28%	**70.22%**	**89.11%**	63.06%	76.56%	88.50%	72.67%	63.72%	65.67%	**75.33%**
Mean	62.95%	56.01%	60.03%	**65.65%**	76.63%	61.07%	69.34%	**81.79%**	69.77%	59.56%	63.83%	**75.71%**

Table 2 presents a summary of DA metric in each scenario with the two models of proposed neural networks. The best results were obtained with the Multilayer Perceptron (MLP). In scenario 1, the mean across all subjects is $65.65\% \pm 7.73\%$, while TPR and TNR are 68.35% and 64.51%, respectively. For scenario 2, in the average DA is $81.79\% \pm 4.59\%$ across all subjects, and 83.74% and 83.07% values can be observed for TPR and TNR. In scenario 3, the average of DA is $75.71\% \pm 6.93\%$, and the specific values for TPR and TNR are 76.43% and 77.62%. Note that the mean for all scenarios are greater than DA_{sig}, and

Table 2. Summary of DA, TPR and TNR computed across all subjects.

	Metrics (%)									
	DMNN					MLP				
	DA	Min	Max	TPR	TNR	DA	Min	Max	TPR	TNR
Relax vs intention	60.03 ± 4.34	50.94	67.22	60.69 ± 4.27	60.15 ± 5.99	$\mathbf{65.65 \pm 7.73}$	47.28	76.67	68.35 ± 9.54	64.51 ± 11.5
Relax vs execution	69.34 ± 5.55	58.50	78.22	71.52 ± 6.2	69.28 ± 5.19	$\mathbf{81.79 \pm 4.59}$	71.17	88.5	83.74 ± 5.24	83.07 ± 5.48
Intention vs execution	63.83 ± 6.93	51.17	81.39	65.26 ± 7.51	64.37 ± 7.41	$\mathbf{75.71 \pm 6.93}$	59.28	92.17	76.43 ± 7.91	77.62 ± 7.73

Table 3. Summary of precision, recall and f1 computed across all subjects.

	Metrics					
	DMNN			MLP		
	Precision	Recall	f1	Precision	Recall	f1
Relax vs intention	0.60 ± 0.04	0.61 ± 0.04	0.59 ± 0.03	0.63 ± 0.07	0.68 ± 0.09	$\mathbf{0.63 \pm 0.07}$
Relax vs execution	0.67 ± 0.05	0.72 ± 0.06	0.68 ± 0.06	0.81 ± 0.07	0.84 ± 0.05	$\mathbf{0.81 \pm 0.06}$
Intention vs execution	0.63 ± 0.07	0.65 ± 0.07	0.63 ± 0.07	0.76 ± 0.08	0.76 ± 0.08	$\mathbf{0.74 \pm 0.08}$

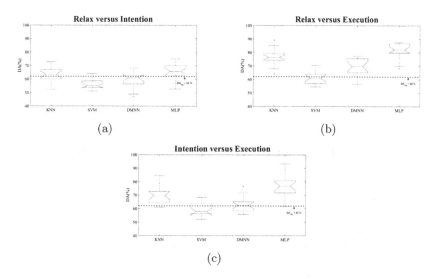

(a)   (b)

(c)

Fig. 6. Distributions of DA computed for each participant in the three classification scenarios with the four proposed classification methods.

particularly the second bi-class classification scenario shows the better results, even the minimal value 71.17% is higher than the significant chance level. The f1 metric considers both precision and recall measures to compute the score. These values are presented in Table 3. For the three different classification scenarios, the best f1 values obtained were 0.63, 0.81, and 0.74 were obtained.

The across all participants distribution of DA is presented in Fig. 6. In the three classification scenarios, the median of the distributions of DA is significantly different and higher than DA_{sig}. Importantly, this shows the identification of movement intention versus relax and movement execution.

6   Conclusions and Further Work

An essential element in a brain–computer interface (BCI) based on electroencephalogram (EEG) is the recognition of the patterns contained in the recorded brain activity that is induced by the mental task performed by the user [1]. To accomplish this, a BCI employs a classification algorithm which discriminates between a discrete number of mental tasks [20]. Although several classification techniques have been successfully applied, they have reached an upper limit in performance. This work proposes the use of ANN to detect motor information directly from brain signals recorded with the electroencephalogram (EEG) in brain–computer interface (BCI) experiments.

The study applied ANN to recognize the intention to move limbs using EEG signals that were recorded in a more challenging BCI protocol. Experiments were conducted with the aid of several healthy volunteers, and the EEG and EMG activity was recorded while they performed self-paced reaching movements. The performance shown by the two types of Artificial Neural Network was also emphasized when comparing this with the typical machine learning algorithms, these showed the best results in most cases.

These results show the possibility of recognizing the intention of movement either from the rest state or the movement execution. The detection of movement information before the execution of a movement provided by the ANN could be used to trigger robotic devices in BCI settings. This will eliminate the time required to collect brain signals while the mental motor task is performed by the user, thus allowing to move from synchronous to asynchronous BCIs. The main objective of this document is to motivate the development of more research on the classification of patterns in brain signals in order to implement Brain-Computer Interfaces.

Acknowledgements. We would like to express our sincere appreciation to the Instituto Politécnico Nacional and the Secretaria de Investigación y Posgrado for the economic support provided to carry out this research. This project was supported economically by SIP-IPN (numbers 20180730 and 20180943) and the National Council of Science and Technology of Mexico (CONACyT) (65 Frontiers of Science, numbers 268958 and PN2015-873).

References

1. Guger, C., Allison, B.Z., Edlinger, G. (eds.): Brain-Computer Interface Research. Springer, Heidelberg (2013). https://doi.org/10.1007/978-3-642-36083-1
2. Antelis, J.M., Gudiño-Mendoza, B., Falcón, L.E., Sanchez-Ante, G., Sossa, H.: Dendrite morphological neural networks for motor task recognition from electroencephalographic signals. Biomed. Signal Process. Control **44**, 12–24 (2018). https://doi.org/10.1016/j.bspc.2018.03.010

3. Asensio Cubero, J., Gan, J.Q., Palaniappan, R.: Extracting optimal tempo-spatial features using local discriminant bases and common spatial patterns for brain computer interfacing. Biomed. Signal Process. Control **8**(6), 772–778 (2013). https://doi.org/10.1016/j.bspc.2013.07.004
4. Bayliss, J.D.: Use of the evoked potential P3 component for control in a virtual apartment. IEEE Trans. Neural Syst. Rehabil. Eng. **11**(2), 113–116 (2003). https://doi.org/10.1109/TNSRE.2003.814438
5. Belhadj, S.A., Benmoussat, N., Krachai, M.D.: CSP features extraction and FLDA classification of EEG-based motor imagery for brain-computer interaction. In: 2015 4th International Conference on Electrical Engineering, ICEE 2015, pp. 3–8 (2016). https://doi.org/10.1109/INTEE.2015.7416697
6. Combrisson, E., Jerbi, K.: Exceeding chance level by chance: the caveat of theoretical chance levels in brain signal classification and statistical assessment of decoding accuracy. J. Neurosci. Methods **250**, 126–136 (2015). https://doi.org/10.1016/j.jneumeth.2015.01.010
7. Donchin, E., Spencer, K.M., Wijesinghe, R.: The mental prosthesis: assessing the speed of a P300-based brain-computer interface. IEEE Trans. Rehabil. Eng. **8**(2), 174–179 (2000). https://doi.org/10.1109/86.847808
8. Goutte, C., Gaussier, E.: A probabilistic interpretation of precision, recall and f-score, with implication for. Evaluation **3408**, 345–359 (2005). https://doi.org/10.1007/978-3-540-31865-125
9. Gudiño-Mendoza, B., Sossa, H., Sanchez-Ante, G., Antelis, J.M.: Classification of motor states from brain rhythms using lattice neural networks. In: Martínez-Trinidad, J.F., Carrasco-Ochoa, J.A., Ayala-Ramírez, V., Olvera-López, J.A., Jiang, X. (eds.) MCPR 2016. LNCS, vol. 9703, pp. 303–312. Springer, Cham (2016). https://doi.org/10.1007/978-3-319-39393-3_30
10. Han, R.X., Wei, Q.G.: Feature extraction by combining wavelet packet transform and common spatial pattern in brain-computer interfaces. Appl. Mech. Mater. **239**, 974–979 (2013). https://doi.org/10.4028/www.scientific.net/AMM.239-240.974
11. Hosni, S.M., Gadallah, M.E., Bahgat, S.F., AbdelWahab, M.S.: Classification of EEG signals using different feature extraction techniques for mental-taskBCI. In: 2007 International Conference on Computer Engineering Systems, pp. 220–226 (2007). https://doi.org/10.1109/ICCES.2007.4447052
12. Iturrate, I., Antelis, J.M., Andrea, K., Minguez, J.: A noninvasive brain-actuated wheelchair based on a P300 neurophysiological protocol and automated navigation. IEEE Trans. Robot. **25**(3), 614–627 (2009)
13. Katona, J., Kovari, A.: EEG-based computer control interface for brain-machine interaction. Int. J. Online Eng. **11**(6), 43–48 (2015). https://doi.org/10.3991/ijoe.v11i6.5119
14. Li, M., Li, W., Zhao, J., Meng, Q., Zeng, M., Chen, G.: A P300 model for cerebot – a mind-controlled humanoid robot. In: Kim, J.-H., Matson, E.T., Myung, H., Xu, P., Karray, F. (eds.) Robot Intelligence Technology and Applications 2. AISC, vol. 274, pp. 495–502. Springer, Cham (2014). https://doi.org/10.1007/978-3-319-05582-4_43
15. Li, Y., et al.: An EEG-based BCI system for 2-D cursor control by combining Mu/Beta rhythm and P300 potential. IEEE Trans. Biomed. Eng. **57**(10 PART 1), 2495–2505 (2010). https://doi.org/10.1109/TBME.2010.2055564
16. Ma, Y., Ding, X., She, Q., Luo, Z., Potter, T., Zhang, Y.: Classification of motor imagery EEG signals with support vector machines and particle swarm optimization. Comput. Math. Methods Med. **2016**(5), 667–677 (2016). https://doi.org/10.1155/2016/4941235

17. Purves, D., et al.: Neuroscience, vol. 3 (2004). ISBN 978-0878937257
18. Ritter, G.X., Sussner, P.: An introduction to morphological neural networks. In: Proceedings - International Conference on Pattern Recognition, vol. 4, pp. 709–717 (1996). https://doi.org/10.1109/ICPR.1996.547657
19. Sossa, H., Guevara, E.: Efficient training for dendrite morphological neural networks. Neurocomputing **131**, 132–142 (2014). https://doi.org/10.1016/j.neucom.2013.10.031
20. Wolpaw, J.R., Birbaumer, N., McFarland, D.J., Pfurscheller, G., Vaughan, T.M.: Brain-computer interfaces for communication and control. Clin. Neurophysiol. **113**, 767–791 (2002)
21. Wolpaw, J.R., McFarland, D.J., Neat, G.W., Forneris, C.A.: An EEG-based brain-computer interface for cursor control. Electroencephalogr. Clin. Neurophysiol. **78**(3), 252–259 (1991). https://doi.org/10.1016/0013-4694(91)90040-B, http://www.sciencedirect.com/science/article/pii/001346949190040B
22. Zhang, Y., Zhou, G., Jin, J., Wang, X., Cichocki, A.: Optimizing spatial patterns with sparse filter bands for motor-imagery based brain-computer interface. J. Neurosci. Methods **255**, 85–91 (2015). https://doi.org/10.1016/j.jneumeth.2015.08.004

Classification of Motor Imagery EEG Signals with CSP Filtering Through Neural Networks Models

Carlos Daniel Virgilio Gonzalez[1]([✉]), Juan Humberto Sossa Azuela[1,2],
Elsa Rubio Espino[1], and Victor H. Ponce Ponce[1]

[1] Centro de Investigación en Computación – Instituto Politécnico Nacional, Av. Juan de Dios Bátiz and M. Othón de Mendizabal, 07738 Mexico City, Mexico
danielvg92@gmail.com, {hsossa,erubio,vponce}@cic.ipn.mx
[2] Tecnológico de Monterrey, Escuela de Ingeniería y Ciencias, Av. General Ramón Corona 2514, Zapopan, Jalisco, Mexico

Abstract. The paper reports the development and evaluation of brain signals classifiers. The proposal consisted of three main stages: organization of EEG signals, feature extraction and execution of classification algorithms. The EEG signals used, represent four motor actions: Left Hand, Right Hand, Tongue and Foot movements; in the frame of the Motor Imagery Paradigm. These EEG signals were obtained from a database provided by the Technological University of Graz. From this dataset, only the EEG signals of two healthy subjects were used to carry out the proposed work. The feature extraction stage was carried out by applying an algorithm known as Common Spatial Pattern, in addition to the statistical method called Root Mean Square. The classification algorithms used were: K-Nearest Neighbors, Support Vector Machine, Multilayer Perceptron and Dendrite Morphological Neural Networks. This algorithms was evaluated with two studies. The first one aimed to evaluate the performance in the recognition between two classes of Motor Imagery tasks; Left Hand vs. Right Hand, Left Hand vs. Tongue, Left Hand vs. Foot, Right Hand vs. Tongue, Right Hand vs. Foot and Tongue vs. Foot. The second study aimed to employ the same algorithms in the recognition between four classes of Motor Imagery tasks; Subject 1 - 93.9% ± 3.9% and Subject 2 - 68.7% ± 7%.

Keywords: EEG signals · Motor Imagery · Common Spatial Pattern
RMS · One vs Rest · Pair-Wise
Dendrite Morphological Neural Network · Multilayer Perceptron

1 Introduction

Brain-Computer Interfaces (BCI) is a promising research field which intends to use non-invasive techniques, allowing communication between humans and computers by analyzing electrical brain activity, recorded at the surface of the scalp

© Springer Nature Switzerland AG 2018
I. Batyrshin et al. (Eds.): MICAI 2018, LNAI 11288, pp. 123–135, 2018.
https://doi.org/10.1007/978-3-030-04491-6_10

with electroencephalography. The main purpose for BCI is to enable communication for people with severe disabilities. In a BCI system, electrical activity within the cerebral cortex is detected through the use of an electrode cap, followed by a EEG signals conversion process in order to obtain instructions to control external devices. In recent years, BCI has been a major focus in the field of brain science and biomedical engineering and it has been developed into a new multidisciplinary cross technology. The key component of a BCI system is how it extracts EEG features to improve the recognition accuracy. At present, the most commonly used pattern recognition methods include the Linear Discriminant Analysis (LDA), K-Nearest Neighbor (KNN) classification algorithm, Artificial Neural Network (ANN) and Naive Bayes (NB).

BCI are commonly used in different applications, such as: spelling system [7], control of a humanoid robot (NAO) [13], control of a cursor on a screen [14], lighting control [12], control of an object in a virtual environment [4], including the control of a wheelchair [11].

In the topic of *signal processing* there are also different ways of approaching the problem, for example in [20] they use frequency analysis of signals acquired from electrodes placed in the central groove of the brain, this method depended on the amplitude of the brain signal. For large amplitudes the cursor moves upwards and for the smallest amplitudes it moves downwards. Another method widely used in the processing of brain signals is the Wavelet Transform, in addition to the Common Spatial Pattern filtering method to classify patterns of 2 kinds of *Motor Imagery* signals [8]. Other extraction methods are used, such as the Autoregressive Model of order 6 [10], Power Band [10], Independent Component Analysis (ICA), Principal Component Analysis (PCA), among others.

The CSP (Common Spatial Pattern) filtering has already been used in the classification of brain signals and it is currently very popular as it improves the performance of them [3,5,9,15,22]. There are different paradigms to address when handling brain signals, in this work the Motor Imagery paradigm was used which will be explained below:

1.1 Motor Imagery

The human being has the ability to imagine anything if he has had any experience related to, be this, an action, a scenario or some object. People can imagine actions although they do not have the capacity to physically develop them. Motor Imagery (MI) is a cognitive process in which the person imagines that he is developing a motor action without moving his muscles. This process requires the activation of the conscious way of the specific areas dedicated to the preparation and execution of the movements [16]. Currently, it is used as therapy for the recovery of patients with neurological injuries due to the positive effects achieved. In some studies, it has been shown that therapy of pain reduction with MI improves up to 40% [21].

1.2   BCI Competition III : Data Set IIIa (4-Class EEG Data)

The database used in this work was provided by the Brain-Computer Interface Laboratory of the Graz University of Technology, in the event called "BCI Competition III" (http://www.bbci.de/competition/iii/). This database was recorded with an electroencephalography amplifier with 64 acquisition channels, taking only 60 in the file of the GDF (General Data Format for Biomedical Signals) format. The EEG signals were acquired with a sampling frequency of 250 Hz and a filter on the frequency band 1–50 Hz. The placement of the 60 electrodes used to record brain signals, it follow the 10/20 standard. From this dataset, only the EEG signals of two healthy subjects were used (*K3B* and *L1B*), the data have 180 (*K3B*) and 120 (*L1B*) patterns of the four different kinds of MI mental task: Left Hand (LH), Right Hand (RH), Tongue (T) and Foot (F) movements. Figure 1 shows how each trial of MI is conducted.

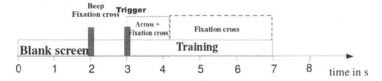

Fig. 1. Sequence for pattern acquisition. During the first two seconds the screen remains blank. At $t = 2\,\mathrm{s}$ an stimulus is heard which indicates the beginning of the test, at the same time a cross appears on the screen to indicate to the person that he must prepare himself. At $t = 3\,\mathrm{s}$ on the screen will appear an arrow to left, right, up or down for a second, at the same time the person should imagine the movement of the left hand, right hand, tongue or foot, respectively until the cross disappear at $t = 7\,\mathrm{s}$.

2   Feature Extraction

In this work, the spatial filtering method was applied to brain signals sampled on the time. As mentioned above, each trial is composed of 60 channels, which were filtered to obtain 60 channels different from the original ones. Later, the calculation of the Root Mean Square (RMS) of each channel was carried out. Finally, a set of 60 features for each trial was obtained and a stage of feature reduction was carried out using the Principal Component Analysis (PCA) algorithm which allowed us to perform the subsequent classification. In this work, the spatial filter is calculated using the Common Spatial Pattern method, which will be explained below.

2.1   Common Spatial Pattern (CSP)

The spatial filtering is an important part of detecting neuromodulatory changes over the motor cortex. The most used and successful method of discriminating such changes is Common Spatial Pattern (CSP). It has been mainly used in face recognition, object recognition, and EEG anomalies detection. It has been successfully applied to Brain-Computer Interfaces. The CSP algorithm is a kind of multidimensional statistics which has often been applied to EEG signals feature extraction and analysis in two-class multichannel methods. CSP finds optimal spatial filters that maximize the ratio of average variances between two different classes. Computationally, CSP is solved by simultaneous diagonalization of the covariance matrices of the two classes [5]. Figure 2 shows an example of the effect of applying spatial filtering on brain signals. On the left side (Channel 19 and 23), there are two EEG channels referring to the mental tasks of imagining the movement of the left hand (Ch 19 and 23 - Up) and right hand (Ch 19 and 23 - Down). As shown in the image, there is no clear difference between the signals of both mental tasks. However, new signals are obtained when CSP filtering is applied (CSP1 and CSP2). Which ones show a greater difference: In CSP1 the mental task of imagining the movement of the right hand shows a greater variance while the task mental of the left hand shows a smaller variance. The opposite occurs in CSP2.

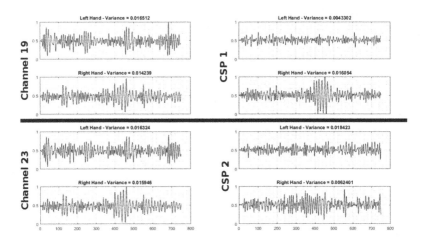

Fig. 2. Effect of CSP filter on patterns.

2.2   Approximation of the CSP Filter to Four Classes

As mentioned above, the CSP filtering is limited to only two classes, due to this two approximation method were used [17].

One vs Rest Approach: It consists in the comparison of each class against the rest of the classes. With each combination, a CSP filter will be obtained: {LH}VS{RH,T,F}, {RH}VS{LH,T,F}, {T}VS{LH,RH,F} and

{F}VS{LH,RH,T}. Subsequently, each trial will be filtered with each of the calculated filters, in the end a total of 240 features will be obtained. Due to the large number of features the PCA method is applied to reduce the amount of these (Fig. 3).

Pair-Wise(PW) Approach: It consists in the comparison of each pair of classes. With each combination, a CSP filter will be obtained: {LH,RH}VS{T,F}, {LH,T}VS{RH,F} and {LH,F}VS{RH,T}. Subsequently, each trial will be filtered with each of the calculated filters, in the end a total of 180 features will be obtained. Due to the large number of features the PCA method is applied to reduce the amount of these.

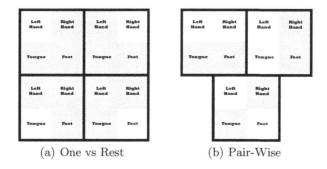

(a) One vs Rest   (b) Pair-Wise

Fig. 3. Approximation of CSP method

3   Classification

Four different classification algorithms were employed in this work, K-Nearest Neighbors (KNN), Support Vector Machines (SVM), Multilayer Perceptron (MLP) and Dendrite Morphological Neural Network (DMNN). Below is explained how the DMNN works, it is a method that is not commonly used with brain signals. For each classification scenario, the performance was assessed by a 5-fold cross-validation process, which was applied independently for the set of trials of each participant. In each case, the set of trials was randomly partitioned into ten subsets which were used to build mutually exclusive training and test sets. Four of the subsets were used to train the classifier while the remaining set was used to measure performance. This process was repeated until the five combinations of train and test sets were exhausted.

3.1   Multilayer Perceptron (MLP)

The multilayer perceptron is an artificial neural network formed by multiple layers, this allows it to solve problems that are not linearly separable, which is the main limitation of the perceptron, it can be totally or locally connected. It is considered a neural network consisting of L layers, without counting the input layer, by taking a random layer called ℓ which has m_ℓ neurons $Y_1^{(\ell)}$, $Y_2^{(\ell)}$, $\ldots$, $Y_{m^{(\ell)}}^{(\ell)}$, each with a transfer function called $f^{(\ell)}$. This transfer function can be different in each layer and when using the delta rule as a base, this function must be differentiable without having to be linear (Fig. 4).

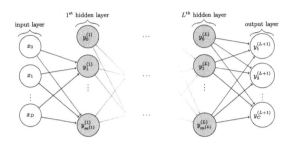

Fig. 4. Network graph of a $(L+1)$-layer perceptron with D input units and C output units. The lth hidden layer contains $m^{(l)}$ hidden units.

3.2   Dendrite Morphological Neural Networks (DMNN)

In order to produce closed separation surfaces to discriminate data from different classes, artificial neural networks (ANN) such as Dendrite Morphological Neural Networks (DMNN) have been proposed by [18]. The difference in these models of neural networks is the incorporation of a computational structure in the dendrites of the neurons. The use of dendrites has several advantages [19]; (i) increases the computational power of the neural model thus no hidden layers are required; (ii) allows the capability of multiclass discrimination; (iii) produces closed separation surfaces between classes.

The architecture of a DMNN is illustrated in Fig. 5. The model consists of n input neurons (number of attributes), m class neurons (number of classes) and a selection unit (final output). All input neurons are connected to each class neuron through d dendrites $D_1, ..., D_d$. Note that each input neuron has at most two weighted connections on a given dendrite, one excitatory and other inhibitory, which are represented as black and white dots, respectively. The weights between neuron N_i and dendrite D_k in the class neuron M_j are denoted by w_{ijk}^l, where $l = 1$ represents excitatory input while $l = 0$ represents inhibitory input. The value for these weights are unknown and have to be learned from a training dataset. The output of dendrite D_k in the class neuron M_j is $\tau_k^j(x)$, which depends on the vector of attributes and the weights. Each class neuron provides an output value $j(x)$ which is computed from the values of all dendrites. Finally, a selection unit determines the class label y from all the values provided by the class neurons (Fig. 5).

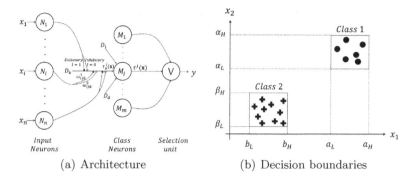

(a) Architecture   (b) Decision boundaries

Fig. 5. Dendrite Morphological Neural Networks.

The training algorithm for this network was reported in [19], which uses an approximation of type divide and conquer to generate hypercubes.

4   Recognition of MI Tasks from EEG Signals

4.1   Study 1: Classification of Two Classes of MI Tasks

The binary classification process was implemented with the four mentioned classifiers, in all possible pairs using the four mental tasks. In this stage an MLP with two hidden layers [Inputs - 10 - 10 - 2] was used. In addition, it was made the variation of the number of features used (from 2 to 10). A summary of the results obtained is shown in the Table 1 (K3B Subject) and Table 2 (L1B Subject).

Table 1 shows the accuracy values obtained with the EEG signals of the K3B Subject for each pair of classes in the four classification methods. In most of the classification scenarios an accuracy above 85% was obtained, the best results was observed when using an MLP. In four of the six different pairs of classes an accuracy higher than 96% was obtained with the MLP, highlighting the Right Hand VS Tongue scenario as the highest with a value of 97.8%±2.1.

For the second test subject (L1B), there was more variation in the classification results. Of the six combinations of pairs of classes, the MLP presents the best result in four combinations (Left H. VS Right H., Left H. VS Foot, Right H. VS Tongue and Right H. VS Foot), while the method KNN presents better results in two combinations (Left H. VS Tongue and Tongue VS Foot). The combination with the highest percentage of accuracy was Left H. VS Foot with 96.7%±1.7.

One aspect to highlight is that state of the art just report the classification of two specific mental tasks: Left Hand VS. Right Hand (Table 3). So the binary classification of each possible pair with the four mental tasks is considered a contribution of this project.

In most cases the classifiers used in this project show better results. However, the subject of the test was not the same in the works mentioned in the state of

Table 1. K3B (Test Trials = 36)

	Accuracy (%)							
	KNN				SVM			
Classes \ Features No.	3	5	7	10	3	5	7	10
Left H. VS Right H.	88.9 ± 2.5	87.5 ± 2.8	90.6 ± 5.7	90.6 ± 2.8	83.9 ± 8.5	87.8 ± 12.6	87.8 ± 12.6	87.8 ± 12.6
Left H. VS Tongue	93.3 ± 4.5	93.9 ± 4.1	93.3 ± 4.5	93.9 ± 4.4	88.3 ± 4.8	86.1 ± 5.3	87.2 ± 4.5	87.2 ± 4.5
Left H. VS Foot	95.6 ± 3.3	96.1 ± 3.8	94.4 ± 4.3	96.1 ± 5.2	92.2 ± 6.4	91.7 ± 5.6	91.7 ± 5.6	91.7 ± 5.6
Right H. VS Tongue	96.7 ± 2.7	97.8 ± 3.2	96.7 ± 3.2	96.1 ± 3.3	90.6 ± 5.2	87.2 ± 6.7	87.8 ± 7.2	87.8 ± 7.2
Right H. VS Foot	96.7 ± 2.7	97.2 ± 3	96.7 ± 4.1	96.1 ± 3.8	93.9 ± 7.1	92.8 ± 6.7	92.8 ± 6.7	92.8 ± 6.7
Tongue VS Foot	83.9 ± 8.5	90 ± 2.8	90 ± 2.8	89.4 ± 3.7	73.9 ± 13.3	75.6 ± 12.5	76.1 ± 12.7	75.6 ± 12.5
	MLP				DMNN			
Classes \ Features No.	3	5	7	10	3	5	7	10
Left H. VS Right H.	90.6 ± 3.8	92.8 ± 3.8	92.8 ± 4.5	**93.3±3.8**	86.7 ± 7.3	87.2 ± 2.8	83.9 ± 6.9	79.4 ± 6.5
Left H. VS Tongue	96.1 ± 3.8	96.1 ± 3.8	96.7 ± 4.1	**96.7±3.2**	93.3 ± 4.5	92.2 ± 4.1	88.3 ± 2.7	88.9 ± 3.5
Left H. VS Foot	95 ± 3.7	96.7 ± 2.1	96.1 ± 2.8	**97.2±1.8**	93.9 ± 3.7	93.3 ± 2.2	92.2 ± 5.4	91.7 ± 6.8
Right H. VS Tongue	**97.8±2.1**	97.8 ± 2.1	97.2 ± 2.5	97.2 ± 1.8	94.4 ± 1.8	93.9 ± 2.1	93.3 ± 2.2	90.6 ± 5.2
Right H. VS Foot	97.2 ± 3	96.7 ± 3.2	**97.2±2.5**	97.2 ± 2.5	94.4 ± 3	93.3 ± 3.8	91.7 ± 6.3	91.7 ± 3
Tongue VS Foot	84.4 ± 5.2	**93.9±3.7**	91.7 ± 2.5	91.1 ± 3.2	79.4 ± 6	87.2 ± 3.3	85.6 ± 1.1	83.9 ± 3.7

Table 2. L1B (Test Trials = 24)

Classes\Features No.	Accuracy (%)							
	KNN				SVM			
	3	5	7	10	3	5	7	10
Left H. VS Right H.	87.5 ± 5.3	88.3 ± 4.9	89.2 ± 6.2	89.2 ± 6.2	77.5 ± 5.7	79.2 ± 5.9	79.2 ± 5.9	79.2 ± 5.9
Left H. VS Tongue	86.7 ± 4.9	**88.3±4.9**	85.8 ± 3.3	85.8 ± 5	73.3 ± 9	73.3 ± 7.3	74.2 ± 7.2	73.3 ± 7.7
Left H. VS Foot	87.5 ± 4.6	91.7 ± 2.6	92.5 ± 3.1	90 ± 2	79.2 ± 4.6	75.8 ± 7.2	75.8 ± 7.2	75.8 ± 7.2
Right H. VS Tongue	85 ± 6.8	85.8 ± 3.3	85 ± 3.3	86.7 ± 4.9	65.8 ± 7.2	66.7 ± 7	65.8 ± 7.2	65.8 ± 7.2
Right H. VS Foot	87.5 ± 2.6	85.8 ± 5.7	85.8 ± 6.2	86.7 ± 6.1	80 ± 7.2	80 ± 7.2	79.2 ± 7	79.2 ± 7
Tongue VS Foot	69.2 ± 11.4	**71.7±7.2**	69.2 ± 9	70 ± 7.6	50 ± 2.6	50 ± 2.6	50 ± 2.6	50 ± 2.6
Classes\Features No.	MLP				DMNN			
	3	5	7	10	3	5	7	10
Left H. VS Right H.	**91.7±5.9**	86.7 ± 6.1	83.3 ± 7	85 ± 6.8	80.8 ± 10.4	75.8 ± 10	72.5 ± 9.7	71.7 ± 8.1
Left H. VS Tongue	86.7 ± 8.5	85 ± 6.2	86.7 ± 4.1	85.8 ± 5.7	83.3 ± 4.6	85 ± 4.2	83.3 ± 5.9	82.5 ± 4.1
Left H. VS Foot	84.2 ± 7.2	95.8 ± 3.7	94.2 ± 2	**96.7±1.7**	81.7 ± 6.2	85 ± 7.3	85 ± 6.8	85 ± 8.6
Right H. VS Tongue	**89.2±7.3**	88.3 ± 5.5	85.8 ± 3.3	85.8 ± 5.7	80 ± 5.5	76.7 ± 10.1	75 ± 8.7	72.5 ± 7.7
Right H. VS Foot	**90±5.7**	87.5 ± 5.3	88.3 ± 4.1	88.3 ± 4.1	81.7 ± 4.2	79.2 ± 12.6	78.3 ± 9.6	78.3 ± 12.2
Tongue VS Foot	70.8 ± 8.7	66.7 ± 7	63.3 ± 6.1	68.3 ± 8.6	60 ± 7.7	66.7 ± 7.9	66.7 ± 8.3	62.5 ± 5.9

the art, so it is not possible to make a direct comparison of the performance of the proposed methods. The same thing happens with the number of tests used to train and evaluate. In addition, in the development of this project, it was observed that the results may vary according to the subject of the test.

4.2   Study 2: Classification of Four Classes of MI Tasks

To carry out the classification of the four mental tasks, the aforementioned classification methods were implemented. This stage was developed using the approach methods previously explained: One versus Rest and Pair Wise. In this

Table 3. State of art (Motor Imagery)

	Left Hand vs Right Hand		
	Feature extraction	Classifier	Accuracy (%)
Ahangi [1]	Wavelet decomposition	KNN	84.28%
		Naive Bayes	68.75%
		MLP	74%
		LDA	87.86%
		SVM	88.57%
Han [8]	Wavelet + CSP (10 channels)	FLDA	93%
		SVM	90.9%
		KNN	92.9%
Asensio Cubero [3]	LDB + CSP	FLDA	75%
		DBI	63%
	LDB + LCT	FLDA	64%
		DBI	71%
Belhadj [5]	CSP (2 features)	FLDA	89.4%
	CSP (10 features)		89.4%
Ma [15]	RCSP	Decision Tree	79.8%
		KNN	92.5%
		LDA	**95.4%**
		PSO - SVM	**97%**
Proposal methods	**CSP**	KNN	S1: 90.6% S2: 89.2%
		SVM	S1: 87.8% S2: 79.2%
		MLP	**S1: 93.3% S2: 91.7%**
		DMNN	S1: 87.2% S2: 80.8%

LDA: Linear Discriminant Analysis
LCT: Local Cosine Transform
LDB: Local Discriminant Bases
FLDA: Fisher Linear Discriminant Analysis
DBO: Davies-Bauldin Index
RCSP: Regularized Common Spatial Pattern
PSO-SVM: Particle Swarm Optimization SVM

stage an MLP with three hidden layers [Inputs - 100 - 100 - 100 - 4] was used. In addition, it was made the variation of the number of features used (from 2 to 25) except for the DMNN in which only was performed in a range of 2–10, because this method is affected by increasing the size of the patterns.

A summary of the results obtained is shown in the Table 4 (K3B subject) and Table 5 (L1B subject). As in the Study 1, the best classifier in both subjects was the MLP, which obtained an accuracy equal to 93.9%±3.9% for K3B subject and 68.7%±7.1% for L1B subject; however, the method to approximate four classes was not the same in both cases, for the K3B subject the best results was obtained with the Pair-Wise method, while for the L1B subject it was the One versus Rest method.

In the same way, in the state of the art a work was found using the same database. In this the classification of the same four mental tasks was carried out, reporting the following results (Table 6):

The application of the neural network of dendritic processing for the classification of brain signals has only been presented in an article of the state of the art [2], here the classification of three mental tasks was performed: relaxation,

Table 4. K3B (Test Trials = 72)

Accuracy (%)								
Classifier\ Features No.	One vs Rest				Pair-Wise			
	5	10	20	25	5	10	20	25
KNN	61.4 ± 5.9	80 ± 4	79.7 ± 5.3	80.6 ± 5.8	65 ± 4.4	**82.2±3.6**	81.1 ± 3.2	80.6 ± 3.6
SVM	58.3 ± 3.5	**60.8±7.4**	59.7 ± 6.5	59.7 ± 6.5	44.4 ± 2.3	53.9 ± 6.2	54.7 ± 5.2	590.64.7 ± 5.4
MLP	62.2 ± 7.3	89.2 ± 2.2	90 ± 1.8	90.8 ± 2.4	67.8 ± 3.7	90 ± 2.4	93.3 ± 3.1	**93.9±3.9**
DMNN	53.6 ± 7.9	**69.4 ± 5.3**	-	-	55.5 ± 4.4	67.5 ± 6.4	-	-

Table 5. L1B (Test Trials = 48)

Accuracy (%)								
Classifier\ Features No.	One vs Rest				Pair-Wise			
	5	10	20	25	5	10	20	25
KNN	62.9 ± 6.6	62.9 ± 6.5	**65 ± 6.6**	63.7 ± 5.4	57.1 ± 7.4	58.7 ± 4.8	60.8 ± 5.2	60 ± 4.2
SVM	47.1 ± 5.4	**50 ± 3.2**	49.6 ± 3.6	49.6 ± 3.6	46.2 ± 6.2	47.5 ± 6.9	47.5 ± 7.4	47.9 ± 6.7
MLP	**68.7 ± 7.1**	62.9 ± 4.4	67.5 ± 5	66.7 ± 2.3	63.3 ± 7.5	60 ± 3.1	66.7 ± 2.3	62.9 ± 4.8
DMNN	57.5 ± 7	51.7 ± 4.6	-	-	**57.5 ± 2.8**	48.3 ± 3.3	-	-

Table 6. State of art (Motor Imagery 4 classes)

Left Hand vs Right Hand vs Tongue vs Foot			
	Feature extraction	Classifier	Accuracy (%)
Chin2009 [6]	FBCSP - One Versus Rest	NBPW	77%
Proposal methods	**CSP - One Versus Rest**	KNN	S1: 80.6% S2: 65%
		SVM	S1: 60.8% S2: 50%
		MLP	S1: 90.8% S2: **68.7 %**
		DMNN	S1: 69.4% S2: 51.7%
	CSP - Pair Wise	KNN	S1: 82.2% S2: 60.8%
		SVM	S1: 54.7% S2: 47.9%
		MLP	**S1: 93.9 %** S2: 66.7%
		DMNN	S1: 67.5% S2: 57.5%

FBCSP: Filter Bank Common Spatial Pattern
NBPW: Naive Bayesian Parzen Window

intention and execution of the hand movement. In this paper, the classification of specific mental tasks of the motor type was carried out, for this reason it is considered as an important contribution. Likewise, it is emphasized that the models of artificial neural networks showed the best results in almost all cases, the neural network of dendritic processing showed a performance that equaled in most cases the results shown by the classic classifiers (KNN, SVM).

5   Conclusions and Further Work

This paper provides an approach to perform the classification of four different mental tasks, showing the binary discrimination between each pair of classes. In the same way, the classification of the four mental tasks is carried out at the same time, showing better results compared to what is found in the state of the art. The performance shown by the two types of Artificial Neural Network was also emphasized when comparing this with the typical machine learning algorithms, these showed the best results in most cases. However, limitations were observed in the proposed method: The protocol of acquisition of the EEG signals requires a total time of seven seconds for each trial, if this proposal were implemented in real time it would generate significant delay. One of the main objectives of BCI is to help people with disabilities, therefore the response of the BCI should be as accurate as possible, making the subject perceive this feedback as natural as possible. This proposal allows to distinguish between four mental tasks of motor imagery type, however, it would be appropriate to train a fifth class in which a state of relaxation of the patient is characterized.

Some possible paths are, for future research on this area, the implementation of a different method for the extraction of characteristics or the use of classifiers with greater computational power. As mentioned in the introduction, this type of research has been applied to control different devices, therefore, it is proposed to control cars, wheelchairs, prostheses, even combine this line of research with others, such as autonomous navigation, virtual reality or augmented reality.

Acknowledgements. We would like to express our sincere appreciation to the Instituto Politécnico Nacional and the Secretaria de Investigación y Posgrado for the economic support provided to carry out this research. This project was supported economically by SIP-IPN (numbers 20180730, 20180943 and 20180846) and CONACYT (65 Frontiers of Science).

References

1. Ahangi, A., Karamnejad, M., Mohammadi, N., Ebrahimpour, R., Bagheri, N.: Multiple classifier system for EEG signal classification with application to brain-computer interfaces. Neural Comput. Appl. **23**(5), 1319–1327 (2013). https://doi.org/10.1007/s00521-012-1074-3
2. Antelis, J.M., Gudiño-Mendoza, B., Falcón, L.E., Sanchez-Ante, G., Sossa, H.: Dendrite morphological neural networks for motor task recognition from electroencephalographic signals. Biomed. Signal Process. Control. **44**, 12–24 (2018). https://doi.org/10.1016/j.bspc.2018.03.010
3. Asensio Cubero, J., Gan, J.Q., Palaniappan, R.: Extracting optimal tempo-spatial features using local discriminant bases and common spatial patterns for brain computer interfacing. Biomed. Signal Process. Control. **8**(6), 772–778 (2013). https://doi.org/10.1016/j.bspc.2013.07.004
4. Bayliss, J.D.: Use of the evoked potential P3 component for control in a virtual apartment. IEEE Trans. Neural Syst. Rehabil. Eng. **11**(2), 113–116 (2003). https://doi.org/10.1109/TNSRE.2003.814438
5. Belhadj, S.A., Benmoussat, N., Krachai, M.D.: CSP features extraction and FLDA classification of EEG-based motor imagery for brain-computer interaction. In: 2015 4th International Conference on Electrical Engineering, ICEE 2015, pp. 3–8 (2016). https://doi.org/10.1109/INTEE.2015.7416697

6. Chin, Z.Y., Ang, K.K., Wang, C., Guan, C., Zhang, H.: Multi-class filter bank common spatial pattern for four-class motor imagery BCI. In: Proceedings of the 31st Annual International Conference of the IEEE Engineering in Medicine and Biology Society: Engineering the Future of Biomedicine, EMBC 2009, vol. 138632, pp. 571–574 (2009). https://doi.org/10.1109/IEMBS.2009.5332383

7. Donchin, E., Spencer, K.M., Wijesinghe, R.: The mental prosthesis: assessing the speed of a P300-based brain- computer interface. IEEE Trans. Rehabil. Eng. **8**(2), 174–179 (2000). https://doi.org/10.1109/86.847808

8. Han, R.X., Wei, Q.G.: Feature extraction by combining wavelet packet transform and common spatial pattern in brain-computer interfaces. Appl. Mech. Mater. **239**, 974–979 (2013). https://doi.org/10.4028/www.scientific.net/AMM.239-240.974

9. Higashi, H., Tanaka, T.: Simultaneous design of FIR filter banks and spatial patterns for EEG signal classification. IEEE Trans. Biomed. Eng. **60**(4), 1100–1110 (2013). https://doi.org/10.1109/TBME.2012.2215960

10. Hosni, S.M., Gadallah, M.E., Bahgat, S.F., AbdelWahab, M.S.: Classification of EEG signals using different feature extraction techniques for mental-task BCI. In: 2007 International Conference on Computer Engineering Systems, pp. 220–226 (2007). https://doi.org/10.1109/ICCES.2007.4447052, http://ieeexplore.ieee.org/lpdocs/epic03/wrapper.htm?arnumber=4447052

11. Iturrate, I., Antelis, J.M., Andrea, K., Minguez, J.: A noninvasive brain-actuated wheelchair based on a P300 neurophysiological protocol and automated navigation. IEEE Trans. Robot. **25**(3), 614–627 (2009)

12. Katona, J., Kovari, A.: EEG-based computer control interface for brain-machine interaction. Int. J. Online Eng. **11**(6), 43–48 (2015). https://doi.org/10.3991/ijoe.v11i6.5119

13. Li, M., Li, W., Zhao, J., Meng, Q., Zeng, M., Chen, G.: A P300 model for cerebot – a mind-controlled humanoid robot. In: Kim, J.-H., Matson, E.T., Myung, H., Xu, P., Karray, F. (eds.) Robot Intelligence Technology and Applications 2. AISC, vol. 274, pp. 495–502. Springer, Cham (2014). https://doi.org/10.1007/978-3-319-05582-4_43

14. Li, Y., et al.: An EEG-based BCI system for 2-D cursor control by combining Mu/Beta rhythm and P300 potential. IEEE Trans. Biomed. Eng. **57**(10 PART 1), 2495–2505 (2010). https://doi.org/10.1109/TBME.2010.2055564

15. Ma, Y., Ding, X., She, Q., Luo, Z., Potter, T., Zhang, Y.: Classification of motor imagery EEG signals with support vector machines and particle swarm optimization. Comput. Math. Methods Med. **2016**(5), 667–677 (2016). https://doi.org/10.1155/2016/4941235

16. Mulder, T.: Motor imagery and action observation: cognitive tools for rehabilitation. J. Neural Transm. **114**(10), 1265–1278 (2007). https://doi.org/10.1007/s00702-007-0763-z

17. Muller-Gerking, J., Pfurtscheller, G., Flyvbjerg, H., Müller-Gerking, J., Pfurtscheller, G., Flyvbjerg, H.: Designing optimal spatial fiters for single-trial EEG classification in a movement task. Clin. Neurophysiol. **110**(5), 787–798 (1999). https://doi.org/10.1016/S1388-2457(98)00038-8. http://www.sciencedirect.com/science/article/pii/S1388245798000388

18. Ritter, G.X., Sussner, P.: An introduction to morphological neural networks. In: Proceedings - International Conference on Pattern Recognition, vol. 4, pp. 709–717 (1996). https://doi.org/10.1109/ICPR.1996.547657

19. Sossa, H., Guevara, E.: Efficient training for dendrite morphological neural networks. Neurocomputing **131**, 132–142 (2014). https://doi.org/10.1016/j.neucom.2013.10.031

20. Wolpaw, J.R., McFarland, D.J., Neat, G.W., Forneris, C.A.: An EEG-based brain-computer interface for cursor control. Electroencephalogr. Clin. Neurophysiol. **78**(3), 252–259 (1991). https://doi.org/10.1016/0013-4694(91)90040-B, http://www.sciencedirect.com/science/article/pii/001346949190040B
21. Zeidan, F., Martucci, K.T.: Brain mechanisms supporting modulation of pain by mindfulness meditation. J. Neurosci.: Off. J. Soc. Neurosci. **31**(14), 5540–5548 (2011). https://doi.org/10.1523/JNEUROSCI.5791-10.2011.Brain
22. Zhang, Y., Zhou, G., Jin, J., Wang, X., Cichocki, A.: Optimizing spatial patterns with sparse filter bands for motor-imagery based brain-computer interface. J. Neurosci. Methods **255**, 85–91 (2015). https://doi.org/10.1016/j.jneumeth.2015.08.004

Efficiency Analysis of Particle Tracking with Synthetic PIV Using SOM

Rubén Hernández-Pérez[⊠], Ruslan Gabbasov, Joel Suárez-Cansino,
Virgilio López-Morales, and Anilú Franco-Árcega

Intelligent Computing Research Group, Information and Systems Technologies
Research Center, Engineering and Basic Sciences Institute,
Autonomous University of the State of Hidalgo,
42184 Mineral de la Reforma, Hidalgo, Mexico
`joel.suarez937@gmail.com`

Abstract. To identify the field of velocities of a fluid, the postprocessing stage in the analysis of fluids using PIV images associates tracers in two consecutive images. Statistical methods have been used to perform this task and investigations have reported models of artificial neural networks, as well. The Self-Organized Map (SOM) model stands out for its simplicity and effectiveness, additionally to presenting areas of opportunity for exploring. The SOM model is efficient in the correlation of tracers detected in consecutive PIV images; however, the necessary operations are computationally expensive. This paper discusses the implementation of these operations on GPU to reduce the time complexity. Furthermore, the function calculating the learning factor of the original network model is too simple, and it is advisable to use one that can better adapt to the characteristics of the fluid's motion. Thus, a proposed 3PL learning factor function modifies the original model for good, because of its greater flexibility due to the presence of three parameters. The results show that this 3PL modification overcomes the efficiency of the original model and one of its variants, in addition to decreasing the computational cost.

Keywords: GPU · Parallel algorithms · PIV technique · SOM
Artificial neural networks

1 Introduction

The experimental data acquisition of the dynamical behavior of fluids becomes very important in some areas of engineering. Turbulent and chaotic behavior are two very well known examples of dynamical fluid regimes where this importance appears. There do exist several techniques to detect the presence of these fluid's behaviors. Usually, the solution of differential equations or the use of

This research project has been supported by the Mexican Public Education Bureau (PRODEP).

I. Batyrshin et al. (Eds.): MICAI 2018, LNAI 11288, pp. 136–153, 2018.
https://doi.org/10.1007/978-3-030-04491-6_11

reported data in the literature, such as Reynold's number, provide the analytical or numerical possibility of doing this.

However, in some cases, these numerical techniques are not necessarily required, and the fluid's behavior visualization can give more useful and sufficient information. The detection of any of these kinds of behaviors is a difficult task indeed. To start with the real fact that the definitions of vortex and turbulence need of an explanation.

For example, the concept of a vortex lies in the so-called vortex's nucleus and a normal plane to some axis of this nucleus. The presence of circular patterns on the plane is related to the visual detection of flow lines. The axes crossing the nucleus define the axes of rotation of the circular patterns [14]. On the other hand, the turbulent flow relates to a list of symptoms, rather than using a formal description as in the case of the vortex [15]. This list includes disorder, irreproducibility of experimental results, efficiency in the fluid mixing, and the presence of vortices with irregular distribution.

However, the lack of definitions for vortex and turbulence has not been a serious obstacle to develop several algorithms to identify, or detect, these fluid characteristics. The algorithms called Maximum Vorticity, Predictor-Corrector and Streamline, are just a few examples of the relatively recent trends in this area of research [14]. The algorithms are classified depending on how the vortex is defined, if the vortex depends on the time or do not, and the nature of the identification process [14].

Even though there has been some progress in the successful detection of vortices, the majority of the algorithms are not able to adequately solve unique vortices phenomena of turbulent flows appearing in different application scenarios [14].

Of course, the algorithms apply to a set of experimental data that specific experimental settings produce. This experimental setting can be invasive or non-invasive. Particle Imaging Velocimetry, or PIV for short, is one of the least intrusive methods for measuring flow fields [16], and its experimental setting has the advantage of making global measurements over the fluid dynamic, through negligible intrusive elements called tracers. The tracers make an observable fluid, which means that the fluid's structures and dynamics can be more visually measurable and testable [16].

A set of consecutive two-dimensional images in discrete time are what the PIV experimental setting produces, and these images also show the tracers in the fluid. A tracer displacement analysis permits to compute the tracer velocity vector over the complete area of two sequential images. Thus, the access to useful information related to fluid structure and dynamic requires a previous analysis and processing of images. However, the analysis and processing are challenging, indeed. The solution requires to apply proper techniques in dealing with the problems of tracer saturations in the images of fluid's velocity.

1.1 Problem's Statement

This paper assumes that a preprocessing stage of the data that PIV provides has already detected particles or tracers in a specific image. Thus the analysis makes the emphasis on the postprocessing phase, where the problem of determining the tracer velocity vectors needs to be solved. Fortunately, Self Organizing Maps artificial neural networks, or SOM for short, which learn in an unsupervised manner, provide an alternative of a solution to the problem. However, some parameters defining the SOM structure have an undesirable influence in the network performance and the results that it produces. In particular, the way the learning factor changes along the training process has considerable effects, sometimes negative ones, on the tracers path detection and the correct identification of the field of velocities, which also has an important consequence on vortex detection. With the additional inconvenience of wasting execution time and perhaps money.

This paper argues that a more general, proper and more flexible definition of the learning law improves the ratios related to the successful detection and identification of vortices in the fluid. Furthermore, the natural parallelism that the SOM architecture involves maps into relatively new computer architectures based on High-Performance Computing. The successful implementation of the proposed learning law will have an impact on more reliable vortices identification, and in the execution time of the complete process.

The rest of the paper organizes as follows. The Section named **State of the Art** introduces the main aspects of the topic and the required technology to study the phenomenon. The Section **Theoretical Framework** deals with the notation, explains the form the SOM works and its architecture and learning factor concepts. The Section **Experimental Results** discusses the experimental setting, the design of the experiment and the results. Finally, the Section **Conclusions** makes some comments about the obtained results and the future work.

2 State of the Art

The fluid flow visualization techniques are used to study the fluid motion and to allow the analysis of complex phenomena that this motion produces, like turbulence for example. Particle Tracking Velocimetry is an alternative to make the fluid flow analysis through images obtained with an experimental setting, which can be numerical or physical. One variation of this technique gives quantitative bidimensional information about the field of velocities of the flow, with the help of the motion of tiny tracers inside the fluid of interest. Images of the fluid at different instants of time are used to define the positions of the lightened tracers. Afterward, this information permits to compute more information related to the flow, like the presence of vorticity or the form of the flow velocity fields, for example [1, 17, 19, 20, 22].

Some authors propose artificial neural networks as efficient tools to reach some levels of precision in the detection of tracer motion in order to make an acceptable particle tracking when the number of tracers is large enough (saturated fields) [18]. Their reported conclusions show excellent results for this type of experimental setting [2,3]. In particular, the use of Self Organizing Map (SOM) [4–6] gives good results that, with the unsupervised learning characteristic, allows to apply this technique to identify the motion of any fluid without requiring previous knowledge of this.

On the other hand, other authors introduce a scheme where the computation of the learning factor, related to the adaptive capability of the neurons in the SOM, depends on the distance between each neuron and its neighbors [7], and another one that includes the rule Delta-Bar-Delta, with the principal purpose of reducing the time complexity of the SOM implementation [8]. Their results show a noticeable improvement of the SOM algorithm applied to PIV or PTV by some of the authors already mentioned in the previous paragraph.

The existent algorithms are not optimized for several computer architectures and, therefore, they do not get the maximum performance. This work reduces the problem of poor performance optimization by implementing the algorithms over parallel architectures, where the concept of GPU plays an important role. This technology offers multi-core functionality, with high flexibility and the capability of executing the arithmetic operation to a lower cost and as an alternative to traditional processors. The number of available computation cores is the main characteristic of this architecture, which is much bigger than any other commercial microprocessor [9].

3 Theoretical Framework

The literature describes the complete theoretical framework about how the SOM is defined [4–7]. However, some essential points need to be clarified. Given an image, there are in fact several SOM architectures requiring to be defined. The number of SOM's depends on the number of detected tracers, while the number of neurons in every SOM architecture relates to a neighborhood or ball $B_r(\boldsymbol{x}_j)$, being $\boldsymbol{x}_j$ the vector position of the tracer j in the image, and r is the neighborhood radius.

Figure 1 considers two sequential images at times t and $t + \Delta t$ for a particular situation. White circles compose the image at time t and indicate the tracer location at the time t. One white circle, at the vector position $\boldsymbol{x}_j$, represents the center of the ball $B_r(\boldsymbol{x}_j)$. The Figure also shows the image at time $t + \Delta t$ defined by black circles inside or outside the ball $B_r(\boldsymbol{x}_j)$.

The SOM architecture is defined by the tracers inside $B_r(\boldsymbol{x}_j)$ at times t and $t + \Delta t$. In fact, the input layer of the SOM is given by the tracers inside $B_r(\boldsymbol{x}_j)$ at time t, while the output layer is defined by the tracers inside $B_r(\boldsymbol{x}_j)$ at time $t + \Delta t$. Therefore, two subnets define the architecture of this specific SOM, such as the Fig. 2 shows.

The input matrix $\boldsymbol{P}$ at any time is defined as follows,

$$\boldsymbol{P} = \begin{bmatrix} x_{11} \ x_{12} \ \dots \ x_{1N} \\ x_{21} \ x_{22} \ \dots \ x_{2N} \end{bmatrix}$$

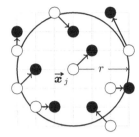

Fig. 1. Ball, $B_r(\boldsymbol{x}_j)$, with center at $\boldsymbol{x}_j$ and radius r. The three white tracers inside the neighbourhood define the first subnet at time t, while the five black tracers inside the same neighbourhood define the second subnet at time $t + \Delta t$.

Fig. 2. Architecture of the SOM, the input layer is defined by the white tracers at time t and the output layer by the tracers at time $t + \Delta t$.

where N denotes the number of detected tracers at time t, the second subindex in every matrix element refers to the detected vector position of the tracer, and the first subindex to the component of this vector.

Every column j of the matrix $\boldsymbol{P}$ connects to every neuron i in the second subnet, and the two-dimensional vector $\boldsymbol{\omega}_{ij}$ weights its contribution so that for the neuron i in the second subnet a matrix $\boldsymbol{W}_i$ can be defined as follows

$$\boldsymbol{W}_i = \begin{bmatrix} \omega_{i1}^{(1)} & \omega_{i2}^{(1)} & \cdots & \omega_{iM}^{(1)} \\ \omega_{i1}^{(2)} & \omega_{i2}^{(2)} & \cdots & \omega_{iM}^{(2)} \end{bmatrix}$$

where i runs from 1 to M. M gives the number of neurons in the second subnet. The second subindex in every matrix element denotes the detected vector position of the tracer at time $t + \Delta t$, while the superindex refers to the component of this vector.

Since the SOM learns in an unsupervised manner, there is enough information to start with the learning process, which is defined by the concept "the winner takes all". In this case, the winner neuron in the output layer will give the position of the tracer, at time $t + \Delta t$, initially located at position $\boldsymbol{x}_j$ at time t. Therefore, $\boldsymbol{x}_j$ itself defines the input to the neuron at position $\boldsymbol{x}_j$ while the positions $\boldsymbol{y}_i$ of the tracers inside $B_r(\boldsymbol{x}_j)$ at time $t + \Delta t$ define the weight's seed.

The process goes backward by considering now that the second subnet becomes the first one, and that the first subnet becomes the second one.

3.1   SOM's Learning Algorithm

Thus, in a first step, two-dimensional input and weight vectors are assigned to every neuron in the two created subnets in accordance to the following equations,

$$
\begin{aligned}
\boldsymbol{p}_j &= \boldsymbol{x}_j, \ \forall j = 1, \ldots, M \\
\boldsymbol{\Omega}_i &= \boldsymbol{y}_i, \ \forall i = 1, \ldots, N
\end{aligned}
\tag{1}
$$

where $\boldsymbol{y}_i$ is the vector position of the neuron i in the second subnet.

Afterward, the working of the SOM model implies a controlled sequence of iterations between the two subnets, where the SOM tries to associate every element of the set of input vectors to the members of the set of weights, and vice versa. At the end of the iterations, a simple closest neighbor search identifies the associated neurons in every subnet (which are in fact related to the tracers in the fluid).

The search produces a list of pair of tracers (one neuron or tracer in the pair belonging to one subnet and the other to the second subnet) that defines a field of vectors such as the Fig. 1 shows.

Next, the neuron's weights in the second layer are updated in accordance to the learning rule given by the Eq. (2)

$$
\Delta \boldsymbol{\Omega}_{ij} = \alpha_i (\boldsymbol{p}_j - \boldsymbol{\Omega}_i), \ \forall i = 1, \ldots, M
\tag{2}
$$

where $\alpha_i \in (0, 1)$ is the SOM's learning factor, which is defined by a simple step function, such as the Fig. 3 illustrates. In other words,

$$
\alpha_i(\boldsymbol{\Omega}_i) = \alpha \chi_{B_r(\boldsymbol{x}_j)}(\boldsymbol{\Omega}_i)
\tag{3}
$$

where $\alpha \in (0, 1)$ and $\chi_A(x)$ is the usual characteristic or indicator function.

In case of using the Ohmi variant, then the learning factor α_j changes and is now a function given by the Eq. (4)

$$
\alpha_i = \alpha \begin{cases} 1, & \text{if } \boldsymbol{\Omega}_i \in B_a(\boldsymbol{p}_j) \\ \exp\left(-\left(\frac{\|\boldsymbol{p}_j - \boldsymbol{\Omega}_i\|}{r} - 1\right)^2\right), & \text{otherwise} \end{cases}
\tag{4}
$$

so that, when the distance between the two points is larger than the radius r, then the learning factor is computed by a Gaussian function [5,7], whose behavior is illustrated by the Fig. 3.

Every time that the first subnet presents the associated set of inputs as a stimulus to the second subnet, then the second subnet updates the corresponding set of weights. After completing this step, the subnets invert their roles to repeat the previous process with the new scenario.

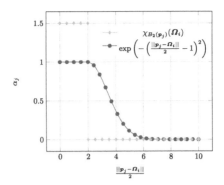

Fig. 3. The figure shows particular examples of the Step and Gaussian functions. The first one was proposed to calculate the learning factor of the SOM model, but this function is too simple to reflect the real dynamic of a fluid [4]. The Gaussian function is used to compute the learning factor and improves the performance of the SOM to represent the actual dynamics of a fluid.

At each step, the radius r of the neighborhood decreases, while the amplitude of the constant α, related to the learning factor, increases with the help of Eq. (5) and the Eq. (6), respectively.

$$r \leftarrow \beta r, \quad 0 < \beta < 1 \qquad (5)$$
$$\alpha \leftarrow \alpha/\beta \qquad (6)$$

The previous steps execute until the radius r reaches a defined value r_f, which must be small enough to ensure that the ball $B_{r_f}(x_0)$ has just one winner neuron. The value of β is a real number in the open interval $(0, 1)$. In the end, the association of the neurons of the first subnet with those neurons of the second subnet is made again with a closest neighbor search algorithm, but now using a radius ϵ that is smaller than r.

The structure of the Eq. (4), computing the learning factor in an innovative way [7], improves the matching quotient of the pairwise tracer identification in two consecutive images [5]. So that, the study of the SOM capabilities involves the analysis of different structures related to the way the learning factor modifies during the adaptive process.

Hence, this work proposes a modification of the structure of the function given by the Eq. (4), and this modification removes the piecewise definition by introducing a single logistic function, with the property or characteristic of a higher flexibility in terms of a set of parameters able to adapt better and faster to the dynamic of the studied fluids [10].

So that the proposed logistic function, which is called 3PL because of the number of parameters involved in the structure [11], must have an acceptable asymptotic behavior in the limits $\frac{\|p_j - \Omega_i\|}{r} \to 0$ (learning factor is α) and

$\frac{||p_j - \Omega_i||}{r} \rightarrow +\infty$ (learning factor is 0). Therefore, the Eq. (7)

$$f(r, b, c, g) = \frac{\alpha}{\left(1 + \left(\frac{||w_c - w_j||}{cr}\right)^b\right)^g} \tag{7}$$

defines the proposed 3PL model, and Fig. 4 gives a particular example of this function.

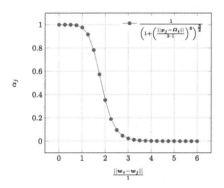

Fig. 4. A particular example of the 3PL function that substitutes the Step and Gaussian functions [7], the general proposed function gives more adaptive flexibility along the training process.

3.2   SOM Implementation

The complexity of the algorithm for tracers detection suggests the use of High-Performance Computing, or HPC for short, through the programming of Graphical Processing Units (GPU for short). The existence of two subnets, the constant interaction between these subnets, and the presence of as many neurons as tracers detected along the stage of particle detection support the use of HPC.

The operations that every neuron x_j needs to detect a neuron y_i in the complementary subnet, and vice versa, imply an exponential time in the execution of several steps. Computationally speaking, this situation becomes worse because of the required number of iterations by the two subnets, which also depends on the compression factor β and the limits r and r_f, previously described.

The performance improvement of the algorithms considers different aspects such as, for example, parallelization of serial code, memory control and the application of very well known parallel programming methods [10].

4   Experimental Results

Even though the phenomenon is three dimensional [21], the fluid flow simulation of a binary mixture of substances passing from laminar to turbulent dynamics

[12] produced a set of frames or images. The images become a useful tool for the analysis of the performance of the SOM when looking for a kind of validation of the proper working of the artificial neural network. The validation test uses the three variants of the SOM [4,7,10] following a previous experimental design where some frames of the simulation play an important role. By considering the characteristics of the best network behavior in every scenario, the analysis of the experimental results is carried out.

4.1   Experimental Setting

The Fig. 5 shows a cause-effect diagram as a guide for the experimental design and the major primary and secondary causes to successfully determine the proper pair matching of tracers inside the two subnets. In particular, the final results of the analysis of the SOM performance consider the followings aspects, such as the Fig. 5 shows,

1. Particles' density. The total amount of particles in the analyzed frame, such as the Fig. 6 shows. As long as the amount of particles increases inside the frame, more complicated the correct detection of the pair of matching tracers is, and the computation time acquires a higher value, as well.
2. Frames' speed. The removal of a predefined set of frames between the current frame and the next frames of interest gives the separation or displacement among the particles in consecutive frames. So that, if one is the number of frames to consider, then there is not an intermediate frame for removal between the current frame and the next one. On the other hand, if three is the number of frames to consider, then there are two intermediate frames for

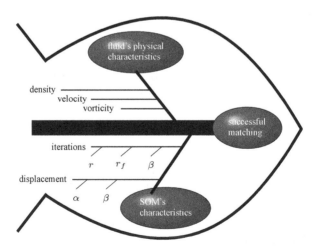

Fig. 5. Cause and effect diagram that shows the independent variables at the postprocessing stage and the relations that they have with the false-positive ratios.

removal between the current one and the next one, and so on. The Fig. 7 shows the effects for different experimental frames' speed, mainly on the velocity vectors of every tracer.

3. Vorticity. Disorder in the particles' path and the presence of vortices in the frames.

The experiment takes two samples from the original image to make the SOM validation test. The first sample considers only the one percent of particles of the original image, which corresponds to 1, 200 elements (tracers), and the size of the second sample is the ten percent of particles of the original image, which corresponds to 12, 100 elements. The Fig. 6 illustrates the visual difference between the first and second samples.

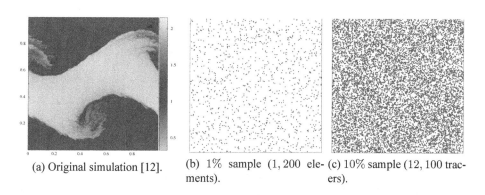

(a) Original simulation [12].   (b) 1% sample (1, 200 elements).   (c) 10% sample (12, 100 tracers).

Fig. 6. The top image shows the result of the original simulation. The bottom pictures are just samples of this image. The left side picture at the bottom considers a 0.01 proportion (1, 210 elements) while the right side picture considers 10% (12,100 tracers).

Furthermore, 1, 000 frames define the simulation and represent the time evolution of the flow of two mixed fluids, such as Fig. 8 shows. The Figure illustrates three instantaneous shoots of the simulation where, at different time stages, the flow initially presents a laminar dynamic and starts to show vortices just at the boundaries or zone of interaction between the two fluids. At the end of the simulation sequence, the velocity of the fluid considerably decreases.

Finally, the experimentation with the velocity takes the original simulation frames, which are increasingly ordered with the corresponding simulation time. Hereafter, and in an attempt to simulate the control of the fluid's velocity, some frames between two frames of interest are discarded, such as it was previously discussed. Figure 7 illustrates how the frame separations increase the vectors of velocity, making more difficult the processing in a real scenario.

The ranges of the experimental parameters are defined to generate the testing conditions through the Fig. 5. Table 1 shows the range of values for the experimental parameters.

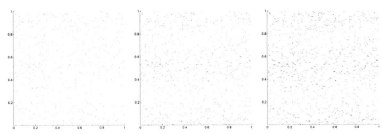

(a) None frame removed between pictures of interest.

(b) One frame removed between pictures of interest.

(c) Two frames removed between pictures of interest.

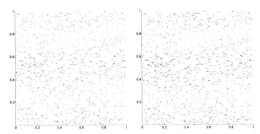

(d) Three frames removed between pictures of interest.

(e) Four frames removed between pictures of interest.

Fig. 7. The main body of this paper discusses the way of obtaining the pair of frames to make several tests to analyze the effect that the velocity of the particles has on the matching results. To get the effect of increasing velocity the sequence of frames is conveniently modified. For example, the removal of none, one, two, three or four frames that follow and precede the first and second images under analysis, respectively, gives the effect of increasing velocity. The images (a), (b), (c), (d) and (e) illustrate the velocities field.

Table 1. The parameters take their values from those suggested by the literature on the topic [4–7].

Variable	Minimum	Maximum	Interval	No. of values
β	0.700	0.900	0.200	2
α	0.005	0.050	0.005	10
r	0.050	0.450	0.050	9

The simulation method is called Smooth Particle Hydrodynamics (or SPH for short) [13] and the structure called Close Packing clusters the simulated particles. The simulation also uses the Navier-Stokes equations, and involves simple aspects about the fluids; including the velocity, density, and pressure, such as the Eq. (8) shows,

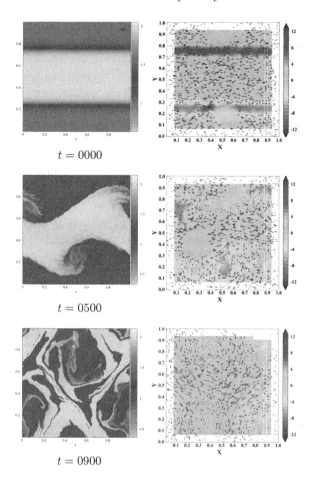

Fig. 8. The simulated fluid dynamic changes from frame to frame, the images represent three different physical configurations of the velocities field, starting with the laminar flow at the top, going through an intermediate vortices configuration in the middle, and ending as a turbulent fluid (images at the bottom). The vertical axes at the right side of the picture give the density values.

$$\frac{\partial \rho}{\partial t} + \rho \boldsymbol{\nabla} \cdot \boldsymbol{v} = 0$$
$$\frac{\partial \boldsymbol{v}}{\partial t} = -\frac{\boldsymbol{\nabla} P}{\rho}$$
$$P = (\gamma - 1)\rho u \qquad (8)$$
$$V_y = A * \sin(4\pi)$$
$$A = 0.01$$

where ρ is the fluid density, P the pressure, and $\boldsymbol{v}$ the velocity, t is the time, u is the internal energy and γ is the ratio of specific heats. A senoidal perturbation, V_y, affects the particle's motion along the y axis.

The column "No. of Values" of the Table 1 permits to compute the total number of possible tests that can be made with the parameters' values and this is equal to 180. If the experiment includes the two samples of particle density, then the number of tests increases to 360. But, the three different configurations of the field of velocities increase this number to 1, 080. Finally, the five separations between frames and the three variants of the SOM's learning factors provide a total of 16, 200 tests over the SOM, and the Table 2 summarizes the partial contributions to the total number of tests.

Table 2. The combination of partial contributions coming from five factors permits to analyze the SOM's behavior. The first three factors refer to fluids' physical characteristics while the SOM characteristics define the other two.

Factor	No. of samples
Density samples	2
Vorticity samples	3
Velocity samples	5
SOM's parameters combinations	180
Proposed SOM's models	3

4.2 Discussion of Results

The way the simulation controls the flow velocity has been already discussed. Next, the best results for every variant of the SOM's factor learning function, taking a frame separation of four, are shown. The simple nearest neighbor search algorithm, which is the first step of the SOM's unsupervised learning, is taken as a basis for the comparisons of results. The efficiency of the nearest neighbor search algorithm decreases in a linear form, which should be expected when the frame separation grows (the particles' velocity increases).

Figure 9a shows the obtained highest efficiency for the Step function [4] and the comparison of this function with the other variants of the SOM's learning factor. It is worth to mention that, even though in this case the Step function reports better results than the others, this does not occur in all the scenarios. A large number of required iterations to get this result is the main objection, because of the number of involved operations.

The main lesson of the results in Fig. 9a says that the Step function works more efficiently, and this occurs when the weights update in a smooth manner, which the value of the factor α (0.005) reflects, with the inconvenience of a large number of iterations (58). This scenario, of course, is not the most adequate, because of the need for reducing the time complexity and, therefore, the computational cost. Even though the results given by the Step function are acceptable

($\approx 72\%$), the required number of iterations makes impractical this function as a solution to the problem of tracers matching.

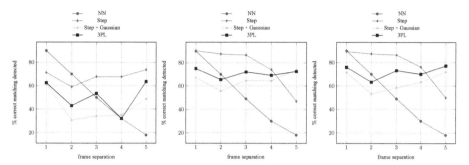

(a) If four frames of the simulation separate the two matching pictures, then the Step function [4] produces the best results. In this case, a small learning factor ($\alpha = 0.005$) is the main characteristic, which implies finer weight updates; however, the number of iterations is large.

(b) If four frames of the simulation separate the two matching pictures, then the Step + Gaussian function [7] produces the best results. Unlike Figure 9a, the number of iterations is considerably reduced, with a larger learning factor ($\alpha = 0.04$).

(c) The 3PL function [10] gives the best result when four simulation frames separate the two images of interest. The obtained result is better than those for the Step and Step+Gaussian functions, under the same experimental conditions. Unlike Figures 9a and 9b, the number of iterations is equal to 13 and the learning factor $\alpha = 0.045$. In fact, a higher learning factor makes the 3PL working better.

Fig. 9. The figures compare the matching efficiency of the Step, the Step plus Gaussian, and the 3PL learning functions when none (1), one (2), two (3), three (4) or four (5) frames separate two images of interest thus giving the appearance of increasing the fluid velocity.

Figure 9b shows the obtained highest efficiency for the Step + Gaussian function [7] and the comparison of this function with the other variants of the SOM's learning factor. The results are similar to those of the Fig. 9a, and they reach an efficiency of $\approx 72\%$, but with a lower number of iterations (13). The number of iterations makes that the processing speed, involving the Step + Gaussian function, becomes four times greater than that involving to the Step function.

The value of the parameter related to the learning factor is another difference worth to mention. First of all, the Step function requires a considerably small value ($\alpha = 0.005$), while the Step + Gaussian function needs a larger one ($\alpha = 0.04$). This fact has a positive impact on the number of iterations that the weight update requires, the convergence becomes faster than that involving to the Step function.

Figure 9c shows the obtained highest efficiency for the 3PL function [10] and the comparison of this function with the other variants of the SOM's learning factor. This result improves those values that the others functions give, reaching 76% of efficiency and using the same number of iterations previously reported by the Step + Gaussian function.

The difference between the learning factor parameters related to the Step + Gaussian and the 3PL functions is another major point. In the first case,

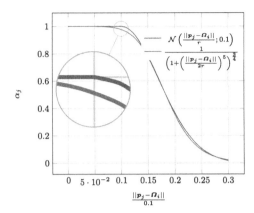

Fig. 10. Graphically, the Step + Gaussian and the 3PL functions are quite similar. The more relevant visual difference is on the top of the figure, where the Step function meets the Gaussian function. The 3PL function can be adapted to have lower values than those of the Step + Gaussian function, and this has a positive effect on the tracers matching.

Table 3. Summary of the parameters giving the best results for the variants of the SOM model to correlate the PIV images. An experiment with images having 1200 elements gives the best results.

Model	Separation	r	β	α	Iterations	e
Labonte	1	0.05	0.9	0.020	38	94%
Ohmi		0.05	0.7	0.005	11	93%
3PL		0.05	0.7	0.005	11	93%
Labonte	2	0.05	0.9	0.050	38	91%
Ohmi		0.05	0.9	0.005	38	87%
3PL		0.05	0.7	0.025	11	88%
Labonte	3	0.10	0.7	0.050	13	87%
Ohmi		0.05	0.9	0.010	38	83%
3PL		0.05	0.7	0.050	11	86%
Labonte	4	0.10	0.9	0.030	44	78%
Ohmi		0.05	0.9	0.030	38	81%
3PL		0.05	0.9	0.030	38	81%
Labonte	5	0.30	0.9	0.005	58	74%
Ohmi		0.10	0.7	0.040	13	74%
3PL		0.10	0.7	0.045	13	79%

$\alpha = 0.04$ and $\alpha = 0.045$ in the second one. This result, at first sight, contradicts the fact that a smooth weight update gives the best results, even at the cost of a large number of iterations.

To fully understand this result and the reasons it appears when the 3PL function is involved, the test parameters permit a closer inspection of these functions. Even though the selected parameters are very close, the real fact confirms by direct observation that they are not equal. Figure 10 shows the local difference between the two functions.

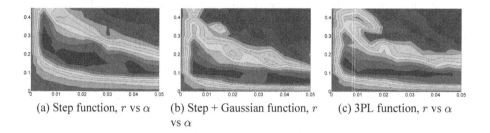

(a) Step function, r vs α (b) Step + Gaussian function, r vs α (c) 3PL function, r vs α

Fig. 11. The images show the level curves of the best results obtained (darker zones) through the three SOM's learning factor models and their relations with the more significative parameters, α, r.

The 3PL function changes smoothly in the region where the Gaussian function meets the non-null value of the Step function, which is in agreement with the fact of a higher learning factor parameter α. In some way, the 3PL smoothness compensates the high value of the learning factor parameter, making less drastic the weight updating process (Fig. 11).

On the other hand, the analysis of the number of iterations involves the factors r, r_f, β, such as Fig. 5 shows. The best experimental results are obtained when $\beta = 0.7$; however, the Step function shows acceptable results when $\beta = 0.9$ (Table 3).

5   Conclusions and Future Work

During the postprocessing phase, the 3PL function provides the best performance of the SOM when computing the learning factor function. The following are the most important results and contributions.

First of all, the 3PL learning factor function provides flexibility to the working of the SOM. The 3PL learning factor function allows flexible adaptation and obtains better results in the tracers matching efficiency, and the relative response times as observed in the tests.

On the other hand, the 3PL learning factor function improves the efficiency and response times as long as the fluid's velocity increases. Under normal conditions of simulation, the obtained results for correct particle matching ($\approx 93\%$) allow getting the velocities field with no variations at all. As soon as the fluid's velocity increases in the way already described lines above, the SOM reaches a 79% of efficiency when the unsupervised training process uses the 3PL learning factor function; in contrast with the other two variants, which only have a 74% of efficiency.

Finally, a relationship between the parameter r of the SOM and the average particles' separation exists that the literature does not report. This fact is quite important since it establishes somehow the initial parameters to start with the execution of the SOM iterations. The equation $r \approx 3D_{\mathrm{prom}}$ gives this relation, where r refers to the radius of the neighborhood related to the SOM, and D_{prom} to the average distance of the particles in the PIV images. This result is quite useful to get suitable particles matching.

5.1   Future Work

Currently, the learning factor functions related to the SOM model consider a radial symmetric neighborhood, which implies isotropy in the field of velocities. The introduction of additional parameters permits the redefinition of the 3PL learning factor function. However, these parameters must not be arbitraries, and they need to provide an elliptical shape to the neighborhood, giving thus a preferred orientation to the local field of velocities. In this case, one expects a better efficiency of the SOM.

References

1. Westerweel, J.: Digital Particle Velocimetry, Theory and Application. Delft University Press, Netherlands (1993)
2. Grant, I., Pan, X.: An investigation of the performance of multilayer neural networks applied to the analysis of PIV images. Exp. Fluids **19**, 159–166 (1995)
3. Grant, I., Pan, X.: The use of neural techniques in PIV and PTV. Meas. Sci. Technol. **8**, 1399–1405 (1997)
4. Labonte, G.: A SOM neural network that reveals continuous displacement fields. In: IEEE World Congress on Computational Intelligence, Neural Networks Proceedings, vol. 2, pp. 880–884 (1998)
5. Labonte, G.: A new neural network for particle tracking velocimetry. Exp. Fluids **26**, 340–346 (1999)
6. Labonte, G.: New neural network reconstruction of fluid flows from tracer-particle displacements. Exp. Fluids **30**, 399–409 (2001)
7. Ohmi, K.: Neural network PIV using a self-organizing maps method. In: Proceedings of 4th Pacific Symposium Flow Visualization and Image Processing, F-4006 (2003)
8. Joshi, S.R.: Improvement of algorithm in the particle tracking velocimetry using self-organizing maps. J. Inst. Eng. **7**, 6–23 (2009)
9. Verber, D.: Chapter 13. Implementation of massive artificial neural networks with CUDA. In: Volosencu, C. (ed.) Cutting Edge Research in New Technologies, INTECH, pp. 277–302 (2012)
10. Hernández-Pérez, R., Gabbasov, R., Suárez-Cansino, J.: Improving performance of particle tracking velocimetry analysis with artificial neural networks and graphics processing units. Res. Comput. Sci. **104**, 71–79 (2015)
11. Gottschalk, P.G., Dunn, J.R.: The five-parameter logistic: a characterization and comparison with the four-parameter logistic. Anal. Biochem. **343**, 54–65 (2005)

12. Gabbasov, R.F., Klapp, J., Suárez-Cansino, J., Sigalotti, L.D.G.: Numerical simulations of the Kelvin-Helmholtz instability with the gadget-2 SPH code. In: Experimental and Computational Fluid Mechanics with Applications to Physics, Enginnering and the Environment (2014). arXiv:1310.3859. [astro-ph.IM]
13. Monaghan, J.J.: Smoothed particle hidrodynamics. Rep. Prog. Phys. **68**(8), 1703–1759 (2005)
14. Jiang, M., Machiraju, R., Thompson, D.: Detection and visualization of vortices. The Visualization Handbook, pp. 296–309. Academic Press, Cambridge (2005)
15. Stewart, R.W.: Turbulence. In: Encyclopaedia Britannica Educational Corporation Film (1969)
16. Unsworth, C.A.: Chapter 3. Section 4. Particle Imaging velocimetry. In: Geomorphological Techniques, British Society for Geomorphology (2015)
17. Shi, B., Wei, J., Pang, M.: A modified cross-correlation algorithm for PIV image processing of particle-fluid two-phase flow. In: Flow Measurement and Instrumentation, October 2015, vol. 45, pp. 105–117 (2015)
18. Rabault, J., Kolaas, J., Jensen, A.: Performing particle image velocimetry using artificial neural networks: a proof-of-concept. In: Measurement Science and Technology, vol. 28, no. 12 (2017)
19. Rossi, R., Malizia, A., Poggi, L.A., Ciparisse, J.-F., Peluso, E., Gaudio, P.: Flow motion and dust tracking software for PIV and dust PTV. J. Fail. Anal. Prev. **16**(6), 951–962 (2016)
20. Jiang, C., Dong, Z., Wang, X.: An improved particle tracking velocimetry (PTV) technique to evaluate the velocity field of saltating particles. J. Arid Land **9**(5), 727–742 (2017)
21. Elhimer, M., Praud, O., Marchal, M., Cazin, S., Bazile, R.: Simultaneous PIV/PTV velocimetry technique in a turbulent particle-laden flow. J. Vis. **20**(2), 289–304 (2017)
22. Dal Sasso, S.F., Pizarro, A., Samela, C., Mita, L., Manfreda, S.: Exploring the optimal experimental setup for surface flow velocity measurements using PTV. Environ. Monit. Assess. **190**, 460 (2018)

Machine Learning

Transforming Mixed Data Bases for Machine Learning: A Case Study

Angel Kuri-Morales[(✉)]

Instituto Tecnológico Autónomo de México,
Río Hondo no. 1, 01000 D.F. Mexico, Mexico
`akuri@itam.mx`

Abstract. Structured Data Bases which include both numerical and categorical attributes (Mixed Databases or MD) ought to be adequately pre-processed so that machine learning algorithms may be applied to their analysis and further processing. Of primordial importance is that the instances of all the categorical attributes be encoded so that the patterns embedded in the MD be preserved. We discuss CESAMO, an algorithm that achieves this by statistically sampling the space of possible codes. CESAMO's implementation requires the determination of the moment when the codes distribute normally. It also requires the approximation of an encoded attribute as a function of other attributes such that the best code assignment may be identified. The MD's categorical attributes are thusly mapped into purely numerical ones. The resulting numerical database (ND) is then accessible to supervised and non-supervised learning algorithms. We discuss CESAMO, normality assessment and functional approximation. A case study of the US census database is described. Data is made strictly numerical using CESAMO. Neural Networks and Self-Organized Maps are then applied. Our results are compared to classical analysis. We show that CESAMO's application yields better results.

Keywords: Machine Learning · Mixed Databases · Non-linear regression
Goodness-of-fit

1 Introduction

"Machine Learning" (ML) has been defined as the field of computer science (a branch of artificial intelligence) whose main purpose is to develop techniques which allow Computers to learn. More concretely, its aim is to create programs able to generalize behaviors derived from the information fed as examples. It is, therefore, a process of induction from knowledge. In many cases the field of application of ML overlaps with computational statistics since both disciplines are based on data analysis. However, ML is also centered in (and needs of) the computational complexity of the problems. Many problems correspond to the NP-hard type. Hence, a substantial parte of the research of ML is focused on the design of feasible solutions to such problems. ML can be seen as an attempt to automate some parts of the scientific method via mathematical methodologies. This has given raise to, for instance, (a) K-Means, (b) Fuzzy C-Means, (c) Self-Organizing Maps, (d) Fuzzy Learning Vector Quantization [1], or (e) Multi-

© Springer Nature Switzerland AG 2018
I. Batyrshin et al. (Eds.): MICAI 2018, LNAI 11288, pp. 157–170, 2018.
https://doi.org/10.1007/978-3-030-04491-6_12

Layered Perceptron Networks, (f) Support Vector Machines [15]. All of these methods have been designed to tackle the analysis of strictly numerical databases, i.e. those in which all the attributes are directly expressible as numbers.

If any of the attributes is non-numerical (i.e. categorical) none of the methods in the list is applicable. Analysis of categorical attributes (i.e., attributes whose domain is not numeric) is a difficult, yet important task: many fields, from statistics to psychology deal with categorical data. In spite of its importance, the task of categorical ML has received relatively scant attention. Much of the published algorithms to apply ML to categorical data rely on the usage of a distance metric that captures the separation between two vectors of categorical attributes, such as the Jaccard coefficient [2]. An interesting alternative is explored in [3] where COOLCAT, a method which uses the notion of entropy to group records, is presented. It is based on information loss minimization. Another reason for the limited exploration of categorical ML techniques is its inherent difficulty.

In [4] a different approach is taken by (a) Preserving the patterns embedded in the database and (b) Pinpointing the codes which preserve such patterns. These two steps result in the correct identification of a set of PPCs. The resulting algorithm is called CENG (Categorical Encoding with Neural Networks and Genetic Algorithms) and its parallelized version ParCENG [5]. However, this approach is computationally very demanding and, to boost its efficiency, it ought to be tackled in ensembles of multiple CPUs. Even so, when the number of CAs and/or the number of category's instances is large, execution time may grow exponentially.

Along the same line, in [6] a new methodology and its basic algorithm, called CESAMO, were introduced. By replacing an exhaustive search of the possible codes by statistical sampling the processing time is considerably reduced. Once the categorical (i.e. non-numerical) instances are replaced by the more adequate codes the preservation of the embedded patterns is achieved. However, it is important to note that the PPCs are NOT to be assumed as an instance applicable to DBs other than the original one. That is to say: a set of PPCs (say PPC1) obtained from a DB (say DB1) is not applicable to a different DB (say DB2) even if DB1 and DB2 are structurally identical. In other words, PPC1 $\neq$ PPC2 for the same DB when the tuples of such DB are different,

The rest of the paper is organized as follows. In Sect. 2 we briefly describe CESAMO. In Sect. 3 we discuss the way the quality of the PPCs is calculated. In Sect. 4 discuss the way in which the universal validity of the PPCs is statistically ascertained. In Sect. 5 we present a case of study where a MD is tackled after preprocessing the data with CESAMO. Finally, in Sect. 6 we offer our conclusions.

2   Encoding Mixed Databases

2.1   Pseudo-binary Encoding

In what follows we denote the i instances of categorical variable c as ci; the number of categorical variables with c; the number of all attributes by n.

A common choice is to replace every CA variable by a set of binary variables, each corresponding to the *ci*s. The CAs in the MD are replaced by numerical ones where every categorical variable is replaced by a set of *ci* binary numerical codes. An MD will be replaced by an ND with $n - c + c \cdot ci$ variables. This approach suffers from the following limitations:

(a) The number of attributes of ND will be larger than that of MD. In many cases this leads to unwieldy databases which are more difficult to store and handle.
(b) The type of coding system selected implies an *a priori* choice since all pseudo-binary variables may be assigned any two values (typically "0" denotes "absence"; "1" denotes "presence"). This choice is subjective. Any two different values are possible. Nevertheless, the mathematical properties of ND will vary with the different choices, thus leading to clusters which depend on the way in which "presence" or "absence" is encoded.
(c) Finally, with this sort of scheme the pseudo-binary variables do no longer reflect the essence of the idea conveyed by a category. A variable corresponding to the *i-th* instance of the category reflects the way a tuple is "affected" by belonging to the *i-th* categorical value, which is correct. But now the original issue "How does the behavior of the individuals change according to the category?" is replaced by "How does the behavior of the individuals change when the category's value is the *i-th*?" The two questions are not interchangeable.

2.2   Pattern Preserving Codes

As stated in the introduction, the basic idea is to apply ML algorithms designed for strictly numerical databases (ND) to MDs by encoding the instances of categorical variables with a number. This is by no means a new concept. MDs, however, offer a particular challenge when clustering is attempted because it is, in principle, impossible to impose a metric on CAs. There is no way in which numerical codes may be assigned to the CAs in general. An alternative goal is to assign codes (which we call Pattern Preserving Codes or PPCs) to each and all the instances of every class (category) which will preserve the patterns present for a given MD.

Consider a set of n-dimensional tuples (say U) whose cardinality is m. Assume there are n unknown functions of $n - 1$ variables each, which we denote with

$$f_k(v_1, \ldots, v_{k-1}, v_{k+1}, \ldots, v_n); \; k = 1, \ldots, n$$

Let us also assume that there is a method which allows us to approximate f_k (from the tuples) with F_k. Denote the resulting n functions of $n - 1$ independent variables with F_i, thus

$$F_k \approx f(v_1, \ldots, v_{k-1}, v_{k+1}, \ldots, v_n); \; k = 1, \ldots, n \tag{1}$$

The difference between f_k and F_k will be denoted with ε_k such that, for attribute k and the m tuples in the database

$$\varepsilon_k = max[abs(f_{ki} - F_{ki})]; \ i = 1, \ldots, m \qquad (2)$$

Our contention is that the PPCs are the ones which minimize ε_k for all k. This is so because only those codes which retain the relationships between variable k and the remaining $n - 1$ variables AND do this for ALL variables in the ensemble will preserve the whole set of relations (i.e. patterns) present in the data base, as in (3).

$$\varXi = min[max \ (\varepsilon_k; \ k = 1, \ldots, n)] \qquad (3)$$

Notice that this is a multi-objective optimization problem because complying with condition k in (2) for any given value of k may induce the non-compliance for a different possible k. Using the min-max expression of (3) equates to selecting a particular point in the Pareto's front [7].

To achieve the purported goal we must have a tool which is capable of identifying the F_k's in (1) and the codes which attain the minimization of (3). This is possible using NN[1]s and GA[2]s. Theoretical considerations (see, for instance, [8–11]) ensure the effectiveness of the method. But, as stated, the use of such ensemble implies a possibly exponential number of floating point operations. An alternative which yields similar but more economical process is CESAMO, which we describe in what follows.

2.3   The CESAMO Algorithm

We denote the number of tuples in the DB by t and the number of categorical attributes by c; the number of numerical attributes by n; the i-th categorical variable by vi; the value obtained for variable i as a function of variable j by $yi(j)$.

We will sample the codes yielding yi as a function of a sought for relationship. This relationship and the model of the population it implies, will be selected so as to preserve the behavioral patterns embedded in the DB.

Two issues are of primordial importance in the proposed methodology:

(a) How to define the function which will preserve the patterns.
(b) How to determine the number of codes to sample.

Regarding (a), we use a mathematical model considering high order relations, as will be discussed below. Regarding (b), we know that, independently of the distribution of the yi's, the distribution of the means of the samples of yi (yi_{AVG}) will become Gaussian. Once the distribution of the yi_{AVG} becomes Gaussian, we will have achieved statistical stability, in the sense that further sampling of the yi's will not significantly modify the characterization of the population.

In essence, therefore, what we propose is to sample enough codes to guarantee the statistical stability of the values calculated from $yi \leftarrow f(vj)$. If $f(vj)$ is adequately chosen the codes corresponding to the best approximation will be those inserted in MD. Furthermore, CESAMO relies on a double level sampling: only pairs of variables are

---

[1] Artificial Neural Networks
[2] Genetic Algorithms

considered and every pair is, in itself, sampling the multivariate space. This avoids the need to explicitly solve the multi-objective optimization underlying problem. The clustering problem may be, then, numerically tackled.

2.4   CESAMO's Pseudo-code

The general algorithm for CESAMO is as follows:

A - Specify the mixed database MD.

B - Specify the sample size (ss)[3]

C - MD is analyzed to determine n, t and $ci(i)$ for $i = 1,...,c$.

D - The numerical data are assumed to have been mapped into [0, 1). Therefore, every ci will be, likewise, in [0, 1).

```
1. for i ← 1 to c
2.       Do until the distribution of yiAVG is Gaussian
3.              Randomly select variable j (j ≠i)
4.              Assign random values to all instances of vi.
5.              yiAVG ← 0
6.              For k ←1 to ss
7.                      yi ← f(vj)
8.                      yiAVG ← yiAVG+yi
9.              endfor
10.             yiAVG ← yiAVG/ss
11.      enddo
12.      Select the codes corresponding to the best value of yi
13. endfor
```

Notice that vj may be, itself, categorical. In that cases every categorical instance of vj is replaced by random codes so that we may calculate $f(vj)$.

3   Finding the PPCs

Step 7 of the pseudo-code implies that, given a table of the form of the one shown in Table 1, we will be able to find an algebraic[4] expression of y as a function of x. The values in Table 1 are assumed to be instances of experimental[5] data.

Innumerable approximation algorithms are known which satisfy this criterion. Three multivariate of these are analyzed in [12]. However, one important feature present in one of them., the so-called Fast Ascent Algorithm (FAA), is that, to

---

[3] For which see Sect. 4.

[4] Several possible forms for the approximant are possible. We selected a polynomial form due to the well known Weierstrass approximation theorem, which states that every continuous function defined on a closed interval [a, b] can be uniformly approximated as closely as desired by a polynomial function.

[5] One of the requirements for the generality of the algorithm is that the characteristics of the data are not known in advance. Assuming them to be experimental underlines this fact.

Table 1. A table of numerical data.

x	y
x_1	y_1
x_2	y_2
...	...
x_n	y_n

implement it, there is no need to load into the CPU's main memory all of the data tuples; a fact that is of utmost relevance where (as is the case for real-world applications) the data bases may consist of thousands or even millions of tuples. Furthermore, with FAA the form of the approximant may be arbitrarily selected.

3.1   The Fast Ascent Algorithm

1. Input the data vectors (call them **D**).
2. Input the degrees of each of the variables of the approximating polynomial.
3. Map the original data vectors into the powers of the selected monomials (call them **P**).
4. Stabilize the vectors of **P** by randomly disturbing the original values (call the resulting data **S**).
5. Select a subset of size M from **S**. Call it **I**. Call the remaining vectors **E**.

BOOTSTRAP

6. Obtain the minimax signs (call the matrix incorporating the σ's **A**).
7. Obtain the inverse of **A** (call it **B**).

LOOP

8. Calculate the coefficients **C** = **f B**. The maximum internal ε_θ error is also calculated.
9. Calculate the maximum external error ε_ϕ from **C** and **E**. Call its index I_E
10. $\varepsilon_\theta \geq \varepsilon_\phi$?
    YES: Stop; the coefficients of **C** are those of the minimax polynomial for the **D** vectors.
11. Calculate the λ vector from $\lambda = A^{I_E} B$
12. Calculate the β vector which maximizes $\sigma_{I_E} \frac{\lambda_j}{B_j}$. Call its index I_I.
13. Interchange vectors I_E and I_I.
14. Calculate the new inverse $\overline{B}$. Make $B \leftarrow \overline{B}$.
15. Go to step 8.

□

In [13] it was shown that the logistic function $1/(1 + e^{-x})$ may be closely approximated by a polynomial of the form

$$y = c_0 + c_1 x + c_3 x^3 + c_5 x^5 + c_7 x^7 + c_9 x^9 + c_{11} x^{11} \tag{4}$$

and that continuous data may be approximated (and its main components retained) from (4). Terms of degree higher than 11 are of little consequence when, as in this case, $0 \leq x < 1$. FAA yields coefficients for the best L_∞ approximation of y given the x as in Table 1. From these CESAMO calculates the RMS approximation error. Therefore, **Step 7** is fulfilled by applying Eq. (4). The code which achieves the smallest mean approximation error (ε_S) once normality is attained will be the PPC.

4 Ascertaining Normality

CESAMO's **Step 2** demands that the number of sampled codes be large enough to guarantee that the distribution of the errors ε_S is normal. To this effect we appeal to the Modified Chi-Squared Distribution (the ЛИ Distribution) which was defined in [14]. In this case we want to compare the distribution of the ε_S to a Gaussian distribution.

The average error codes corresponding to the approximating polynomials of the FFA will become Gaussian from the well known Central Limit Theorem. It states that for a large sample size (rule of thumb: $n \geq 30$), ε_S is approximately normally distributed, regardless of the distribution of the population one samples from, as illustrated in Figs. 1 and 2. μ and σ denote the mean and standard deviation of an arbitrary population. μ_X and σ_X denote the same parameters for the distribution of the means from that same arbitrary distribution.

We will therefore, statistically determine that CESAMO has reached equilibrium (and no further samples are needed) when the ε_S are normally distributed. CESAMO will keep on sampling codes until the ε_S distribute normally.

The problem we address is, therefore, how to know that such is the case. The ЛИ Distribution goodness-of-fit test allows us to determine whether two experimental samples may be said to correspond to the same population.

At least, 5 observations per decile are required. In this case we defined 10 deciles. The expected number of observations in the *i-th* decile (E_i) we expect to find for a Gaussian distribution in every decile may be theoretically determined. The observed number of observations is denoted by O_i.

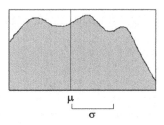

Fig. 1. Arbitrary population

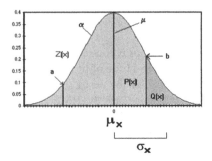

Fig. 2. Gaussian distribution

Then we calculate Л from (5).

$$\text{Л} = \sum_{i=1}^{Q} \frac{(O_i - E_i)^2}{E_i} \wedge [\, O_i \geq 5 \;\; \forall i \,]$$ (5)

For $ss = 30$ (the sample size), 10 deciles ($Q = 10$) and $p\text{-}value = 0.05$ the critical value of Л is 3.2. When Л $= 3.2$ the probability that the ε_S are normally distributed is better than 0.95. In other words: (a) A categorical variable is selected, (b) Random codes are assigned to the instances of the variable ss times and (c) ε_S is calculated. Since there are 10 deciles and 5 observations/decile are required, at least 50 ε_S are calculated. Thereafter Л is calculated for every new sample of ss codes. If Л is less than 3.2 CESAMO stops: the codes corresponding to the smallest ε_S are assigned to the instances of the selected variable.

5 Case Study

From <http://www.census.gov/ftp/pub/DES/www/welcome.html> we extracted a database (USCDB) consisting of 12 attributes: age (N), workclass (C), fnlwgt (N), education (C), education-num (N), marital-status (C), occupation (C), relationship (C), race (C), sex (C), hours-per-week (N), salary (C)[6]. The data base consisted of 29,267 tuples, with a total of 58 different instances for categorical attributes. A very small segment is illustrated in Fig. 3.

CESAMO was run on this data. A small sample of the encoded USCDB is illustrated in Fig. 4.

5.1 Supervised Learning

The original problem was to find the salary (only 2 values) as a function of the remaining variables. This is a classification problem which was tackled with various algorithms, with the results illustrated in Fig. 5.

---

[6] (N) denotes a numerical attribute; (C) denotes a categorical attribute.

Age	Workclass	Fnlwgt	Education	Edunum	Marstat	Occup	Relat	Race	Sex	Hours	Salary
39.0000	State-gov	77516.0000000000	Bachelors	13.000000000000	Never-married	Adm-clerical	Not-in-family	White	Male	40.000000000000	<=50K
50.0000	Self-emp-not-inc	83311.0000000000	Bachelors	13.000000000000	Married-civ-spouse	Exec-managerial	Husband	White	Male	13.000000000000	<=50K
38.0000	Private	215646.000000000	HS-grad	9.000000000000	Divorced	Handlers-cleaners	Not-in-family	White	Male	40.000000000000	<=50K
53.0000	Private	234721.000000000	11th	7.000000000000	Married-civ-spouse	Handlers-cleaners	Husband	Black	Male	40.000000000000	<=50K
28.0000	Private	338409.000000000	Bachelors	13.000000000000	Married-civ-spouse	Prof-specialty	Wife	Black	Female	40.000000000000	<=50K
37.0000	Private	284582.000000000	Masters	14.000000000000	Married-civ-spouse	Exec-managerial	Wife	White	Female	40.000000000000	<=50K
49.0000	Private	160187.000000000	9th	5.000000000000	Married-spouse-absen	Other-service	Not-in-family	Black	Female	16.000000000000	<=50K
52.0000	Self-emp-not-inc	209642.000000000	HS-grad	9.000000000000	Married-civ-spouse	Exec-managerial	Husband	White	Male	45.000000000000	>50K
31.0000	Private	45781.0000000000	Masters	14.000000000000	Never-married	Prof-specialty	Not-in-family	White	Female	50.000000000000	>50K
42.0000	Private	159449.000000000	Bachelors	13.000000000000	Married-civ-spouse	Exec-managerial	Husband	White	Male	40.000000000000	>50K
37.0000	Private	280464.000000000	Some-college	10.000000000000	Married-civ-spouse	Exec-managerial	Husband	Black	Male	80.000000000000	>50K
30.0000	State-gov	141297.000000000	Bachelors	13.000000000000	Married-civ-spouse	Prof-specialty	Husband	Asian-Pac-Islander	Male	40.000000000000	>50K
23.0000	Private	122272.000000000	Bachelors	13.000000000000	Never-married	Adm-clerical	Own-child	White	Female	30.000000000000	<=50K
32.0000	Private	205019.000000000	Assoc-acdm	12.000000000000	Never-married	Sales	Not-in-family	Black	Male	50.000000000000	<=50K

Fig. 3. A small sample of the USA census database.

Age	Workclass	Fnlwgt	Education	Edunum	Marstat	Occup	Relat	Race	Sex	Hours	Salary
0.2602739726030000	0.71293	0.1052997673050000	0.24537	0.5333333333330000	0.78817	0.24414	0.82007	0.26555	0.24873	0.5000000000000000	0.95332
0.1369863013700000	0.44104	0.0949914149280000	0.26996	0.8000000000000000	0.26843	0.13650	0.08753	0.26555	0.70355	0.6020408163270000	0.95332
0.4657534246580000	0.77985	0.0847895926370000	0.47333	0.7333333333330000	0.78817	0.24414	0.82007	0.26555	0.24873	0.6020408163270000	0.95332
0.1780821917810000	0.77985	0.3569917157190000	0.24537	0.5333333333330000	0.78817	0.23989	0.82007	0.26555	0.24873	0.6020408163270000	0.95332
0.0273972602740000	0.71293	0.3753451783650000	0.83202	0.4666666666670000	0.26843	0.23989	0.54264	0.26555	0.24873	0.1938775510200000	0.95332
0.6027397260270000	0.15898	0.0726190525920000	0.22641	1.0000000000000000	0.78817	0.13650	0.82007	0.26555	0.24873	0.3979591836730000	0.84631
0.2465753424660000	0.71293	0.1580352927910000	0.24537	0.5333333333330000	0.15566	0.63638	0.54264	0.26555	0.24873	0.3979591836730000	0.95332
0.2054794520550000	0.71293	0.0680748133630000	0.24537	0.5333333333330000	0.78817	0.66028	0.82007	0.26555	0.24873	0.3979591836730000	0.95332
0.2739726027400000	0.71293	0.1631497813000000	0.78443	0.2666666666670000	0.78817	0.63638	0.84983	0.51542	0.70355	0.4795918367350000	0.95332
0.1780821917810000	0.71293	0.2484386031610000	0.26996	0.8000000000000000	0.78817	0.13650	0.84983	0.26555	0.70355	0.3979591836730000	0.95332
0.3150684931510000	0.71293	0.1338842957940000	0.24537	0.5333333333330000	0.78817	0.63638	0.82007	0.26555	0.24873	0.3979591836730000	0.95332
0.5205479452050000	0.44104	0.0172192305600000	0.26996	0.8000000000000000	0.13077	0.13650	0.08753	0.26555	0.70355	0.3979591836730000	0.84631

Fig. 4. CESAMO encoded USCDB (sample).

	Algorithm	Error
1	C4.5	15.54
2	C4.5-auto	14.46
3	C4.5 rules	14.94
4	Voted ID3 (0.6)	15.64
5	Voted ID3 (0.8)	16.47
6	T2	16.84
7	1R	19.54
8	NBTree	14.10
9	CN2	16.00
10	HOODG	14.82
11	FSS Naive Bayes	14.05
12	IDTM (Decision table)	14.46
13	Naive-Bayes	16.12
14	Nearest-neighbor (1)	21.42
15	Nearest-neighbor (3)	20.35
16	OC1	15.04
17	Pebls	Crashed

Fig. 5. Reported performance for USCDB tackled with traditional algorithms

Once the data was encoded, we trained a NN with 11 input neurons, one output neuron and one neuron in the hidden layer. Notice that the output neuron corresponded to the class "salary" and that, therefore, perfect classification is achieved if the maximum approximation error is ≤ 0.5. Remember that all variables are encoded into the $[0, 1)$ interval.

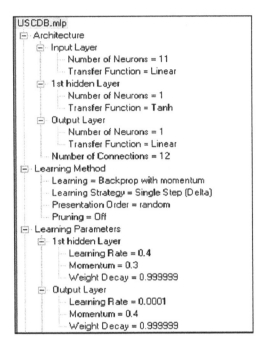

Fig. 6. Neural network parameters

For only two classes, as in this case, a maximum error of 0.5 will allow us to distinguish between the 2 classes unequivocally The parameters are illustrated in Fig. 6. The maximum error for this NN was 0.0824. The results and learning curve are illustrated in Fig. 7. The achieved maximum training error guarantees perfect classification. None of the reported traditional methods achieves error-free classification.

5.2   Non-supervised Learning

We also performed a clustering exercise using the same data. In this case we trained several SOMs (for 2, 3,…, 7 clusters). From these we determined that the number of clusters was 5, as illustrated in Fig. 8. "MAX DELTA" refers to the relative increment of the maximum approximation error difference between consecutive numbers of clusters. Bezdek's criterion [16] suggests that the number of clusters lies within the region of steepest change.

Lack of space disallows a detailed analysis of the results. As an example, however, we show (Fig. 9) the way age is clustered. Age is numerical. A final example is presented in Fig. 10, for the attribute "Sex". The codes found by CESAMO for "Sex" and "Education" are shown next,

Sex:

Female   0.7035457848105580
Male   0.2487332848832010

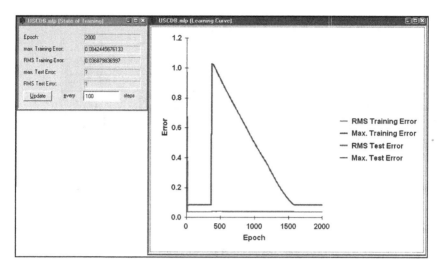

Fig. 7. Training results for the NN with CESAMO encoded USCDB

CLUSTERS	MEAN DISTANCE	MEAN DELTA	MAX DISTANCE	MAX DELTA
2	0.524	0.025	0.997	0.001
3	0.499	0.011	0.996	0.009
4	0.488	0.055	0.987	0.016
5	0.433	0.008	0.971	0.020
6	0.425	0.014	0.951	0.017
7	0.411	0.411	0.968	0.968

Fig. 8. Determination of the number of clusters for USCDB

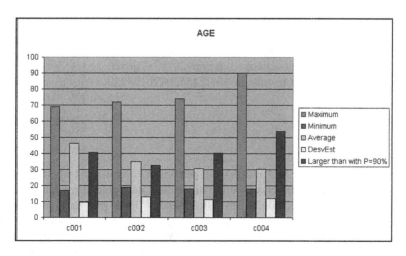

Fig. 9. Distribution of age in 4 clusters.

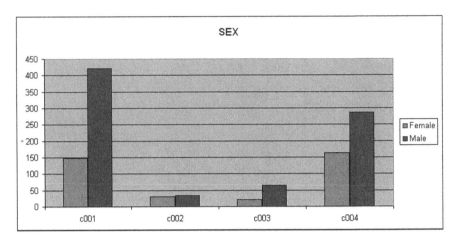

Fig. 10. Distribution of sex in 4 clusters

Education:

10th	0.9491241523064670
11th	0.8457170522306110
12th	0.8320180689916020
1st-4th	0.7622485130559650
5th-6th	0.5937019358389080
7th-8th	0.5366302013862880
9th	0.7844320349395280
Assoc-acdm	0.4793326088692990
Assoc-voc	0.3429927569814030
Bachelors	0.2699594611767680
Doctorate	0.2264134744182230
HS-grad	0.2453737098257990
Masters	0.2603427632711830
Preschool	0.9723846812266860
Prof-school	0.3469207938760520
Some-college	0.2745520470198240

6 Conclusions

We have shown that CESAMO allows us to perform a generalized analysis of MDs. It outperforms other traditional classification algorithms and generalizes the application of Machine Learning Techniques to all kinds of data (be they numerical or non/numerical). CESAMO allows us to treat large numbers of attributes with large numbers of instances for the categorical ones. We were able to solve the two-class classification problem with no errors by finding the PPCs in USCDB. The typical pseudo-binary approach mentioned in Sect. 2.1 would have yielded a total of 52 variables. Considering USCDB's size this would have implied an increment of roughly 1,200,000 additional records.

On the other hand, the clustering analysis with SOMs would have been simply impossible without CESAMO. From the clusters, furthermore, simple classification rules may be found to make the results quite intuitive by identifying the attributes which outstand in every cluster.

At present CESAMO is being applied to the data base of Chikongunya in Mexico, with a database from the Secretaría de Salud. We expect to publish in this regard in the very near future.

References

1. Goebel, M., Gruenwald, L.: A survey of data mining and knowledge discovery software tools. ACM SIGKDD Explor. Newsl. **1**(1), 20–33 (1999)
2. Sokal, R.R.: The principles of numerical taxonomy: twenty-five years later. Comput.-Assist. Bacterial Syst. **15**, 1 (1985)
3. Barbará, D., Li, Y., Couto, J.: COOLCAT: an entropy-based algorithm for categorical clustering. In: Proceedings of the Eleventh International Conference on Information and Knowledge Management, pp. 582–589. ACM (2002)
4. Kuri-Morales, A.F.: Categorical encoding with neural networks and genetic algorithms. In: Zhuang, X., Guarnaccia, C. (eds.) WSEAS Proceedings of the 6th International Conference on Applied Informatics and. Computing Theory, pp. 167–175, 01 Jul 2015. ISBN 9781618043139, ISSN 1790-5109
5. Kuri-Morales, A., Sagastuy-Breña, J.: A parallel genetic algorithm for pattern recognition in mixed databases. In: Carrasco-Ochoa, J.A., Martínez-Trinidad, J.F., Olvera-López, J.A. (eds.) MCPR 2017. LNCS, vol. 10267, pp. 13–21. Springer, Cham (2017). https://doi.org/10.1007/978-3-319-59226-8_2
6. Kuri-Morales, A.: Pattern discovery in mixed data bases. In: Martínez-Trinidad, J.F., Carrasco-Ochoa, J.A., Olvera-López, J.A., Sarkar, S. (eds.) MCPR 2018. LNCS, vol. 10880, pp. 178–188. Springer, Cham (2018). https://doi.org/10.1007/978-3-319-92198-3_18
7. Deb, K., Agrawal, S., Pratap, A., Meyarivan, T.: A fast elitist non-dominated sorting genetic algorithm for multi-objective optimization: NSGA-II. In: Schoenauer, M., et al. (eds.) PPSN 2000. LNCS, vol. 1917, pp. 849–858. Springer, Heidelberg (2000). https://doi.org/10.1007/3-540-45356-3_83
8. Cybenko, G.: Approximation by superpositions of a sigmoidal function. Math. Control Sig. Syst. **2**(4), 303–314 (1989)
9. Rudolph, G.: Convergence analysis of canonical genetic algorithms. IEEE Trans. Neural Netw. **5**(1), 96–101 (1994)
10. Kuri-Morales, A.F., Aldana-Bobadilla, E., López-Peña, I.: The best genetic algorithm II. In: Castro, F., Gelbukh, A., González, M. (eds.) MICAI 2013. LNCS (LNAI), vol. 8266, pp. 16–29. Springer, Heidelberg (2013). https://doi.org/10.1007/978-3-642-45111-9_2
11. Widrow, B., Lehr, M.A.: 30 years of adaptive neural networks: perceptron, madaline, and backpropagation. Proc. IEEE **78**(9), 1415–1442 (1990)
12. Lopez-Peña, I., Kuri-Morales, A.: Multivariate approximation methods using polynomial models: a comparative study. In: 2015 Fourteenth Mexican International Conference on Artificial Intelligence (MICAI). IEEE (2015)

13. Kuri-Morales, A., Cartas-Ayala, A.: Polynomial multivariate approximation with genetic algorithms. In: Sokolova, M., van Beek, P. (eds.) AI 2014. LNCS (LNAI), vol. 8436, pp. 307–312. Springer, Cham (2014). https://doi.org/10.1007/978-3-319-06483-3_30

14. Kuri-Morales, A.F., López-Peña, I.: Normality from monte carlo simulation for statistical validation of computer intensive algorithms. In: Pichardo-Lagunas, O., Miranda-Jiménez, S. (eds.) MICAI 2016. LNCS (LNAI), vol. 10062, pp. 3–14. Springer, Cham (2017). https://doi.org/10.1007/978-3-319-62428-0_1

15. Haykin, S.: Neural Networks: A Comprehensive Foundation. Prentice Hall PTR, Upper Saddle River (1994)

16. Kwon, S.H.: Cluster validity index for fuzzy clustering. Electron. Lett. **34**(22), 2176–2177 (1998)

Full Model Selection in Huge Datasets and for Proxy Models Construction

Angel Díaz-Pacheco$^{(\boxtimes)}$ and Carlos Alberto Reyes-García

Computer Science Department, Instituto Nacional de Astrofísica,
Óptica y Electrónica (INAOE), Luis Enrique Erro No.1, Santa María Tonantzintla,
72840 San Andrés Cholula, Puebla, Mexico
diazpacheco@inaoep.mx

Abstract. Full Model Selection is a technique for improving the accuracy of machine learning algorithms through the search of the most adequate combination on each dataset of feature selection, data preparation, a machine learning algorithm and its hyper-parameters tuning. With the increasingly larger quantities of information generated in the world, the emergence of the paradigm known as Big Data has made possible the analysis of gigantic datasets in order to obtain useful information for science and business. Though Full Model Selection is a powerful tool, it has been poorly explored in the Big Data context, due to the vast search space and the elevated number of fitness evaluations of candidate models. In order to overcome this obstacle, we propose the use of proxy models in order to reduce the number of expensive fitness functions evaluations and also the use of the Full Model Selection paradigm in the construction of such proxy models.

Keywords: Big Data · Model Selection · Machine learning

1 Introduction

Data can be considered as an expenditure in storage or a valuable asset, this valuation relies on the analysis made to such information. One of the main tendencies to analyze data is the use of machine learning techniques though, to choose an adequate learning algorithm to a specific dataset is not a trivial task. This process requires to find the combination of algorithms together with their hyper-parameters to achieve the lowest misclassification rate in a wide search space [17]. Other factors that have a major impact in the generalization capacities of a classification algorithm are: feature selection and data-preparation. These factors, in combination with the selection of a classification algorithm integrates the Full Model Selection paradigm (FMS). Under this paradigm, every factor combination represents a set of data transformations and performing the learning process over the training set [6]. Although this paradigm is useful, has

---

The first author is grateful for the support from CONACyT scholarship no. 428581.

I. Batyrshin et al. (Eds.): MICAI 2018, LNAI 11288, pp. 171–182, 2018.
https://doi.org/10.1007/978-3-030-04491-6_13

been poorly explored in the Big Data context due the huge search space and the time every transformation and learning process takes in a dataset of this context. FMS has been addressed as an optimization problem varying the search technique employed. As an example, in [11] a hybrid method based on grid search and the theoretic hyper-parameter decision technique (ThD) of Cherkassy and Ma for the algorithm SVR (Support Vector Regression) was proposed. In [12] a genetic algorithm was employed for the hyper-parameter tuning of the SVM algorithm. In [6] they tackled and defined the full model selection problem with the use of a particle swarm optimization algorithm (PSO), in [9] a PSO algorithm was also used but just for hyper-parameter optimization for the ls-SVM algorithm meanwhile in [2] was proposed the use of the bat algorithm for solving FMS. Similar problems where the computing time is prohibitive, have been approached in the literature by means of proxy models also known as surrogate functions. A proxy model is a computationally inexpensive alternative to a full numerical simulation and can be defined as mathematically, statistically, or data-driven model defined function that replicates the simulation model output for selected input parameters [1]. Some of the main approaches analyzed can be grouped as follows: **(a) The surrogate function is based on a single regression algorithm to predict the fitness in the objective or objectives to optimize**. In [5] a neural network based on fuzzy granules was employed as a proxy model. In [15] a multi-objective genetic algorithm assisted by surrogate functions was proposed for model selection. A neural network with the parameters selected by hand was employed in order to predict each objective. Every generation, the surrogate predicts the performance of the individuals and the promissory ones were evaluated by the complete fitness function and the neural networks were re-trained with the new samples. **(b) Model selection is employed in order to increase the quality of the surrogate functions**. An island model genetic algorithm is employed in [4] for model selection and hyper-parameter optimization. The GA is combined with the Expected Improvement algorithm (EI) for the selection of the interest data points that can improve the performance of the surrogate model. In [14] the model selection is performed among several algorithms with their hyper-parameters previously configured by hand. From all the analyzed works we can see two fundamental aspects for the development of this work. The first one is that, the full model selection problem could be benefited by the use of surrogate models in order to reduce the high number of fitness evaluations during the search step. The second one is that, proxy models can be used as a way to reduce the fitness evaluations and also guide the search process as a compass, therefore, a wise decision is to built the best compass possible and a way to do that is through the FMS paradigm. This work has the following organization. In Sect. 2 we present some background on Big Data and MapReduce. Section 3 describes our proposed algorithm. Section 4 shows the experiments performed to test the validity of our proposal. Finally, Sect. 5 presents the conclusions.

2   Big Data and the MapReduce Programming Model

MapReduce was introduced by Dean and Ghemawat in 2004 with the goal of enabling the parallelization and distribution of big scale computation required to analyse the large datasets. This programming model was designed to work over computing clusters and it works under the master-slave communication model. In MapReduce a computing task is specified as a sequence of stages: map, shuffle and reduce that works on a dataset $X = \{x_1, x_2, ..., x_n\}$. The map step applies a function μ to each value x_i to produce a finite set of key-value pairs (k, v). To allow for parallel execution, the computation of function $\mu(x_i)$, must depend only on x_i. The shuffle step collects all the key-value pairs produced in the previous map step, and produces a set of lists, $L_k = (k; v_1, v_2, ..., v_n)$ where each of such lists consists of all values v_i, such that $k_i = k$ for a key k assigned in the map step. The reduce stage applies a function ρ to each list $L_k = (k; v_1, v_2, ..., v_n)$, created during the shuffle step, to produce a set of values $y_1, y_2, ..., y_n$. The reduce function ρ is defined to work sequentially on L_k but should be independent of other lists L_k, where $k' \neq k$ [7]. A widespread definition of Big Data describes this concept in terms of three characteristics of information in this field: Volume, Velocity and Variety [18] referring to the huge quantity, the high speed of generation and the different formats of the information. This definition does not provide rules to identify a dataset that belongs to Big Data and the justification of using a dataset in the literature only considers its size. The size of a dataset is relative to resources available to manage it and, in a country like Mexico (where this research was done), the availability of specialized hardware is an important limitation. Taking these factors into account, we propose an alternative definition of Big Data relative to the FMS problem. We propose that a huge dataset for the model selection problem must to accomplish two rules: (1) The dataset size is big enough that at least one of the considered classification algorithms in their sequential version cannot process it. (2) The dataset size is defined by their file size considering the number of instances (I) and features (F) as long as $I \gg F$.

3   PSMS for FMS and Proxy Model Construction

One of the most popular and successful algorithms to perform the FMS analysis is the PSMS algorithm proposed by [6]. This algorithm is based on the PSO algorithm which is a population-based search inspired by the behavior of biological communities that exhibit both individual and social behavior [6]. PSMS is faster and easy to implement because relies in just one operator unlike the evolutionary algorithms. Due to the aforementioned reasons the PSMS algorithm was chosen and adapted to MapReduce in order to deal with datasets of any size.

3.1   Codification and Functioning

The solutions encoded in PSMS needs to be codified in a vector called particle. Each particle $x_i^t = [x_{i,1}^t, x_{i,2}^t, ..., x_{i,16}^t,]$ is encoded as follows: In position 1 the fitness of the potential models is stored. Position 2 allows to determine which operation will be done first: data-preparation or feature selection. Position 3 indicates if the data-preparation step will be done. Positions 4 to 6 are parameters for the data-preparation step (method identifier, parameter 1 and parameter 2). Position 7 determines if the feature selection step will be done. Positions 8 and 9 are for the feature selection step (Method identifier and number of features to be selected respectively). Positions 10 to 16 are for the machine learning algorithm construction. The range of values that every element in the vector can take is as follows: $[0{-}100]$; $[0, 1]$, $[0, 1]$, $[1, 30]$, $[1, NF]$, $[1, 50]$, $[0, 1]$, $[1, 5]$, $[1, NF]$, $[1, 6]$, $[1, 2]$, $[1, 4]$, $[1, 100]$, $[1, 60]$, $[1, 400]$, $[-20, 20]$ with $NF = $ Number of Features. At each time t, each particle, i, has a position in the search space. A set of particles $S = \{x_1^t, x_2^t, ..., x_m^t\}$ is called a swarm. Every particle has a related velocity value that is used to explore the search space and the velocity of such particle at time t is as follows $V_i^t = [v_{i,1}^t, v_{i,2}^t, ..., v_{i,16}^t]$ where $v_{i,k}^t$ is the velocity for dimension k of the particle i at time t. The search trajectories are adjusted employing the following equations:

$$v_{i,j}^{t+1} = W \times v_{i,j}^t + c1 \times r1 \times (p_{i,j} - x_{i,j}^t) + c2 \times r2 \times (p_{g,j} - x_{i,j}^t) \quad (1)$$

$$x_{i,j}^{t+1} = x_{i,j}^t + v_{i,j}^{t+1} \quad (2)$$

from the previous equations $p_{i,j}$ is the value in dimension j of the best solution found so far, also called personal best. $p_{g,j}$ is the value in dimension j of the best particle found so far in the swarm. Regarding to $c1, c2 \in \mathbb{R}$, are constants weighting the influence of local and global best solutions, and $r1, r2 \sim U[0, 1]$ are values that introduce randomness into the search process. The inertia weight W controls the impact of the past velocity of a particle over the current one, influencing the local and global exploration. As in the original paper, the inertia weight is adaptive and specified by the triplet $W = (w_{start}, w_f, w_{end})$; where w_{start} and w_{end} are the initial values of W, w_f indicates the fraction of iterations in which W is decreased. W is decreased by $W = W - w_{dec}$ from the first iteration where $W = W_{start}$ to the last iteration where $W = w_{end}$ and $w_{dec} = \frac{w_{start} - w_{end}}{Number\_of\_iterations}$ [6].

3.2   Models Evaluation

This version of PSMS was developed under Apache Spark 1.6.0, that is based on MapReduce. Apache Spark was selected because of its enhanced capacity to deal with iterative algorithms and the possibility to perform data processing in main memory (if memory capacity allows it). An analysis of the advantages of Spark over traditional MapReduce is out of the scope of this work, but we refer to [19]. The cornerstone of Apache Spark is the RDD or Resilent Distributed Dataset which is a collection of partitioned data elements that can be processed in parallel [8]. As described above, the models evaluation stage is comprised of data preparation, feature selection, and training of a classification algorithm, in the following algorithms this process is described.

Algorithm 1. Data preparation

1: **procedure** DATAPREP(DataSet,particle)
2:     Return(DataSet.map(row → row.toArray.map(col → Transform(col,particle))))
3:     ▷ The Transform function is applied to every column of each row in the RDD according to the parameters encoded in the particle
4: **end procedure**

Algorithm 2. Feature Selection

1: **procedure** FEATSELECTION(DataSet,particle)
2:     numFeat = particle(9)
3:     rankRDD = DataSet.map(row → RankingCalculation(row))
4:     ▷ The RankingCalculation function obtains the ranking of the features of the dataset
5:     reducedRDD = rankRDD.map(row → getF(row,numFeat))
6:     ▷ The function getF is applied to every row in rankRDD and returns a reduced dataset
7:     Return(reducedRDD)
8: **end procedure**

Algorithm 3. Classification

1: **procedure** CLASSIFICATION(DataSet,particle)
2:     NumFolds = 2
3:     kFolds = createFold(DataSet,NumFolds)
4:     ▷ The createFold function creates an RDD for k-Fold Cross validation
5:     error=kFolds.map {
6:       **case**(Training,Validation)
7:     ▷ The dataset is separated in Training and Validation partitions
8:         model = createModel(Training, particle)
9:     ▷ The createModel function create a model using the parameters codified in the particle
10:        PredictedTargets = Validation.map(Instance → model.predict(Instance.features))
11:    ▷ Performs the predictions in the validation set
12:        accuracy= getAcc(PredictedTargets,Validation.targets)
13:    ▷ Obtains the accuracy in each fold
14:        error = 100-accuracy
15:        Return(error)
16:      }
17:      meanError=error.sum/error.length
18:      Return(meanError)
19: **end procedure**

In the proposed version of PSMS, the mean error over the 2-fold cross validation is used in order to evaluate the performance of every potential model. During the test stage in the development of the algorithm, different number of folds were evaluated (2, ...,10) without significant differences, but adding to the computing time factor, the 2-fold cross validation was the best choice. As to choose a single final model is not a trivial decision, another major change in our PSMS version was that the final model is a weighted ensemble of the best models found during the search process.

3.3   Proxy Model Construction Through the FMS Paradigm

As mentioned earlier, the search space of the FMS problem is huge even if restrictions are imposed to hyper-parameter values. The bio-inspired search methods have proved a high capacity to deal with this kind of problems and, particularly, PSMS has shown that is capable to solve the FMS problem. It is not easey to asses the complexity of the FMS problem, which varies from linear to quadratic or higher orders. The fitness function in a normal execution of PSMS should be evaluated $\rho = m \times (I + 1)$ times, where m is the swarm size and I the number of iterations. Supposing that the complexity of a model λ is bounded by λ_0, the complexity of PSMS will be bounded by $\rho \times \lambda_0$. The computing time of FMS is related to the dimensionality and size of the dataset, and when high volume datasets are analyzed, MapReduce allows to divide the load among MapReduce nodes. Although the complexity is the same ($\lambda_0 \times \rho$), the computing time is reduced to manageable levels. An excellent alternative to reduce the computing time, is through the use of proxy models in order to reduce the value of ρ and to guide the search in a similar fashion of a compass. An effective way to built the best compass possible is through PSMS, to automate the selection of the best combination of regression algorithm, feature selection and data preparation of the meta-dataset. In this work, the meta-dataset, was built employing the particles evaluated by the PSMS algorithm for FMS in huge datasets. All the evaluated particles, are vectors that describe combinations of factors that represent a full model along with the particle performance, therefore, the meta-dataset constitutes a regression problem in order to predict the performance of new not analyzed particles and just to evaluate those that are promising. For the PSMS algorithm assisted by proxy models constructed with the FMS paradigm (onwards FMSProxy-PSMS), we used the regression and data-preparation algorithms available in the machine learning tool WEKA [10]. As feature selection algorithm the Principal Components Analysis (PCA) was employed because the available algorithms for FS need discrete classes in order to perform an analysis and of course the problem at hand is a regression problem with a continuous target. Finally the process of the FMSProxy-PSMS algorithm can be described as follows: (1) Evaluate all the particles in the first "N" iterations of the algorithm with the costly fitness function, (2) use this particles to build a first proxy model that will be used during the next "N" iterations, (3) evaluate the promising particles with the costly fitness function and discard the rest, (4) after complete "N" iterations build a new proxy model, (5) iterate until the termination criteria is met. All the particles evaluated with the costly fitness function are stored in a separated file used as meta-dataset.

4 Experiments and Results

With the purpose to evaluate the proposed algorithm performance, we experimented with the datasets shown in Table 1. The datasets "Synthetic 1" and "Synthetic 2" were created using the tool for synthetic datasets generation in the context of ordinal regression: "Synthetic Datasets Nspheres" provided in [16]. Despite of have been developed for ordinal regression, the tool can be properly adjusted for traditional binary or multi-class problems and provides the mechanism to control the overlaps and classes balance. Another major feature of the aforementioned datasets is its intrinsic dimension. The intrinsic dimension (ID) is the minimum number of parameters needed to represent the data without information loss [13]. The id of the employed datasets was estimated with the "Minimum neighbor distance estimator" (MNDE) [13] and the "Dimensionality from angle and norm concentration" (DANCO) estimator [3]. The importance of the estimation of the "id" of each dataset is to ensure that each dataset represent a different computational problem and, therefore, that proposed algorithm have the capability to deal with a wide range of problems and in the context of this work also with datasets of different sizes. In the Table 2 the calculated intrinsic dimension using the aforementioned estimators is shown.

Table 1. Datasets used in the experiments.

Datasets	Data points	Attributes	Samples by class	Type of variables	File size
RLCP	5749111	11	(5728197;20915)	Real	261.6 MB
KDD	4856150	41	(972780;3883369)	Categorical	653 MB
Synthetic 1	200000000	3	(100000000;100000000)	Real	5.5 GB
Higgs	11000000	28	(5170877;5829123)	Real	7.5 GB
Synthetic 2	49000002	30	(24500001;24500001)	Real	12.7 GB
Epsilon	500, 000	2000	(249778;250222)	Real	15.6 GB

Table 2. Intrinsic dimension of the datasets.

Datasets	MNDE	DANCO
RLCP	2	2
KDD	1	1
Synthetic 1	3	3
Higgs	12	15
Synthetic 2	22	28
Epsilon	160	78

The performance of the proposed approach FMSProxy-PSMS was contrasted against the complete search (PSMS) and a surrogate-assisted version of PSMS based on a Multi-layer perceptron (onwards MLP-PSMS) as in [15] whose hyperparameters were determined with a grid search at the beginning of the process and during the rest of the search was just re-trained with the new samples of the meta-dataset. The termination criteria of PSMS was to complete 50 iterations, this in conjunction with a swarm size of 30 particles means that the search explored 1500 possible models before to build the final model. As mentioned earlier the proxy model approach can be though as a reduction of the expensive fitness functions or as a compass that can guide the search and, therefore, the best model must be find earlier in the search. With this in mind, as a termination criteria of the surrogate-assisted searches, the evaluation of a certain number of models was employed. In this case, the termination criteria was set to complete 500 evaluations with the expensive fitness function with no differences in the increment of this value (500, 1000, 1500 evaluations). In order to obtain an statistical power of 90% in an ANOVA test, 20 replications for every dataset were performed. Each replication was performed with a particular random sample of the data points with different random samples among replications. The dataset was divided in two disjoint datasets with 60% of the data samples for the training set and 40% for the test set. In Table 3, the mean error of the contrasted methods is shown.

Table 3. Mean classification error obtained in the test dataset by FMSProxy-PSMS, MLP-PSMS and PSMS, over 20 replications. The best results are in **bold**

Dataset	FMSProxy-PSMS	MLP-PSMS	PSMS
RLCP	0.059 ± 0.123	0.426 ± 1.593	$\mathbf{0.052 \pm 0.001}$
KDD	$\mathbf{0.079 \pm 0.003}$	2.556 ± 7.638	0.156 ± 0.134
Synthetic 1	15.865 ± 0.005	15.863 ± 0.004	$\mathbf{15.862 \pm 0.004}$
Higgs	30.193 ± 0.685	29.491 ± 2.220	$\mathbf{28.299 \pm 0.057}$
Synthetic 2	6.682 ± 0.003	6.682 ± 0.003	$\mathbf{6.681 \pm 0.005}$
Epsilon	$\mathbf{28.816 \pm 2.762}$	32.223 ± 5.397	54.008 ± 0.926

From previous table, the best method was PSMS, obtaining the lowest errors in the datasets RLCP, Synthetic 1, Higgs and Synthetic 2. The second best in the comparative was FMSProxy-PSMS in the datasets KDD and Epsilon. As was mentioned before, the complete search PSMS, explores 1500 models before to find the best model, and the surrogate-based methods only explores 500 models. In order to understand the search process of the methods under evaluation, Fig. 1 is provided.

The previous figure shows the approximate number of evaluation models in the "X" axis and the estimated error of the best particle found in the "Y" axis in Higgs dataset. It can be seen that PSMS got the lowest error with 39.2%, the second best was FMSProxy-PSMS with 42% and finally MLP-PSMS with 42.2%. Regarding to the number of evaluated models, MLP-PSMS was the fastest with 68 models evaluated until build the final model, FMSProxy-PSMS with 120 models and finally PSMS with 1220 evaluated models. Though, PSMS was the best in the comparative with a mean error of 28.299%, the surrogate-based searches were pretty near with 29.491% for MLP-PSMS and 30.193% for FMSProxy-PSMS with an smaller amount of models evaluations. Regarding to computing time factor, in Fig. 2, average execution times are shown as a bar chart in each dataset. The experiments were performed in a workstation of 12 threads with a Intel(R) Xeon(R) CPU E5-2695 at 2.40 GHz and 30 GB in RAM.

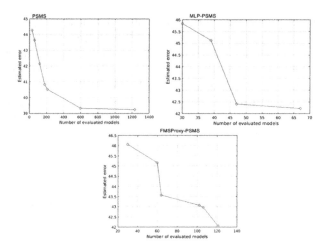

Fig. 1. Evolution of the search of the proposed methods in dataset Higgs. In X axis are the number of evaluated models (as a way to point out the progress of the search) and in Y axis the estimated error.

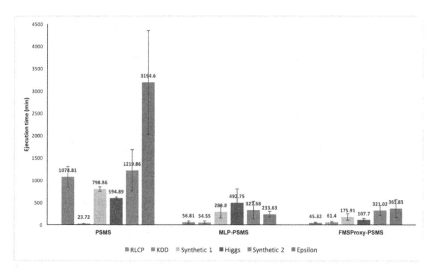

Fig. 2. Bar chart with the average execution time (in minutes) performed in 20 replications in each dataset. Each color represent a different dataset and bars are grouped by algorithm. The standard deviation is depicted as a solid black line on the bars. (Color figure online)

On Fig. 2 it can be appreciated the average execution times of contrasted algorithms with the aforementioned hardware configuration. With no surprise, highest execution times are for PSMS (the complete search), while surrogate-assisted approaches are considerably faster and specially for Epsilon dataset. Due its random nature (the initial swarm is randomly initialized), PSMS has also the largest standard deviation, but, taking into account that all the surrogated versions of PSMS have a random initial swarm too, we can conclude that a proxy model is capable to guide the search in an effective way. Concerning to standard deviation, a simpler visual analysis shows that FMSProxy-PSMS has a smaller standard deviation that traditional approach (MLP-PSMS). Finally, although computing times of FMSProxy-PSMS are no the smallest in all datasets, results in Table 3 shows that this approach is almost as good as the complete search but with moderate to low execution times. In order to find significant differences in the performance of the evaluated methods, an ANOVA test was performed. In the following table the results of the test are shown (Table 4).

Table 4. F-statistic obtained from the ANOVA test and q-values from the Tukey HSD test for performing all possible pairwise comparisons among the proposed strategies for the final model construction. The critical values at the 95% confidence level for the ANOVA test are 3.16 (F(2,57)) for all datasets. The critical values at the 95% confidence level for the Tukey HSD test are 3.44 (57 degrees of freedom). Cases that exceed the critical value are considered as a difference that is statistically significant at the fixed level and are marked with an asterisk (*)

Dataset	ANOVA F	FMSProxy-PSMS vs MLP-PSMS	FMSProxy-PSMS vs PSMS
RLCP	1.076	1.7785	0.035
KDD	2.039	2.511	0.079
Synthetic 1	1.629	2.070	2.328
Higgs	**10.189***	2.338	**6.313**
Synthetic 2	0.486	0.346	1.344
Epsilon	**298.013***	**4.303**	**31.819**

The statistic tests shows that there are almost no differences in the performance of the surrogate-based algorithms except in the Epsilon dataset where the lowest error was obtained by the FMSProxy-PSMS algorithm. Regarding to PSMS it can bee seen that exist significant differences in the Higgs dataset and the Epsilon dataset, with PSMS winning in the Higgs dataset and FMSProxy-PSMS winning in the Epsilon dataset. From Table 2 it can be seen that from the perspective of the intrinsic dimension analysis, the Epsilon dataset is the hardest one in the comparative with an ID of 160/78 against the one of Higgs with 12/15. Considering the previous analysis, we can see that the FMSProxy-PSMS algorithm is almost as good as PSMS in a wide range of dataset sizes and with different intrinsic dimensions. The exploration performed by the FMSProxy-PSMS algorithm is more conservative than the one performed by MLP-PSMS and therefore, the guide provided by the proxy models created under the FMS paradigm promote a better exploration of the search space that leads to better final models than using a single regression algorithm as proxy model. The mean error obtained by FMSProxy-PSMS in the dataset Epsilon shows its capacity to perform a better exploration than the complete search of PSMS with just a third part of the explored models (500 against 1500). Though the creation of a new proxy model based on the FMS paradigm every so often could be think as a major drawback of our approach, the time employed in the process is nothing in comparison to the transformation and training on a big dataset. The time dedicated to the training and transformation of the dataset of just one model commonly take several hours even under MapReduce and considering that a good search explores several potential models, the time of the entire process takes a lot more. A final consideration of our proposed approach shows that although the traditional approach of proxy model makes a quicker search, the FMS approach reduces the time employed and also perform a better exploration.

5   Conclusions

The full model selection paradigm provides a way to improve the predictive accuracy of learning algorithms and to determine the best one for a determined dataset, which is of big help because there is no a thumb rule to chose an algorithm for a dataset. This paradigm is poorly explored in the context of huge datasets due to the higher computing times intrinsically related to the size of the dataset. As a way to perform a reduction of the time employed in the search, the use of proxy models in conjunction with the FMS paradigm was proposed in this work. The use of the FMS paradigm in the construction of proxy models showed an improvement over traditional proxy models based on a single regression algorithm, performing a better exploration of the search space and reducing the number of the explored models to a third part in comparison to the complete search. Though the construction of FMS-based proxy models every so often could be think as a major drawback of our approach, it is not hard to see that the time employed is insignificant in comparison to the time employed in the training in a big dataset of several non promising models that will not be used in the final model construction. Our approach provides a tool comparable to a compass to provide a better guide of the search and reducing the time of the process.

References

1. Alenezi, F., Mohaghegh, S.: A data-driven smart proxy model for a comprehensive reservoir simulation. In: Saudi International Conference on Information Technology (Big Data Analysis) (KACSTIT), pp. 1–6. IEEE (2016)
2. Bansal, B., Sahoo, A.: Full model selection using bat algorithm. In: 2015 International Conference on Cognitive Computing and Information Processing (CCIP), pp. 1–4. IEEE (2015)
3. Ceruti, C., Bassis, S., Rozza, A., Lombardi, G., Casiraghi, E., Campadelli, P.: DANCo: dimensionality from angle and norm concentration. arXiv preprint arXiv:1206.3881 (2012)
4. Couckuyt, I., De Turck, F., Dhaene, T., Gorissen, D.: Automatic surrogate modeltype selection during the optimization of expensive black-box problems. In: Proceedings of the 2011 Winter Simulation Conference (WSC), pp. 4269–4279. IEEE (2011)

5. Cruz-Vega, I., García, C.A.R., Gil, P.G., Cortés, J.M.R., Magdaleno, J.d.J.R.: Genetic algorithms based on a granular surrogate model and fuzzy aptitude functions. In: 2016 IEEE Congress on Evolutionary Computation (CEC), pp. 2122–2128. IEEE (2016)
6. Escalante, H.J., Montes, M., Sucar, L.E.: Particle swarm model selection. J. Mach. Learn. Res. **10**(Feb), 405–440 (2009)
7. Goodrich, M.T., Sitchinava, N., Zhang, Q.: Sorting, searching, and simulation in the mapreduce framework. In: Asano, T., Nakano, S., Okamoto, Y., Watanabe, O. (eds.) ISAAC 2011. LNCS, vol. 7074, pp. 374–383. Springer, Heidelberg (2011). https://doi.org/10.1007/978-3-642-25591-5_39
8. Guller, M.: Big Data Analytics with Spark: A Practitioner's Guide to Using Spark for Large Scale Data Analysis. Apress, New York City (2015). https://www.apress.com/9781484209653
9. Guo, X., Yang, J., Wu, C., Wang, C., Liang, Y.: A novel LS-SVMs hyper-parameter selection based on particle swarm optimization. Neurocomputing **71**(16), 3211–3215 (2008)
10. Hall, M., Frank, E., Holmes, G., Pfahringer, B., Reutemann, P., Witten, I.H.: The WEKA data mining software: an update. SIGKDD Explor. **11**(1), 10–18 (2009)
11. Kaneko, H., Funatsu, K.: Fast optimization of hyperparameters for support vector regression models with highly predictive ability. Chemom. Intell. Lab. Syst. **142**, 64–69 (2015). https://doi.org/10.1016/j.chemolab.2015.01.001, http://linkinghub.elsevier.com/retrieve/pii/S0169743915000039
12. Lessmann, S., Stahlbock, R., Crone, S.F.: Genetic algorithms for support vector machine model selection. In: 2006 International Joint Conference on Neural Networks. IJCNN 2006, pp. 3063–3069. IEEE (2006)
13. Lombardi, G., Rozza, A., Ceruti, C., Casiraghi, E., Campadelli, P.: Minimum neighbor distance estimators of intrinsic dimension. In: Gunopulos, D., Hofmann, T., Malerba, D., Vazirgiannis, M. (eds.) ECML PKDD 2011. LNCS (LNAI), vol. 6912, pp. 374–389. Springer, Heidelberg (2011). https://doi.org/10.1007/978-3-642-23783-6_24
14. Pilat, M., Neruda, R.: Meta-learning and model selection in multi-objective evolutionary algorithms. In: 2012 11th International Conference on Machine Learning and Applications (ICMLA), vol. 1, pp. 433–438. IEEE (2012)
15. Rosales-Pérez, A., Gonzalez, J.A., Coello Coello, C.A., Escalante, H.J., Reyes-Garcia, C.A.: Surrogate-assisted multi-objective model selection for support vector machines. J. Neurocomputing **150**, 163–172 (2015)
16. Sánchez-Monedero, J., Gutiérrez, P.A., Pérez-Ortiz, M., Hervás-Martínez, C.: An n-spheres based synthetic data generator for supervised classification. In: Rojas, I., Joya, G., Gabestany, J. (eds.) IWANN 2013. LNCS, vol. 7902, pp. 613–621. Springer, Heidelberg (2013). https://doi.org/10.1007/978-3-642-38679-4_62
17. Thornton, C., Hutter, F., Hoos, H.H., Leyton-Brown, K.: Auto-WEKA: combinedselection and hyperparameter optimization of classification algorithms. In: Proceedings of the 19th ACM SIGKDD International Conference on Knowledgediscovery and Data Mining, pp. 847–855. ACM (2013)
18. Tlili, M., Hamdani, T.M.: Big data clustering validity. In: 2014 6th International Conference of Soft Computing and Pattern Recognition (SoCPaR), pp. 348–352. IEEE (2014)
19. Zaharia, M., et al.: Apache spark: a unified engine for big data processing. Commun. ACM **59**(11), 56–65 (2016)

Single Imputation Methods Applied to a Global Geothermal Database

Román-Flores Mariana Alelhí[1][(✉)], Santamaría-Bonfil Guillermo[2],
Díaz-González Lorena[3], and Arroyo-Figueroa Gustavo[4]

[1] Posgrado en Optimización y Cómputo Aplicado,
Universidad Autónoma del Estado de Morelos, Avenida Universidad 1001,
Chamilpa, 62209 Cuernavaca, Morelos, Mexico
alheli155@gmail.com
[2] Instituto Nacional de Electricidad y Energías Limpias,
Gerencia de Tecnologías de la Información,
Reforma 113 Col. Palmira, 62490 Cuernavaca, Morelos, Mexico
guillermo.santamaria@ineel.mx
[3] Departamento de Computación, Centro de Investigación en Ciencias,
Instituto de Investigación en Ciencias Básicas Aplicadas,
Universidad Autónoma del Estado de Morelos,
Av. Universidad 1001, Chamilpa, 62209 Cuernavaca, Morelos, Mexico
ldg@uaem.mx
[4] Instituto Nacional de Electricidad y Energías Limpias, Av. Reforma # 113,
Col. Palmira, 62490 Cuernavaca, Morelos, Mexico
garroyo@iie.org.mx

Abstract. In the exploitation stage of a geothermal reservoir, the estimation of the bottomhole temperature (BHT) is essential to know the available energy potential, as well as the viability of its exploitation. This BHT estimate can be measured directly, which is very expensive, therefore, statistical models used as virtual geothermometers are preferred. Geothermometers have been widely used to infer the temperature of deep geothermal reservoirs from the analysis of fluid samples collected at the soil surface from springs and exploration wells. Our procedure is based on an extensive geochemical data base (n = 708) with measurements of BHT and geothermal fluid of eight main element compositions. Unfortunately, the geochemical database has missing data in terms of some compositions of measured principal elements. Therefore, to take advantage of all this information in the BHT estimate, a process of imputation or completion of the values is necessary.

In the present work, we compare the imputations using medium and medium statistics, as well as the stochastic regression and the support vector machine to complete our data set of geochemical components. The results showed that the regression and SVM are superior to the mean and median, especially because these methods obtained the smallest RMSE and MAE errors.

Keywords: Geothermal data · Missing data · Imputation
Stochastic regression

© Springer Nature Switzerland AG 2018
I. Batyrshin et al. (Eds.): MICAI 2018, LNAI 11288, pp. 183–194, 2018.
https://doi.org/10.1007/978-3-030-04491-6_14

1 Introduction

In the exploration stage of a geothermal reservoir, the estimation of bottomhole temperatures is a fundamental activity to estimate the available energy potential and the feasibility of exploiting its resources for the generation of electric power [1]. For this, there are low cost geothermometric tools that allow obtaining an approximate bottomhole temperature based on the chemical composition of the sampled fluids of natural manifestations of geothermal reservoirs (thermal springs, geysers or volcanoes).

Today there are several geothermometric tools reported in the literature, several of which tend to overestimate temperatures, due in large part to the fact that the amount of data available for development is small or its origin is unreliable.

For the development of a geothermometric tool that improves the estimations of bottomhole temperatures, a geochemical data base of n = 708 is available, which contains measured temperatures and concentrations of eight main components of wells producing different parts of the world. Unfortunately, the geochemical database shows absence of data in some variables since they were not reported by the original authors.

The missing data in the geochemical database represents a limitation to attack the problem of estimation of bottomhole temperatures, since incomplete data sets can cause bias due to differences between observed and unobserved data. The most common approach to managing missing values is the analysis of complete cases [2]. However, Allison [3] observed that this approach reduces the sample size and study power. Alternatively, this problem can be solved by means of data imputation, which consists of the replacement of missing data by calculated values [4].

The imputation can be generally classified into statistical techniques and machine learning [5]. This work compares the performance of four statistical techniques for the imputation of missing data [6]: mean, median, stochastic linear regression, and Support Vector Machines (SVM). In the imputation of the geochemical database, using the techniques mentioned above, the data set (n = 150) that did not contain missing data were extracted, which were later split into two groups, a for training and another for testing. The results showed that the stochastic regression and SVM methods estimated more precise missing values than the substitution methods by the mean and median.

The rest of the document is organized as follows: Sect. 2 presents some studies related to the imputation of the mean, median, stochastic regression and SVM. Section 3 describes the mechanisms of missing data, as well as the proposed techniques for imputation of the geochemical database. Section 4 includes the information of the data, the experimental configuration and the results obtained from the evaluation of the performance of each method. Finally, the conclusions of the document and future work are exposed in Sect. 5.

2 Literature Review

Currently, no reported works have been found in the literature in which the imputation to geothermal fluids data is performed. However, there are reports studies in other areas, such as environmental pollution, air quality and medicine. Norazian, et al. [7] and Noor [8] applied the interpolation and imputation of the mean in a set of PM10 concentration data, simulated different percentages of missing data, and concluded that the mean is the best method only when the number of missing values is small. Razak, et al. [9] evaluated the methods of imputation of the mean, hot deck and maximization of expectations (EM) in PM10 concentrations, and concluded that the error of these methods is considerable when the percentage of missing data is very high (e.g., 50%). Junninen, et al. [10] compared the performance of various imputation methods in a set of air quality data, and concluded that multivariate statistical methods (e.g., regression-based imputation) are superior to univariate methods (e.g., linear, spline and nearest neighbor interpolation). Yahaya, et al. [11] compared univariate imputation techniques (e.g., mean, median, nearest neighbor, linear interpolation, spline interpolation and regression) in Weibull distributions, and obtained that no single imputation technique is the best for each sample size and for each percentage of missing values.

On the other hand, the imputation of values in the medical area has also been applied. Jerez, et al. [12] applied several methods of statistical imputation (e.g., mean, hot-deck and multiple imputation), and machine learning techniques (e.g., multi-layer perceptron, self-realization maps and k-nearest neighbor) in an extensive real breast cancer data set, where methods based on machine learning techniques were the most suited for the imputation of missing values. Engels, et al. [13] compared different methods of imputing missing data on depression, weight, cognitive functioning, and self-rated health in a longitudinal cohort of older adults, where the imputations that used no information specific to the person, such as using the sample mean, had the worst performance. In contrast, Shrive, et al. [14] compared different imputation techniques for dealing with missing data in the Zung Self-reported Depression scale, and showed that the individual mean and single regression method produced similar results, when the percent of missing information increased to 30%. Also, Newman [15], Olinsky, et al. [16], Aydilek, et al. [17] reported comparative studies of various imputation techniques such as stochastic regression, fuzzy c-means and SVR. Finally, Wang, et al. [18] demonstrated that the SVR impute method has a powerful estimation ability for DNA microarray gene expression data.

3   Methods to Treat Missing Data

The reasonable way to handle missing data depends on how the data points are missing. In 1976 Rubin [6] classified the data loss into three categories. In your theory, each data point has some probability of missing. The process that governs these probabilities is called the response mechanism or missing data mechanism. To explain these three categories Z is denoted as a variable with missing data, S as a set of complete variables, Rz as a binary variable that has a value of 1 if the data in Z is missing and 0 if observed. The categories of the Rubin classification can be expressed by the following statements:

Missing Completely At Random (MCAR)

$$\Pr(Rz = 1 | S, Z) = \Pr(Rz = 1) \tag{1}$$

That is, the probability of missing a value in Z does not depend either on S or Z and therefore its estimates cannot depend on any variable.

Missing At Random (MAR).

$$\Pr(Rz = 1|S, Z) = \Pr(Rz = 1|S) \tag{2}$$

Where the loss of a value in Z depends on S but not on Z, therefore its estimation can depend on S.

Missing Not At Random (MNAR)

$$\Pr(Rz = 1|S, Z) = \Pr(Rz = 1|Z) \tag{3}$$

The absence of data in Z depends on Z itself, to generate estimates under this assumption, special methods are required.

Rubin's distinction is important, since his theory establishes the conditions under which a method to deal with missing data can provide valid statistical inferences. On several occasions, the assumption that data loss is MAR is acceptable and that a treatment can be resorted to using imputation methods. Unfortunately, the assumptions that are necessary to justify a method of imputation are generally quite strong and often unverifiable [3].

3.1   Proposed Methods

The present work focuses on the imputation of the geochemical database based on the assumption that the loss is MAR. The available information may be used to estimate the missing values. The use of the complete analysis method Schafer [2] was discarded, which consists in ignoring the records that contain missing data, because applying this method is practical only when the number of incomplete records is less than 5% of the total data and the data loss is of the MCAR Buuren type [19]. The geochemical database is incomplete in more than 50% and its MAR type loss is assumed.

The single imputation methods are broadly classified into statistical and machine learning techniques [5]. The most commonly imputed forms of imputation are substitution by means, median and stochastic regression [15]. Within the current machine learning techniques, we can find of SVM. The imputation with statistical techniques provides estimates of lost values by replacing them with the observed data. When the variables are continuous, the simplest statistical parameters based on the mean and median are used.

3.2   Mean and Median Imputation

Substitution by the mean is an imputation technique where the missing data for any variable is completed with the average of the observed value of that variable [20]. The average is obtained by Eq. 4.

$$\bar{x} = \frac{\sum_{i=1}^{n} x_i}{n} \tag{4}$$

On the other hand, impute the median is done from an ordered vector that contains the observed data of an incomplete variable, the missing values of said variable are imputed through Eq. 5.

$$\tilde{x} = X_{\left[\frac{n+1}{2}\right]}, \text{if } n \text{ is even}$$
$$\tilde{x} = \frac{X_{\left[\frac{n+1}{2}\right]} + X_{\left[\frac{n}{2}\right]}}{2}, \text{if } n \text{ is odd} \tag{5}$$

The use of these techniques entails the disadvantage that the variance of the imputed variable is systematically underestimated.

3.3 Imputation Stochastic Regression

A slightly more robust but popular method is stochastic regression, in which the variable with the missing data uses all other variables to produce a regression equation (depending on the complete cases).

$$\hat{Y} = \beta_0 + \beta_1 X1 + \ldots + \beta_p XP + \epsilon \tag{6}$$

where β_0 is the intersection, $\beta_1, \ldots \beta_p$ are the rate of change of $\hat{Y}$ for a unit change in $X1, \ldots, XP$ correspondent $X1, \ldots, XP$ are the predictors and ϵ random noise added to $\hat{Y}$. The random error term is a normal random variant with a mean of zero and a standard deviation equal to the standard error of the estimation of the regression equation [15]. The addition of the random error is a method used to avoid that the variance of the imputed variable is underestimated, and the correlations with the imputed variable are overestimated.

The most important thing when modeling the equation for an incomplete variable is the selection of predictors. The inclusion of as many predictors as possible tends to make the MAR assumption more plausible [3]. The missing values are replaced with predicted values of the regression equation.

One strategy for selecting predictive variables is to inspect their correlations and the response indicator, the latter measures the percentage of observed cases of one variable while there is absence in another. This means that a variable that has good correlation with the target variable must also have a high proportion of observed data to be a predictor.

3.4 SVM Imputation

In supervised learning techniques, imputation is considered a pattern classification task [5]. In them the missing values are the output obtained and the observed values are the inputs used for the training of the models. SVM is one of the machine learning techniques currently used for imputation.

Support Vector Regression (SVR) proposed by Drucker et al. [21], is the regression version of Support Vector Machines [22]. This method fits a hyperplane to a continuous dependent variable y in terms of one or more continuous independent variables, i.e. $y = f(x, \theta)$, where $y \in \mathbb{R}$ is the dependent variable, $x \in \mathbb{R}^N$ is a vector of N independent features, and $\theta \in \mathbb{R}$ is a vector of model parameters. SVR estimates the hyperplane by minimizing the Structural Risk which guarantees a good generalization of the model by controlling its complexity [21]. This control is achieved by wrapping the hyperplane with a margin which (a) constrains the number of points that the function can represent, and (b) obtains a f in terms of a subset of the train sample which is called Support Vectors (SV). Additionally, SVR can handle data noise and non-linearity: the former is achieved by including slack variables into the model's formulation, whereas the latter is achieved by the *Kernel Trick* [23]. Although SVR has been neatly defined elsewhere [21], for the sake of completeness we now provide its formulation:

$$\text{Max.} W(\alpha, \alpha^*)$$
$$= -\tfrac{1}{2} \sum_{i,j=1}^{m} \left(\alpha_i - \alpha_j^* \right) \left(\alpha_i - \alpha_j^* \right) \langle \phi(x_i), \phi(x_j) \rangle - \varepsilon \sum_{i=1}^{m} \left(\alpha_i^* + \alpha_i \right)$$
$$+ \sum_{i=1}^{m} (\alpha_i^* - \alpha_i) y_i \tag{7}$$
$$\text{subject to } \sum_{i=1}^{m} \left(\alpha_i - \alpha_i^* \right) = 0$$
$$0 \le \alpha_i, \alpha_i^* \le C, \forall i = 1, \ldots, m,$$

where C is the complexity penalization term, ε is the width of the margin, ϕ is the kernel function, and α, α^* corresponds to the weights of each element in the train set. Particularly, those $\alpha, \alpha^* \ge 0$ correspond to the SV.

4   Experimentation

The main objective of this work is to evaluate the proposed unique imputation techniques, applied to a set of geochemical data to increase the sample that has and thus allow the development of a geothermometric tool that better estimates the bottomhole temperatures of a geothermal reservoir. To do this, we have a geochemical database with 708 rows, each one represents a well producing geothermal energy, by 9 columns that correspond to the measured temperature (°C) of the well and the chemical concentrations of Li, Na, K, Mg Ca, Cl, SO_4 and HCO_3 given in mg/L. Table 1 shows the descriptive statistics and the total of missing values.

As shown in Table 1, the temperature and the Na and K components have no missing data. However, the components Li, Mg, Ca, Cl, SO_4 and HCO_3 are incomplete. Figure 2 shows the percentages that represent the missing data of each variable (Fig. 1).

Variables such as Li and HCO_3 are incomplete by more than 50%. To avoid discarding possible useful data for the development of a geothermometric tool that improves the bottomhole temperature estimates of a geothermal energy producing well,

Table 1. Statistical information for bottomhole temperatures and compositional database of geochemical fluids.

Variable	Min	Median	Mean	Max	SD	Na
Temperature (°C)	59	230	217	359	69.86	0
Li (mg/L)	0.02	6.40	14.03	215	24.31	452
Na (mg/L)	22	1,416	11,472.20	565,578.60	52,014.33	0
K (mg/L)	0.55	196.50	1,583.50	66,473.40	6,755.31	0
Mg (mg/L)	0.001	0.18	114.60	3,920	512.24	114
Ca (mg/L)	0.06	17	2,302.73	55,600	7,685.21	44
Cl (mg/L)	2	1,714	6,918	52,4690	28,522.53	157
SO_4 (mg/L)	0.60	51.80	140.3	2,500	246.94	191
HCO_3 (mg/L)	0	88.50	349.5	3,074	566.55	412

Minimum value (Min), maximum (Max), mean (Mean), median, standard deviation (SD) and the number of missing data (Na) of each variable contained in the geochemical database.

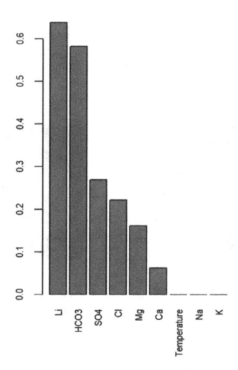

Fig. 1. Histogram of missing data. Shows the percentage of missing data in each variable of the geochemical database. The temperature, Na and K have no missing values, but the rest of the variables there is a percentage of missing data: Li 63%, HCO_3 58%, SO_4 26%, Cl 22%, Mg 16%, Ca 6%.

unique imputation methods were implemented, using statistical techniques such as mean, median and stochastic regression and machine learning. such as SVM.

4.1 Experimental Configuration

The geochemical database was divided into two sets. A complete set with the 150 rows that have all their observed chemical elements and another with the 558 records with missing data. From the complete data set, 120 rows (80%) were taken for training of the models of each variable and 30 rows (20%) for testing. From the training set, the mean and median of each incomplete variable were obtained and the values obtained replaced the missing values of the incomplete set.

For the stochastic regression and SVM imputation, we first analyzed the relationship between the observed data of the variables, by means of the pairwise correlation and the response indicator (described in Sect. 3) to determine the predictor variables that would be included in the model of incomplete variables.

Figure 2a shows the correlation of the observed data for each pair of variables in the geochemical database. Figure 2b shows the percentage of observed data of one variable while the other variable has lost data. For example, to determine the predictive variables of Li, according to Fig. 2a, Li has a correlation above 0.5 with Na, K, Ca and Cl, as well as a correlation very close to 0 with Temperature, Mg, SO_4 and HCO_3. Moreover, according to Fig. 2b, when Li has missing data, the variables Temperature, Na, K and Ca have a proportion of data observed in more than 75%. Therefore, the predictive variables of Li can be Na, K and Ca.

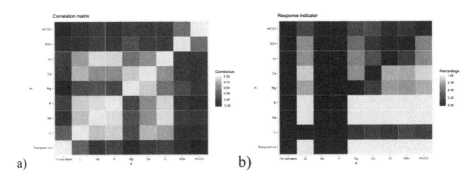

Fig. 2. (a) Correlation matrix shows the correlation of the observed data for each pair of variables; the yellow boxes indicate a correlation of 1, while the blue boxes indicate a close negative correlation to 0. (b) Response indicator indicates the percentage of observed data of one variable while the other variable has lost data, the yellow color indicates 100% and the blue 0%. The Temperature, Na and K have columns in blue totally since these variables do not contain missing data. Both are read from left to right, from bottom to top. (Color figure online)

The same analysis was carried out to select the predictive variables of the rest of the incomplete variables of the geochemical database. It is important to mention that in some variables very low correlations were found with the other variables and at the

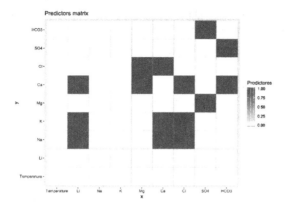

Fig. 3. Predictors matrix indicates in purple the predictors of each incomplete variable. It reads from left to right, from bottom to top.

same time when the variable had missing data, they were also missing in the rest of the variables. Despite this, the variables that obtained the highest values in comparison with the others were selected as predictors. The predictors for each incomplete variable are shown in Fig. 3.

4.2   Method Validation

To quantify the accuracy of the imputation models in the prediction of missing data, the two precision measures detailed below were used:

Root Mean Squared Error (RMSE) which indicates the variance in the estimates, has the same units as the measured and calculated data. The smaller values indicate a better concordance between the true values and the estimated ones.

$$RMSE = \sqrt{\left\langle (x_c - x_m)^2 \right\rangle} \tag{8}$$

Mean Absolute Error (MAE) like the RMSE, smaller values of MAE indicate a better concordance between the true and calculated values. MAE outputs a number that can be directly interpreted since the loss is in the same units of the output variable.

$$MAE = \left\langle |x_c - x_m| \right\rangle \tag{9}$$

4.3   Results and Discussion

The imputation techniques were implemented in the set of test data extracted from the complete set of the database, with this the RMSE and MAE errors could be measured between the predicted and the measured values. Tables 2 and 3 show the results of RMSE and MAE of the estimates of the mean, median, stochastic regression and SVM.

Table 2. Comparison of RMSE values obtained by the imputation methods of the mean, median, stochastic regression and SVM.

Variable	Mean	Median	Stochastic regression	SVM
Li	12	13.21	13.99	8.10
Mg	18.99	15.11	134.28	12.60
Ca	4,329.27	4,363.56	213.83	2,493.21
Cl	18,832.98	19,049	340.43	11,223.85
SO_4	126.43	133.97	127.38	83.14
HCO_3	361.06	375.39	457.45	204.26

Table 3. Comparison of MAE values obtained by the imputation methods of the mean, median, stochastic regression and SVM.

Variable	Mean	Median	Stochastic regression	SVM
Li	10	9.30	7.98	10.69
Mg	17.40	3.96	32.14	57.65
Ca	1,011.15	868.33	96.72	147.21
Cl	5,828.08	5,271.19	145.86	227.34
SO_4	78.50	58.97	52.71	85.92
HCO_3	191.08	135.29	183.73	230.97

The results presented in Tables 2 and 3 show that the imputations by the mean and median have the highest errors in most of the experiments compared to the stochastic regression and SVM methods. With the exception, for the variables Mg (16% of missing data) and HCO_3 (58% of missing data), in which the imputation of the median was the best according to the MAE parameter. On the other hand, the stochastic regression was the best (according to RMSE and MAE) in the imputation of the variables of Ca and Cl, which presented 16% and 22% of missing data, respectively. While, the SVM method obtained the best results (according to RMSE) in the estimation of Li (63% missing data), Mg (16% of missing data), SO_4 (26% missing data) and HCO_3 (58% of Missing data). Finally, with these results it was found that the best methods to estimate the lost values of the variables of Ca and Cl is the stochastic regression; and for the variables Li, Mg, SO_4 and HCO_3 is SVM.

5   Conclusions

In this paper, the unique imputation methods were compared by means, median, stochastic regression and SVM, applying them in a geochemical data set of geothermal fluids. This study is aimed at obtaining a complete and larger geochemical database that allows the development of a geothermometric tool that best estimates the bottomhole temperatures of a geothermal reservoir. From the complete data set, the training (80%) and testing (20%) sets were obtained. From training set, the mean and median values were calculated to replace the missing values, as well as the regression and SVM imputation models were developed.

To evaluate the performance of these methods, two indicators were calculated, the mean absolute error (MAE) and mean square error (RMSE) between the test set and the values estimated by the methods. The results showed that the stochastic regression and SVM are superior to the mean and median. From these performance indicators, it is concluded that the best methods to estimate the lost values of the variables of Ca and Cl are stochastic regression; and for the variables Li, Mg, SO_4 and HCO_3 is SVM. Therefore, both techniques were used for the completion of the geochemical database.

As future work, our task will be to analyze the statistical distribution of the imputation errors for the possible choice of more sophisticated validation parameters not sensitive to the presence of discordant values. On the other hand, the future plans for the project are to carry out a detailed study of the new complete geochemical database to develop a new geothermometric model.

References

1. Díaz-González, L., Santoyo, E., Reyes-Reyes, J.: Tres nuevos geotermómetros mejorados de Na/K usando herramientas computacionales y geoquimiométricas: aplicación a la predicción de temperaturas de sistemas geotérmicos. Revista Mexicana de Ciencias Geológicas **25**(3), 465–482 (2008)
2. Schafer, J.L.: Analysis of Incomplete Multivariate Data. Chapman and Hall/CRC, New York/Boca Raton (1997)
3. Allison, P.D.: Missing Data, vol. 136. Sage Publications, Thousand Oaks (2001)
4. Batista, G.E., Monard, M.C.: An analysis of four missing data treatment methods for supervised learning. Appl. Artif. Intell. **17**(5–6), 519–533 (2003)
5. Tsai, C.F., Li, M.L., Lin, W.C.: A class center based approach for missing value imputation. Knowl.-Based Syst. **151**, 124–135 (2018)
6. Rubin, D.B.: Inference and missing data. Biometrika **63**(3), 581–592 (1976)
7. Norazian, M.N., Shukri, Y.A., Azam, R.N.: Al Bakri, A.M.M.: Estimation of missing values in air pollution data using single imputation techniques. ScienceAsia **34**, 341–345 (2008)
8. Noor, N.M., Abdullah, M.M.A.B., Yahaya, A.S., Ramli, N.A.: Comparison of linear interpolation method and mean method to replace the missing values in environmental data set. Small **5**, 10 (2015)
9. Razak, N.A., Zubairi, Y.Z., Yunus, R.M.: Imputing missing values in modelling the PM10 concentrations. Sains Malays. **43**, 1599–1607 (2014)
10. Junninen, H., Niska, H., Tuppurainen, K., Ruuskanen, J., Kolehmainen, M.: Methods for imputation of missing values in air quality data sets. Atmos. Environ. **38**, 2895–2907 (2004)
11. Yahaya, A.S., Ramli, N.A., Ahmad, F., Mohd, N., Muhammad, N., Bahrim, N.H.: Determination of the best imputation technique for estimating missing values when fitting the weibull distribution. Int. J. Appl. Sci. Technol. (2011)
12. Jerez, J.M., et al.: Missing data imputation using statistical and machine learning methods in a real breast cancer problem. Artif. Intell. Med. **50**, 105–115 (2010)
13. Engels, J.M., Diehr, P.: Imputation of missing longitudinal data: a comparison of methods. J. Clin. Epidemiol. **56**(10), 968–976 (2003)

14. Shrive, F.M., Stuart, H., Quan, H., Ghali, W.A.: Dealing with missing data in a multi-question depression scale: a comparison of imputation methods. BMC Med. Res. Methodol. **6**(1), 57 (2006)
15. Newman, D.A.: Longitudinal modeling with randomly and systematically missing data: a simulation of ad hoc, maximum likelihood, and multiple imputation techniques. Organ. Res. Methods **6**, 328–362 (2003)
16. Olinsky, A., Chen, S., Harlow, L.: The comparative efficacy of imputation methods for missing data in structural equation modeling. Eur. J. Oper. Res. **151**(1), 53–79 (2003)
17. Aydilek, I.B., Arslan, A.: A hybrid method for imputation of missing values using optimized fuzzy c-means with support vector regression and a genetic algorithm. Inf. Sci. **233**, 25–35 (2013)
18. Wang, X., Li, A., Jiang, Z., Feng, H.: Missing value estimation for DNA microarray gene expression data by support vector regression imputation and orthogonal coding scheme. BMC Bioinformatics **7**(1), 32 (2006)
19. Buuren, S.V., Groothuis-Oudshoorn, K.: MICE: multivariate imputation by chained equations in R. J. Stat. Softw. 1–68 (2010)
20. Schafer, J.L., Graham, J.W.: Missing data: our view of the state of the art. Psychol. Methods **7**, 147 (2002)
21. Drucker, H., Burges, C.J., Kaufman, L., Smola, A.J., Vapnik, V.: Support vector regression machines. In: Advances in Neural Information Processing Systems, pp. 155–161 (1997)
22. Cortes, C., Vapnik, V.: Support-vector networks. Mach. Learn. **20**(3), 273–297 (1995)
23. Schölkopf, B., Smola, A.J.: Learning With Kernels: Support Vector Machines, Regularization, Optimization, and Beyond, p. 644. MIT Press, Cambridge (2002)
24. Lakshminarayan, K., Harp, S.A., Samad, T.: Imputation of missing data in industrial databases. Appl. Intell. **11**(3), 259–275 (1999)
25. Baraldi, A.N., Enders, C.K.: An introduction to modern missing data analyses. J. Sch. Psychol. **48**(1), 5–37 (2010)

Feature Selection for Automatic Classification of Gamma-Ray and Background Hadron Events with Different Noise Levels

Andrea Burgos-Madrigal[✉][iD], Ariel Esaú Ortiz-Esquivel[✉][iD],
Raquel Díaz-Hernández[✉], and Leopoldo Altamirano-Robles[✉]

Instituto Nacional de Astrofísica Óptica y Electrónica (INAOE), Puebla, Mexico
{burgosmad,arieleo,raqueld,robles}@inaoep.mx

Abstract. In this paper we present a feature set for Gamma-ray and Background Hadron events automatic classification. We selected the best parameters combination collected by Cherenkov telescopes in order to make a robust Gamma-ray recognition against different signal noise levels using multiple Machine Learning approaches for pattern recognition. We made a comparison of the robustness to noise for four classifiers reaching an accuracy up to 90.14% in high noise level cases.

Keywords: Gamma-ray separation · Relief-F
Machine learning classifiers · Robustness to noise

1 Introduction

When cosmic rays of high-energy *Gamma-ray* enter to the earth atmosphere, they initiate a chain reaction of particles known as atmospheric cascade. Because an atmospheric cascade produces Cherenkov light, the particles can be detected using *Imaging air-Cherenkov telescopes* (IACTs) by collecting Cherenkov photons with telescopes sensitive to ultra violet light. IACTs has a low duty cycle but excellent sensitivity over a narrow field-of-view. Another way to detect atmospheric cascades is by Air-shower arrays to build an array of particle detectors on in the ground and observe particles once they reach the surface of telescopes. Air-shower arrays has comparatively lower instantaneous sensitivity but have a nearly continuous duty cycle and wide field-of-view of the overhead sky. This advantage gives air-shower arrays the ability to make unbiased surveys of the *GeV-TeV Gamma-ray* sky, and will complement IACT observations at higher energies with greater sensitivity for extended sources [9].

The *Gamma-ray* and *Hadron* Background events separation for the low energies (less than 1 TeV Tera electron-volts) is a difficult task for ground based gamma observatories because of the presence of signal noise in atmosphere and other kind of ray events being mostly Hadrons. In this case only a small part of

© Springer Nature Switzerland AG 2018
I. Batyrshin et al. (Eds.): MICAI 2018, LNAI 11288, pp. 195–204, 2018.
https://doi.org/10.1007/978-3-030-04491-6_15

shower reaches the ground and the difference between the topology of hadronic and electromagnetic showers can not be studied effectively [8]. Thus, discrimination between *Gamma-ray* and Background events is key to achieving better sensitivity [10,11]. Moreover, according to [13], the success of current and future air shower arrays for detecting point sources of cosmic rays above 100 GeV (Giga electron-volts) depends crucially on the possibility of finding efficient methods for separating induced *Gamma-ray* events from the overwhelming background of *Hadron* events showers.

The Ground-based arrays of Atmospheric Cherenkov telescopes have been emerged as the most effective way to detect *Gamma-rays* in ranges of about 100 GeV and above. The strengths of these arrays are a very large effective collection area on the order of $10^5 \, \mathrm{m}^2$.

In this work, we gathered a dataset of simulated *Gamma-ray* and *Hadron* events from a Cherenkov telescope and proposed a feature set for *Gamma-ray* and *Hadron* events in order to automatically identify them.

The study of the best feature combination for this problem was done by analyzing the relevance through *Relief-F* feature selection algorithm. Preliminary tests returned that the relation between accuracy and the number of selected features is stable in a range of values. As a result, a set of 12 parameters in data was proposed. Finally, we determined that the best accuracy classification is reached with specified classification models, however, computational models as *Feed-Forward Neural Networks* have showed best *Gamma-ray* precision rates even when the *Noise to Signal* factor in dataset is changed.

This paper is structured as follows: Sect. 2 discusses the most relevant state-of-art approaches for *Gamma-ray* and *Hadron* events automatic separation, describing advantages and disadvantages. In Sect. 3 we describe the followed processes to identify a feature set for best recognition of *Gamma-ray* and Background events using a large data collection from Cherenkov telescopes. Section 4 describes the different Machine Learning used methods for classification and their robustness to noise for each case. We report the obtained accuracy and precision results in Sect. 4 as well the discussion of advantages for using this feature set with different *SNR* levels in Sect. 5.

2   Related Work

The challenge of separating the *Gamma-ray* induced air showers from the Background events has been studied before. [4] summarizes some background models used to identify the non *Gamma-ray* data. Nevertheless, the majority of approaches are based on basic arithmetical data subtraction and have no robustness to noise and external conditions of the used sensors.

In order to separate the *Gamma-ray* from the *Hadron* events, some approaches have been proposed introducing different parameters such as *Surface Brightness* in [3] which depends directly of simulating multiple ray events.

Recently, *Compactness* and *PINCness* features were proposed in [1,2] for identifying particles using a technique called *Standard Cut*. Both parameters depend of spatial distribution in deposited charges across the whole telescopes array during the event capture.

Based on the timing information of the events [6] proposed the *SFCFChi2* and *logmaxPE* parameters, which were selected for their higher separation potential in each *Gamma-ray* and *Hadron* events.

However, these approaches do not study the effect of noise caused by the detectors in multiple events and they are not robust enough to this issue. Moreover, other proposed techniques require simulated data to update parameters.

3 Automatic Classification of Gamma-Ray and Hadron Events

The acquired data from Cherenkov telescopes are a big data collection of *Gamma-ray* and other Background events such as *Hadron* particles. In [7] different techniques have been proposed for manipulate these large data containing a whole set of 189 parameters related to position and recorded electric charge for each detector in array-grounded sensors. A complete list of contained features in such data can be consulted in [7].

According to [6] captured events contain relevant parameters to determine the event type and other important information about that. Some of the most important present features are described as follows:

- **nDect**: Number of detectors of the telescope array activated during the event.
- **nCh**: Number of detector channels hit in the air shower (nCh) as an energy proxy.
- **nCont**: number of steel container densely distributed to observe the particles of atmospheric cascades.
- **PE**$_{SUM}$ and **PE**$_{MAX}$: Sum and maximum of *Photo-Electron* (PE) charge in the detectors during the event respectively.
- **dect.gridID** and **dect.contID**: It contains an array with all the identifiers of the detectors and containers of the array that were activated in the event respectively.
- **dect.xPos**, **dect.yPos** and **dect.zPos**: Contain an array that indicates the position on the X, Y and Z axis of the detectors that were activated in the event.
- **ground.dectID**: The ID of the most important detector in array-ground telescope with largest electric charge during recorded event.

In our study we determined a combination of features that best represents the *Gamma-ray* and Background *Hadron* events through feature selection algorithms. These can include some of the explained features and other parameters extracted from data as well.

Table 1. Instances of each class

Gamma-rays	Hadrons
2229	6009

3.1 Used Dataset

Gamma-rays is the class of interest while the *Hadrons* are part of the Background of events to be discarded. In this work, we collected a whole simulated dataset containing both, Gamma-ray and Hadron Background events in ideal conditions. In the Table 1 the number of classes and instances per class considered is listed.

3.2 Proposed Feature Set

Relief-F algorithm is a feature selector process proposed in [12] aimed to rank a set of descriptors in a classification problem. This is made by an iterative process assigning weights to each feature. The algorithm penalizes the predictors that give different values to neighbors of the same class, and rewards predictors that give different values to neighbors of different classes.

We applied *Relief-F* to the 189 values from Sect. 3 in order to get the most relevant features for this classification problem. According to this process, the features with more relevance for Gamma-ray automatic recognition are ranked in Table 2.

Table 2. Main features and their relevance order according to *ReliefF*

Relevance order	Feature
1	nDect
2	nCh
3	nCont
4	PE_{SUM}
5	PE_{MAX}
6	dect.gridID
7	dect.contID
8	dect.xPos
9	dect.yPos
10	dect.zPos
11	ground.dectID

These features are present in the files according to [7] and are the principal parameters to identify the difference between *Gamma-ray* events and others Background data.

3.3   Added Noise Process

The measures from Cherenkov telescopes are exposed to the presence of noise depending of the aperture and efficiency. The majority of problems are related to the night-sky background and mechanisms of the telescope itself [5]. Adding white noise to the data, the classifiers performances of several *Signal-to Noise Ratio* (SNR) levels were compared to identify their robustness. The Figs. 1, 2 and 3 show the behavior of adding different Gaussian noise levels to the original data collection of *Gamma-ray* and *Hadron* events. In these figures, the most important features **nDect** and **nCh** are plotted for showing the difference between both classes.

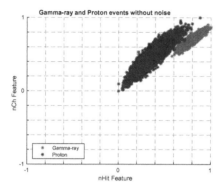

Fig. 1. Original samples of *Gamma-ray* (red) and *Hadron* (blue) events. (Color figure online)

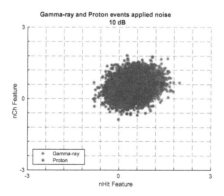

Fig. 2. *Gamma-ray* (red) and *Hadron* (blue) events varying SNR 10 dB. (Color figure online)

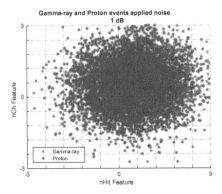

Fig. 3. *Gamma-ray* (red) and *Hadron* (blue) events varying SNR 1 dB. (Color figure online)

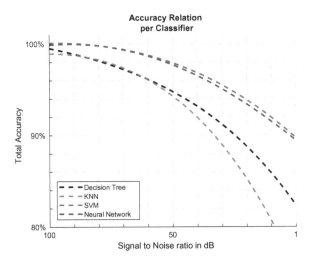

Fig. 4. Accuracy through different SNR values

4   Experiments and Results

For our experiments, we take a large dataset of *Gamma-ray* and *Hadron* data from Cherenkov telescope. These include multiple data for identifying the type of event that belongs to. However the original data is composed of 189 elements per sample. In order to determine a better reduced feature set, we applied *Relief-F* algorithm as explained in Sect. 3.2 getting the 12 most discriminant features.

In first place, we evaluate the total accuracy for four distinct Machine Learning approaches to automatic classification: *Support Vector Machines, K-NN Linear Classifier, Decision Trees* and *Feed-Forward Neural Networks*. Figure 4 shows the relation between total reached accuracy and *SNR* applied to the signals per each classifier.

Table 3. Resulting precision of each classifier through the addition of noise.

Classifier	DT	KNN	SVM	NN
SNR				
No noise	100%	100%	100%	100%
10	97.58%	97.67%	99.73%	99.96%
9	95.07%	94.12%	96.07%	99.87%
8	95.65%	93.09%	98.92%	99.15%
7	91.25%	94.93%	97.44%	97.08%
6	89.86%	92.10%	96.55%	95.93%
5	85.06%	76.13%	93.32%	93.09%
4	80.98%	79.14%	89.91%	89.68%
3	73.76%	64.24%	86.00%	85.73%
2	70.75%	57.34%	81.34%	83.36%
1	51.82%	55.45%	78.29%	78.42%

In Fig. 4 can be observed that the total accuracy decreases in relation with added noise. *SVM* Classifier is the most robust to the higher noise values, reaching up to 90.14% of accuracy for *SNR* values of 1 dB in comparison of others classification approaches.

By other hand, we measured the reached precision for Gamma-ray by each classifier. Table 3 shows the rate for true positive classified samples *Gamma-ray* events sorted by classifier, taking account the *SNR* in each case.

Neural Networks shows better precision rate for *Gamma-ray* events recognition, reaching up to 78.42% in the worst case when the noise is higher.

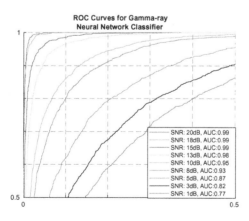

Fig. 5. ROC curve through different SNR values for ANN

4.1   ROC Curves Comparison

The *SNR* was variated from 100 dB to 1 dB in order to compare the classifiers.

In order to study the effectiveness of a classifier for recognize the class of interest *Gamma-ray* events, the precision values must be studied in detail using ROC curve in each case. For this, we compared the *Area Under the Curve* (AUC) varying the *SNR* value per each classifier.

Classification without noise have AUC values of 0.99 for all classifiers which is an original distribution as illustrated in Fig. 1. Then, the maximum area through the different levels of SNR was using SVM. When the noise is highest the AUC is equal to 0.77 units (Fig. 5).

5   Conclusions and Future Work

In this work, we studied the principal features for the *Gamma-ray* and *Hadron* events classification problem taking into account the possibility of having noising scenarios (Fig. 6).

First, we used 189 features from simulated data of *Gamma-ray* and *Hadron* events and studied the most important of them for *Gamma-ray* automatic recognition tasks, resulting in a subset of 12 features.

In second place, we added different Gaussian noise levels to collected dataset in order to evaluate the robustness of multiple machine learning classifiers against difficult conditions and *SNR* rates. *SVM* got the highest results for total accuracy while *Neural Networks* reached the best precision for single class recognition *Gamma-ray* considering the highest noise rates.

Unlike other related approaches, in this work we gathered a large collection of *Gamma-ray* and *Hadron* events with more than 8, 200 samples and considering scenarios with high noise rates as well (Figs. 7 and 8).

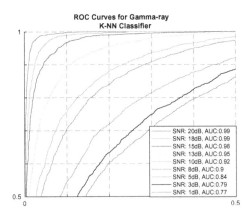

Fig. 6. ROC curve through different SNR values for KNN

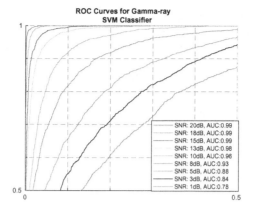

Fig. 7. ROC curve through different SNR values for SVM

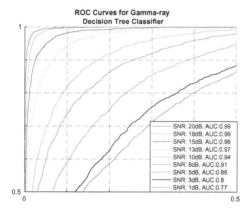

Fig. 8. ROC curve through different SNR values for decision tree

For future work, it is necessary to enlarge our dataset using other sources such as different datasets and no-simulated *Gamma-ray* and *Hadron* Background events from Cherenkov-telescopes in order to improve the obtained classification techniques.

References

1. Abeysekara, A., et al.: Sensitivity of the high altitude water Cherenkov detector to sources of multi-TeV gamma rays. Astropart. Phys. **50**, 26–32 (2013)
2. Atkins, R., et al.: Observation of TeV gamma rays from the Crab nebula with MILAGRO using a new background rejection technique. Astrophys. J. **595**(2), 803 (2003)
3. Badran, H., Weekes, T.: Improvement of gamma-hadron discrimination at TeV energies using a new parameter, image surface brightness. Astropart. Phys. **7**(4), 307–314 (1997)

4. Berge, D., Funk, S., Hinton, J.: Background modelling in very-high-energy γ-ray astronomy. Astron. Astrophys. **466**(3), 1219–1229 (2007)
5. Bernlöhr, K., et al.: Monte Carlo design studies for the Cherenkov telescope array. Astropart. Phys. **43**, 171–188 (2013)
6. Bourbeau, E., Capistrán, T., Torres, I., Moreno, E.: New gamma/hadron separation parameters for a neural network for HAWC. arXiv preprint arXiv:1708.03585 (2017)
7. Brun, R., Rademakers, F.: Root—an object oriented data analysis framework. Nucl. Instrum. Methods Phys. Res. Sect. A: Accel. Spectrometers Detect. Assoc. Equip. **389**(1–2), 81–86 (1997)
8. Grabski, V., Chilingarian, A., Nellen, L.: Gamma/hadron separation study for the HAWC detector on the basis of the multidimensional feature space using non parametric approach (2011)
9. Hampel-Arias, Z., et al.: Gamma hadron separation using pairwise compactness method with HAWC. arXiv preprint arXiv:1508.04047 (2015)
10. Krause, M., Pueschel, E., Maier, G.: Improved γ/hadron separation for the detection of faint γ-ray sources using boosted decision trees. Astropart. Phys. **89**, 1–9 (2017)
11. Reynolds, P., et al.: Survey of candidate gamma-ray sources at TeV energies using a high-resolution cerenkov imaging system-1988-1991. Astrophys. J. **404**, 206–218 (1993)
12. Robnik-Šikonja, M., Kononenko, I.: Theoretical and empirical analysis of ReliefF and RReliefF. Mach. Learn. **53**(1–2), 23–69 (2003)
13. Westerhoff, S., et al.: Separating γ-and hadron-induced cosmic ray air showers with feed-forward neural networks using the charged particle information. Astropart. Phys. **4**(2), 119–132 (1995)

Ranking Based Unsupervised Feature Selection Methods: An Empirical Comparative Study in High Dimensional Datasets

Saúl Solorio-Fernández[(✉)], J. Ariel Carrasco-Ochoa,
and José Fco. Martínez-Trinidad

Computer Sciences Department, Instituto Nacional de Astrofísica,
Óptica y Electrónica, Luis Enrique Erro # 1, Santa María Tonantzintla,
72840 Puebla, Mexico
sausolofer@inaoep.mx

Abstract. Unsupervised Feature Selection methods have raised considerable interest in the scientific community due to their capability of identifying and selecting relevant features in unlabeled data. In this paper, we evaluate and compare seven of the most widely used and outstanding ranking based unsupervised feature selection methods of the state-of-the-art, which belong to the filter approach. Our study was made on 25 high dimensional real-world datasets taken from the ASU Feature Selection Repository. From our experiments, we conclude which methods perform significantly better in terms of quality of selection and runtime.

Keywords: Unsupervised feature selection · Filter methods
Feature ranking

1 Introduction

The aim of Unsupervised Feature Selection (UFS) [1] is to identify and select a relevant and non-redundant feature subset in unlabeled datasets for discovering similar or better cluster structures in the data than those obtained by using the whole set of features. UFS is more challenging than Supervised Feature Selection because, in UFS, there is no information (class labels) that can guide the search for relevant features. However, in spite of this challenging issue, in the last 30 years, several UFS methods (selectors), which can identify and select relevant features in unlabeled data, have been reported in the literature.

Similar to the supervised case, unsupervised feature selection methods can be divided into three main approaches [2]. Filter, wrapper, and hybrid methods. Filter methods [3–10] select the most relevant features through the data itself, i.e., features are evaluated based on intrinsic properties of the data. According to [2], filter UFS methods can be divided into univariate and multivariate. The

© Springer Nature Switzerland AG 2018
I. Batyrshin et al. (Eds.): MICAI 2018, LNAI 11288, pp. 205–218, 2018.
https://doi.org/10.1007/978-3-030-04491-6_16

former, also known as ranking based UFS methods use some criteria to evaluate each feature to get an ordered list (ranking), where the final feature subset is selected according to this ordering. On the other hand, multivariate filter methods evaluate feature subsets rather than individual features. Most multivariate UFS methods, commonly produce a feature subset as a result. In wrapper methods [1,11–14], on the other hand, a specific clustering algorithm is employed to evaluate features; therefore, they have a high computational cost, and they are limited to be used in conjunction with a particular clustering algorithm. Hybrid methods [15–18] perform feature selection by exploiting the advantages of both approaches, trying to have a good compromise between efficiency and effectiveness. In this paper, we will focus our study on those filter UFS methods that produce a feature ranking as a result. These filter methods are extensively used in practice for high dimensional datasets, due to their effectiveness, and scalability.

The rest of this paper is organized as follows, in Sect. 2, the filter unsupervised feature selection methods used in our study will be described. The setup of our experimental study and the results will be presented in Sect. 3. A discussion about the results is given in Sect. 4. Finally, in Sect. 5, we present some conclusions and the future work.

2   Ranking Based Unsupervised Feature Selection Methods

Among the most relevant filter ranking based UFS methods are Variance [19], Laplacian Score (LS) [5], SPEC [7], UDFS [8], SVD-Entropy [6], SUD [3], and USFSM [10]. The output of these methods is a sorted list of features (ranking) from the most to the least relevant (or vise versa) according to their respective relevance evaluation. In this section, we will briefly describe these methods.

Variance [1,19,20] is the most simple univariate filter method for evaluating the importance of a single feature into an unsupervised context. The idea is that features with higher variance are more relevant for building good cluster structures. This method ranks features by their sample variance.

Laplacian Score [5] is a univariate spectral feature selection method that measures the relevancy of a feature based on the analysis of the eigensystem of the Laplacian matrix derived from the similarities among objects. The idea is to weight each feature, measuring how this feature reflects the underlying manifold structure [21], i.e., those features most preserving the pre-defined graph structure represented by the Laplacian matrix are the most relevant (consistent features).

SPEC [7] is a univariate spectral feature selection method that evaluates the relevance of a feature by its consistency with the structure of the graph induced from the similarities among objects. The feature evaluation is performed by measuring the consistency between each feature and those nontrivial eigenvectors of the Laplacian matrix.

UDFS [8] (Unsupervised Discriminative Feature Selection method) is a multivariate filter method that produces a feature ranking as a result. This method weights the features by simultaneously exploiting discriminative information (information contained in the scatter matrices) and feature correlations. UDFS addresses the feature selection problem by taking into account the trace criterion [22] into a constraints regression problem optimized through an efficient algorithm.

SVD-Entropy [6] is a univariate UFS method based on the Singular Value Decomposition (SVD) of the data matrix [23]. The basic idea of this method is to measure the entropy of the data according to its Singular Values; since when the entropy is low, well-formed clusters are generated; by contrast, when the entropy is high the spectrum is uniformly distributed, and well-formed clusters cannot be generated.

SUD [3] (Sequential backward selection method for Unsupervised Data) is a univariate UFS method that weighs features using a measure of entropy of similarities based on distance. The idea is to measure the entropy of the data based on the fact that when every pair of objects is very close or very far, the entropy is low, and it is high if most of the distances between pairs of objects are close to the average distance.

USFSM [10] is an univariate filter feature selection method based on spectral feature selection. The idea of this method is to rank features according to their consistency in the dataset by analyzing the changes in the spectrum distribution (spectral gaps) of the Normalized Laplacian matrix when each feature is excluded from the whole set of features separately.

3   Experimental Study

In this section, we first describe the experimental setup in Sect. 3.1, and later, the results obtained from the evaluation of the analyzed filter UFS methods over high dimensional datasets are presented in Sect. 3.2.

3.1   Experimental Setup

For our study, we selected the 25 datasets with the highest dimensionality from the ASU Feature Selection Repository[1] [24]. All of them are real-world datasets of different types which include text, face images, biological data, among others. Detailed information about the selected datasets is summarized in Table 1.

Following the most common way for assessing the performance of filter UFS methods, in our experiments, an evaluation framework similar to that described in [25] was used, where the UFS methods are evaluated through the accuracy of supervised classifiers. The evaluation was made by using three well-known classifiers: kNN ($k = 3$) [26], SVM [27], and Naive Bayes (NB) [28,29]. These three classifiers were chosen because they represent three quite different approaches for supervised classification, and they are among the most used classifiers for validation of unsupervised feature selection methods. These classifiers were taken from the Weka data mining software tool [30], using their default parameter values.

---

[1] http://featureselection.asu.edu/datasets.php.

Table 1. Dataset description.

#	Dataset	No. of objects	No. of features	No. of classes
1	Isolet	1560	617	26
2	Yale	165	1024	15
3	OLR	400	1024	40
4	WarpAR10P	130	2400	10
5	Colon	62	2000	2
6	WarpPIE10P	2010	2420	10
7	Lung	203	3312	5
8	COIL20	1440	1024	20
9	Lymphoma	96	4026	9
10	GLIOMA	50	4434	4
11	ALLAML	72	7129	2
12	Prostate_GE	102	5966	2
13	TOX_171	171	5748	4
14	Leukemia	72	7070	2
15	Nci9	69	9712	9
16	Carcinom	174	9182	11
17	Arcene	200	10000	2
18	Orlraws10P	100	10304	10
19	Pixraw10P	100	10000	10
20	RELATHE	1427	4322	2
21	PCMAC	1943	3289	2
22	BASEHOCK	1993	4862	2
23	CLL_SUB_111	111	11340	3
24	GLI_85	85	22283	2
25	SMK_CAN_187	187	19993	2

For our experiments, we applied stratified 5-fold cross-validation, and the final classification performance is reported as the average over the five folds. For each fold, each unsupervised feature selection method is first applied on the training set to obtain a feature ranking. Then, after training the classifier using the first r, $r = 20, 40, \ldots, 480, 500$ ranked features, the respective test sets are used for assessing each classifier through its accuracy (ACC). In order to evaluate the overall performance of each ranking based UFS method with the three supervised classifiers (kNN, SVM, and NB), we compute the *aggregated accuracy* [31]. The aggregated accuracy is obtained by averaging the averaged accuracy achieved by the classifiers using the top $20, 40, \ldots, 480, 500$ features ranked by each filter UFS method. It is worth mentioning that, for our experiments, the class labels in all datasets were removed before performing feature selection.

In order to perform a comprehensive comparison among the analyzed UFS methods, the Friedman test [32] was employed to measure the statistical significance of the results. The Friedman test is a non-parametric test used to measure the statistical differences of several methods over multiple datasets. For each dataset, the evaluated methods are ranked separately based on their aggregated accuracies. The method with the highest aggregated accuracy gets rank 1, the second highest result gets rank 2, and so on. In our experiments, the significance level was set to 0.05. If statistical differences are detected, the Holm test [33] is applied to compare every pair of unsupervised feature selection methods.

In all cases except SUD, we used the author's Matlab implementation with the parameters recommended by their respective authors. Meanwhile, SUD was implemented in Matlab based on the description provided by their authors in [3]. For building the similarity matrix between objects, which is required for Laplacian Score, SPEC and USFSM methods, we use the popular Gaussian radial basis function (kernel function) [34]. For USFSM, the Apache Commons Math[2] library and Statistical Machine Intelligence & Learning Engine[3] (SMILE) were used for matrix operations and eigen-system computation, respectively.

All experiments were run in Matlab® R2018a with Java 9.04, using a computer with an Intel Core i7-2600 3.40 GHz × 8 processor with 32 GB DDR4 RAM, running 64-bit Ubuntu 16.04 LTS (GNU/Linux 4.13.0-38 generic) operating system.

3.2   Experimental Results and Comparisons

In Tables 2, 3 and 4, the aggregated accuracy of all evaluated UFS methods using kNN, NB and SVM on the 25 datasets of Table 1 is shown, respectively. In these tables, the last row shows the overall aggregated accuracy obtained on all tested datasets. Additionally, in the last column, the classification accuracy using the whole set of features of each dataset is included. Notice that some results in these tables are marked with the symbol "-", this indicates that after a while (72 h), the UFS method was unable to produce a result for the specified dataset.

As we can see in Table 2, for kNN, among the UFS methods that could process all datasets, the best overall aggregated accuracy was obtained by Variance and USFSM. Laplacian Score and SPEC got slightly worse results on these datasets. Likewise, in Table 3, for the NB classifier, we can observe that the best results were obtained by Variance, followed by USFSM and SPEC. For this classifier, Laplacian Score got the worst results. Finally in Table 4, we show the results for SVM. In this case, we can see that the best overall aggregated accuracy was achieved by USFSM, followed by Variance and Laplacian Score, while SPEC got the worst results.

Table 5 reports the average ranks of the filter UFS methods that could process all datasets using kNN, NB, and SVM classifiers. It should be noticed that these

---

[2] http://commons.apache.org/proper/commons-math/.
[3] http://haifengl.github.io/smile/.

Table 2. Aggregated accuracy of the evaluated UFS methods using kNN ($k = 3$).

Dataset	Variance	LS	SPEC	UDFS	SVD-Entropy	SUD	USFSM	All-features
Isolet	0.814	0.809	0.741	0.721	0.743	0.741	0.791	0.886
Yale	0.518	0.606	0.509	0.564	0.573	0.509	0.552	0.673
ORL	0.846	0.878	0.809	0.911	0.866	0.809	0.885	0.948
WarpAR10P	0.569	0.492	0.739	0.467	0.578	0.739	0.660	0.492
Colon	0.712	0.643	0.728	0.703	0.655	0.681	0.698	0.663
WarpPIE10P	0.827	0.865	0.955	0.967	0.840	0.955	0.980	0.990
Lung	0.918	0.917	0.922	0.932	0.916	0.922	0.900	0.936
COIL20	0.938	0.917	0.856	0.942	0.922	0.856	0.969	0.999
Lymphoma	0.912	0.865	0.816	0.876	0.849	0.816	0.864	0.948
GLIOMA	0.447	0.722	0.782	0.614	0.726	0.766	0.650	0.700
ALLAML	0.737	0.825	0.799	0.731	0.762	0.819	0.783	0.874
Prostate_GE	0.798	0.646	0.763	0.765	0.651	0.762	0.754	0.823
TOX_171	0.826	0.782	0.811	0.818	0.670	0.811	0.747	0.836
Leukemia	0.876	0.767	0.789	0.803	-	0.850	0.809	0.917
Nci9	0.384	0.329	0.389	0.453	-	0.383	0.421	0.450
Carcinom	0.901	0.791	0.770	0.839	-	0.770	0.768	0.862
Arcene	0.757	0.779	0.767	0.659	-	-	0.760	0.865
Orlraws10P	0.828	0.881	0.897	0.868	-	-	0.830	0.930
Pixraw10P	0.927	0.974	0.954	0.982	-	-	0.983	0.990
RELATHE	0.735	0.745	0.697	0.734	-	-	0.666	0.772
PCMAC	0.707	0.626	0.627	0.654	-	-	0.625	0.712
BASEHOCK	0.801	0.664	0.691	0.734	-	-	0.681	0.776
CLL_SUB_111	0.585	0.570	0.544	0.613	-	-	0.568	0.621
GLI_85	0.798	0.826	0.818	-	-	-	0.818	0.882
SMK_CAN_187	0.692	0.628	0.618	-	-	-	0.698	0.647
Overall average	**0.754**	**0.742**	**0.752**	-	-	-	**0.754**	**0.808**

average ranks are computed according to the values in Tables 2, 3 and 4. In this table, we can see that the best ranks considering all classifiers were obtained by Variance, followed by USFSM. While SPEC and Laplacian Score got the third and fourth best average rank, respectively.

Table 6 shows the results of the Friedman test for the comparison among the UFS methods shown in Table 5. From this table, we can see that for kNN and SVM there is no statistical difference among the evaluated methods since the Fisher's statistic (F_F) is less than the corresponding critical value for a significance level of $\alpha = 0.05$. However, for the NB classifier, the differences are statistically significant. Therefore, for this classifier, we proceed with the post-hoc tests in order to detect significant pairwise differences among the evaluated UFS methods.

In Table 7, we show the Holm test for the pairwise comparison of the UFS methods of Table 5 for the NB classifier. Holm's test will reject those hypotheses (the compared selectors perform equally) that have a p-value ≤ 0.01. Thus, the first two hypotheses in Table 7 are rejected, which means that both Variance

Table 3. Aggregated accuracy of the evaluated UFS methods using NB.

Dataset	Variance	LS	SPEC	UDFS	SVD-Entropy	SUD	USFSM	All-features
Isolet	0.812	0.782	0.773	0.738	0.711	0.773	0.797	0.896
Yale	0.437	0.538	0.422	0.548	0.424	0.422	0.437	0.715
ORL	0.645	0.702	0.560	0.774	0.668	0.560	0.733	0.880
WarpAR10P	0.445	0.263	0.436	0.631	0.219	0.436	0.387	0.769
Colon	0.712	0.595	0.691	0.749	0.621	0.654	0.634	0.727
WarpPIE10P	0.485	0.316	0.797	0.879	0.253	0.797	0.876	0.929
Lung	0.911	0.886	0.932	0.936	0.865	0.932	0.914	0.956
COIL20	0.873	0.833	0.762	0.912	0.853	0.762	0.904	0.985
Lymphoma	0.886	0.789	0.718	0.732	0.756	0.718	0.746	0.938
GLIOMA	0.581	0.690	0.693	0.701	0.656	0.686	0.674	0.840
ALLAML	0.867	0.825	0.878	0.828	0.703	0.801	0.857	0.958
Prostate_GE	0.803	0.573	0.778	0.844	0.560	0.777	0.745	0.833
TOX_171	0.574	0.590	0.595	0.621	0.548	0.595	0.617	0.701
Leukemia	0.931	0.736	0.874	0.826	-	0.708	0.793	0.959
nci9	0.208	0.142	0.187	0.185	-	0.150	0.177	0.500
Carcinom	0.906	0.759	0.753	0.854	-	0.753	0.781	0.914
Arcene	0.647	0.695	0.665	0.673	-	-	0.661	0.670
orlraws10P	0.890	0.821	0.904	0.911	-	-	0.832	0.990
pixraw10P	0.896	0.961	0.955	0.988	-	-	0.995	0.980
RELATHE	0.758	0.680	0.623	0.687	-	-	0.648	0.849
PCMAC	0.845	0.543	0.603	0.752	-	-	0.601	0.874
BASEHOCK	0.934	0.592	0.662	0.857	-	-	0.681	0.967
CLL_SUB_111	0.655	0.532	0.548	0.663	-	-	0.599	0.676
GLI_85	0.784	0.790	0.819	-	-	-	0.850	0.859
SMK_CAN_187	0.635	0.634	0.649	-	-	-	0.649	0.647
Overal average	**0.725**	**0.651**	**0.691**	-	-	-	**0.704**	**0.840**

and USFSM are significantly better with a significance level of $\alpha = 0.05$, than Laplacian Score using NB. Moreover, between Variance and USFSM there is not a significant difference, therefore we can conclude that for the largest datasets (among those selectors that finished), both Variance and USFSM are the best.

Performing a similar analysis, but in this case using only the datasets that could be processed by all the evaluated UFS methods (the first 13 datasets of Tables 2, 3 and 4), in Table 8 we report the average ranks of the filter UFS methods using kNN, NB, and SVM classifiers. In this table, we can see that the best ranks considering all classifiers were obtained by UDFS, followed by USFSM and Variance. While Laplacian Score, SPEC, SUD, and SVD-Entropy were the worst ranked methods.

Table 9 shows the results of the Friedman test for the comparison among the evaluated UFS methods of Table 8. From this table, we can see that for kNN and SVM there is no statistical difference among the evaluated methods. However, again, for the NB classifier, the differences among the results of the UFS methods are statistically significant. Therefore, for this classifier, we proceed

Table 4. Aggregated accuracy of the evaluated UFS methods using SVM.

Dataset	Variance	LS	SPEC	UDFS	SVD-Entropy	SUD	USFSM	All-features
Isolet	0.899	0.884	0.843	0.839	0.844	0.843	0.873	0.962
Yale	0.549	0.636	0.586	0.661	0.595	0.586	0.589	0.764
ORL	0.862	0.874	0.828	0.913	0.861	0.828	0.915	0.973
WarpAR10P	0.695	0.588	0.786	0.893	0.603	0.786	0.717	1.000
Colon	0.793	0.750	0.788	0.731	0.724	0.742	0.765	0.788
WarpPIE10P	0.844	0.898	0.872	0.963	0.859	0.872	0.986	1.000
Lung	0.948	0.938	0.951	0.945	0.932	0.951	0.934	0.966
COIL20	0.913	0.904	0.819	0.930	0.888	0.819	0.942	1.000
Lymphoma	0.921	0.892	0.855	0.918	0.873	0.855	0.887	0.958
GLIOMA	0.658	0.789	0.766	0.719	0.753	0.756	0.762	0.780
ALLAML	0.771	0.913	0.831	0.853	0.859	0.847	0.886	0.987
Prostate_GE	0.885	0.804	0.846	0.870	0.765	0.845	0.843	0.892
TOX_171	0.887	0.844	0.801	0.884	0.834	0.801	0.829	0.959
Leukemia	0.936	0.833	0.851	0.849	-	0.889	0.886	0.987
nci9	0.483	0.409	0.443	0.513	-	0.457	0.505	0.583
Carcinom	0.923	0.870	0.852	0.896	-	0.852	0.854	0.925
Arcene	0.709	0.732	0.788	0.668	-	-	0.778	0.855
orlraws10P	0.825	0.856	0.801	0.939	-	-	0.810	0.980
pixraw10P	0.911	0.936	0.936	0.980	-	-	0.981	0.990
RELATHE	0.754	0.739	0.727	0.706	-	-	0.687	0.885
PCMAC	0.801	0.619	0.652	0.723	-	-	0.642	0.850
BASEHOCK	0.892	0.654	0.716	0.796	-	-	0.726	0.946
CLL_SUB_111	0.613	0.614	0.636	0.671	-	-	0.680	0.757
GLI_85	0.848	0.841	0.838	-	-	-	0.854	0.906
SMK_CAN_187	0.703	0.698	0.672	-	-	-	0.714	0.722
Overall average	**0.801**	**0.781**	**0.779**	-	-	-	**0.802**	**0.897**

Table 5. Average ranks of the UFS methods that could process all datasets, based on the aggregated accuracy of kNN, SVM, and NB classifiers.

Classifier	Variance	LS	SPEC	USFSM
kNN	2.36	2.52	2.54	2.58
NB	2.10	3.16	2.50	2.24
SVM	2.24	2.66	2.90	2.20
Average	**2.23**	**2.78**	**2.65**	**2.34**

Table 6. Results of the Friedman test for the comparisons among UFS methods of Table 5.

Classifier	F_F	Critical value	Significant differences
kNN	0.135	2.732	No
NB	3.67	2.732	Yes
SVM	1.769	2.732	No

Table 7. Results (p-values) achieved on post hoc comparisons among the UFS methods of Table 5 for $\alpha = 0.05$ using NB.

i	UFS methods	p	Significant differences
1	Variance vs. LS	0.003697	Yes
2	LS vs. USFSM	0.011751	Yes
3	LS vs. SPEC	0.070687	No
4	Variance vs. SPEC	0.273322	No
5	SPEC vs. USFSM	0.47644	No
6	Variance vs. USFSM	0.701419	No

Table 8. Average ranks of the UFS methods over the first 13 datasets of Tables 2, 3 and 4, based on the aggregated accuracy of kNN, SVM, and NB classifiers.

Classifier	Variance	LS	SPEC	UDFS	SVD-Entropy	SUD	USFSM
kNN	3.69	4.15	3.88	3.46	4.77	4.27	3.77
NB	3.42	4.54	4.04	1.92	5.92	4.73	3.42
SVM	3.69	3.38	4.50	3.15	5.15	4.88	3.23
Average	**3.60**	**4.02**	**4.14**	**2.84**	**5.28**	**4.63**	**3.47**

with the post-hoc tests in order to detect significant pairwise differences among UFS methods.

Table 9. Results of the Friedman test for the comparisons among UFS methods of Table 8.

Classifier	F_F	Critical value	Significant differences
kNN	0.474	2.227	No
NB	6.095	2.227	Yes
SVM	1.990	2.227	No

In Table 10, we show the Holm test for the pairwise comparison of the UFS methods for the NB classifier. In this case, the Holm's test will reject those hypotheses (the compared selectors perform equally) that have a p-value ≤ 0.002778. Thus, the first three hypotheses in Table 10 are rejected, this means that UDFS is significantly better, with a significance level of $\alpha = 0.05$, than SVD-Entropy, SUD, and Laplacian Score using NB; while between UDFS and SPEC, Variance and USFSM respectively, there is not a statistically significant difference.

Finally, Table 11 shows the average runtime over the five folds, spent by the evaluated UFS methods, as well as the total average runtime for those UFS methods that were able to process all datasets. As we can see, the fastest methods are

Table 10. Results (*p*-values) achieved on post hoc comparisons among the UFS methods of Table 8 for $\alpha = 0.05$ using NB.

i	UFS methods	*p*	Significant differences
1	UDFS vs. SVD-Entropy	0.000001	Yes
2	UDFS vs. SUD	0.00033	Yes
3	LS vs. UDFS	0.000472	Yes
4	SPEC vs. UDFS	0.006139	No
5	SVD-Entropy vs. USFSM	0.009363	No
6	Variance vs. SVD-Entropy	0.012279	No
7	Variance vs. UDFS	0.020611	No
8	UDFS vs. USFSM	0.026382	No
9	SPEC vs. SVD-Entropy	0.037635	No
10	SUD vs. USFSM	0.170649	No
11	LS vs. SVD-Entropy	0.185877	No
12	LS vs. USFSM	0.202086	No
13	Variance vs. SUD	0.202086	No
14	SVD-Entropy vs. SUD	0.219303	No
15	Variance vs. LS	0.237548	No
16	SPEC vs. SUD	0.395092	No
17	LS vs. SPEC	0.449692	No
18	SPEC vs. USFSM	0.603272	No
19	Variance vs. SPEC	0.670684	No
20	Variance vs. USFSM	0.924719	No
21	LS vs. SUD	0.924719	No

Variance, Laplacian Score, and SPEC, taking 4.87, 9.68 and 12.74 s on average respectively. USFSM was the fourth method, requiring on average 4187.3 s on average. UDFS, SUD, and SVD-Entropy were the slowest methods, and they could not process all datasets.

4   Discussion

As we can see in Tables 2, 3 and 4, for the largest datasets (among those selectors that finished), USFSM and Variance obtained the best results on average. After performing the Friedman and Holm's statistical tests, no matter the classifier used both Variance and USFSM arise as the best selectors (See Tables 5, 6 and 7). On the other hand, when we included in our study only the first 13 datasets, those datasets where all the selectors produced a result, UDFS outperformed all the others UFS methods on average (see the ranking shown in Table 8). Moreover, after performing the Friedman and Holm's statistical tests, no matter

Table 11. Average runtime (in seconds) of the UFS methods.

Dataset	Variance	LS	SPEC	UDFS	SVD-Entropy	SUD	USFSM
Isolet	1.01	10.60	22.26	71.47	1634.71	7351.46	484.82
Yale	0.25	0.05	0.05	5.68	22.88	95.47	25.94
ORL	0.28	0.08	0.10	6.81	63.13	502.46	179.25
WarpAR10P	0.57	0.09	0.06	80.40	2119.51	411.52	80.23
Colon	0.45	0.04	0.02	45.38	591.71	128.87	15.91
WarpPIE10P	0.58	0.10	0.10	81.54	231.66	837.27	186.72
Lung	0.79	0.14	0.13	208.56	544.47	1618.71	326.89
COIL20	0.40	2.62	2.85	10.99	482.86	6964.91	456.74
Lymphoma	1.69	0.16	0.09	792.33	14772.88	1786.77	182.64
GLIOMA	1.81	0.12	0.06	1063.73	10636.00	1533.86	123.59
ALLAML	1.76	0.18	0.10	2474.12	49267.96	5484.51	438.07
Prostate_GE	4.38	0.37	0.18	2882.82	50547.10	10596.04	974.22
TOX_171	4.04	0.82	0.61	2671.18	10439.65	18580.66	1875.43
leukemia	5.10	0.35	0.15	4572.43	–	5940.42	634.59
nci9	6.95	0.44	0.19	11542.31	–	10805.58	953.46
Carcinom	6.88	1.46	0.89	9408.85	–	46248.05	5022.85
arcene	7.61	1.60	1.24	12138.14	–	–	6989.49
orlraws10P	7.79	0.54	0.29	12882.70	–	–	2660.10
pixraw10P	7.35	0.60	0.29	12419.02	–	–	2444.30
RELATHE	5.56	47.94	74.02	1207.20	–	–	6841.69
PCMAC	5.95	66.74	89.09	611.09	–	–	9047.39
BASEHOCK	7.30	102.72	132.15	1485.95	–	–	18969.03
CLL_SUB_111	8.99	0.80	0.42	17125.59	–	–	3892.56
GLI_85	16.01	1.30	0.53	–	–	–	10370.65
SMK_CAN_187	14.47	3.14	2.07	–	–	–	27803.48
Average	4.87	9.68	12.74	–	–	–	4187.30

the classifier used, UDFS, Variance, and USFSM arise as the best selectors. However, Variance was the fastest (see Table 11), while UDFS appears among the slowest methods. Therefore, both Variance and USFSM become also as the best option.

Finally, in light of the results presented in Tables 2, 3, 4, 5, 6, 7, 8, 9, 10 and 11, some guidelines are suggested:

- For high dimensional datasets, Variance and USFSM are the best ranking based unsupervised feature selection methods.
- For datasets with 6000 features or less, USFSM is the best option, since it gives statistically the same results to the best one (UDFS) but USFSM is much faster than UDFS. While for very high dimensional datasets (more than 6000 features) Variance is clearly the best option in terms of quality and runtime.

5   Conclusions and Future Work

Unsupervised feature selection is an important task in data mining for selecting those features that allow discovering groups and identifying interesting cluster structures in the unsupervised data. In practice, class information is commonly unknown, so only unsupervised feature selection methods can be applied.

In this paper, we evaluated the most widely used and outstanding filter ranking based unsupervised feature selection methods, including Variance [1,19,20], Laplacian Score [5], SPEC [7], UDFS [8], SVD-Entropy [6], SUD [3], and USFSM [10] on high dimensional datasets. The results have shown that, in general, in terms of classification accuracy and runtime the best option for datasets with 6000 or less features is USFSM. While for datasets with more than 6000 features Variance is the best option.

Finally, given that, in this work, we only compared ranking based UFS methods, a comparison against other kinds of UFS methods is part of the future work of this research.

Acknowledgements. The first author gratefully acknowledges to the National Council of Science and Technology of Mexico (CONACyT) for his Ph.D. fellowship, through the scholarship 428478.

References

1. Dy, J.G., Brodley, C.E.: Feature selection for unsupervised learning. J. Mach. Learn. Res. **5**, 845–889 (2004)
2. Alelyani, S., Tang, J., Liu, H.: Feature selection for clustering: a review. Data Clust.: Algorithms Appl. **29**, 110–121 (2013)
3. Dash, M., Liu, H., Yao, J.: Dimensionality reduction of unsupervised data. In: Proceedings Ninth IEEE International Conference on Tools with Artificial Intelligence, pp. 532–539. IEEE Computer Society (1997)
4. Mitra, P., Murthy, C.A., Pal, S.K.: Unsupervised feature selection using feature similarity. IEEE Trans. Pattern Anal. Mach. Intell. PAMI **24**(3), 301–312 (2002)
5. He, X., Cai, D., Niyogi, P.: Laplacian score for feature selection. In: Advances in Neural Information Processing Systems 18, vol. 186, pp. 507–514 (2005)
6. Varshavsky, R., Gottlieb, A., Linial, M., Horn, D.: Novel unsupervised feature filtering of biological data. Bioinformatics **22**(14), e507–e513 (2006)
7. Zhao, Z., Liu, H.: Spectral feature selection for supervised and unsupervised learning. In: Proceedings of the 24th International Conference on Machine learning, pp. 1151–1157. ACM (2007)
8. Yang, Y., Shen, H.T., Ma, Z., Huang, Z., Zhou, X.: L2, 1-norm regularized discriminative feature selection for unsupervised learning. In: IJCAI International Joint Conference on Artificial Intelligence, pp. 1589–1594 (2011)
9. Tabakhi, S., Moradi, P., Akhlaghian, F.: An unsupervised feature selection algorithm based on ant colony optimization. Eng. Appl. Artif. Intell. **32**, 112–123 (2014)
10. Solorio-Fernández, S., Martínez-Trinidad, J.F., Carrasco-Ochoa, J.A.: A new unsupervised spectral feature selection method for mixed data: a filter approach. Pattern Recogn. **72**, 314–326 (2017)

11. Kim, Y., Street, W.N., Menczer, F.: Evolutionary model selection in unsupervised learning. Intell. Data Anal. **6**(6), 531–556 (2002)
12. Law, M.H.C., Figueiredo, M.A.T., Jain, A.K.: Simultaneous feature selection and clustering using mixture models. IEEE Trans. Pattern Anal. Mach. Intell. **26**(9), 1154–1166 (2004)
13. Breaban, M., Luchian, H.: A unifying criterion for unsupervised clustering and feature selection. Pattern Recogn. **44**(4), 854–865 (2011)
14. Dutta, D., Dutta, P., Sil, J.: Simultaneous feature selection and clustering with mixed features by multi objective genetic algorithm. Int. J. Hybrid Intell. Syst. **11**(1), 41–54 (2014)
15. Dash, M., Liu, H.: Feature selection for clustering. In: Terano, T., Liu, H., Chen, A.L.P. (eds.) PAKDD 2000. LNCS (LNAI), vol. 1805, pp. 110–121. Springer, Heidelberg (2000). https://doi.org/10.1007/3-540-45571-X_13
16. Hruschka, E.R., Hruschka, E.R., Covoes, T.F., Ebecken, N.F.F.: Feature selection for clustering problems: a hybrid algorithm that iterates between k-means and a Bayesian filter. In: 2005 Fifth International Conference on Hybrid Intelligent Systems. HIS 2005. IEEE (2005)
17. Li, Y., Lu, B.L., Wu, Z.F.: A hybrid method of unsupervised feature selection based on ranking. In: 18th International Conference on Pattern Recognition. ICPR 2006, Hong Kong, China, pp. 687–690 (2006)
18. Solorio-Fernández, S., Carrasco-Ochoa, J., Martínez-Trinidad, J.: A new hybrid filter-wrapper feature selection method for clustering based on ranking. Neurocomputing **214**, 866–880 (2016)
19. Theodoridis, S., Koutroumbas, K.: Pattern Recognition. Elsevier Science, Amsterdam (2008)
20. Liu, L., Kang, J., Yu, J., Wang, Z.: A comparative study on unsupervised feature selection methods for text clustering. In: Proceedings of 2005 IEEE International Conference on Natural Language Processing and Knowledge Engineering. IEEE NLP-KE 2005, pp. 597–601. IEEE (2005)
21. He, X., Niyogi, P.: Locality preserving projections. In: Advances in Neural Information Processing Systems, pp. 153–160 (2004)
22. Fukunaga, K.: Introduction to Statistical Pattern Recognition, 2nd edn. Academic Press Professional Inc., San Diego (1990)
23. Alter, O., Alter, O.: Singular value decomposition for genome-wide expression data processing and modeling. Proc. Natl. Acad. Sci. U.S.A. **97**(18), 10101–10106 (2000)
24. Li, J., et al.: Feature selection: a data perspective. ACM Comput. Surv. (CSUR) **50**(6), 94 (2017)
25. Zhao, Z., Morstatter, F., Sharma, S., Alelyani, S., Anand, A., Liu, H.: Advancing feature selection research. ASU Feature Selection Repository (2010)
26. Fix, E., Hodges Jr., J.L.: Discriminatory analysis-nonparametric discrimination: consistency properties. Technical report, California University, Berkeley (1951)
27. Cortes, C., Vapnik, V.: Support-vector networks. Mach. Learn. **20**(3), 273–297 (1995)
28. Maron, M.E.: Automatic indexing: an experimental inquiry. J. ACM **8**(3), 404–417 (1961)
29. John, G.H., Langley, P.: Estimating continuous distributions in Bayesian classifiers. In: Proceedings of the Eleventh Conference on Uncertainty in Artificial Intelligence, pp. 338–345. Morgan Kaufmann Publishers Inc. (1995)
30. Hall, M., Frank, E., Holmes, G., Pfahringer, B., Reutemann, P., Witten, I.H.: The WEKA data mining software: an update. SIGKDD Explor. Newsl. **11**(1), 10–18 (2009)

31. Zhao, Z., Wang, L., Liu, H., Ye, J.: On similarity preserving feature selection. IEEE Trans. Knowl. Data Eng. **25**(3), 619–632 (2013)
32. Friedman, M.: The use of ranks to avoid the assumption of normality implicit in the analysis of variance. J. Am. Stat. Assoc. **32**(200), 675–701 (1937)
33. Holm, S.: A simple sequentially rejective multiple test procedure. Scand. J. Stat. **6**(2), 65–70 (1979)
34. Buhmann, M.D.: Radial Basis Functions: Theory and Implementations. Cambridge Monographs on Applied and Computational Mathematics. Cambridge University Press, Cambridge (2003)

Dynamic Selection Feature Extractor
for Trademark Retrieval

Simone B. K. Aires[1]([⊠]) [iD], Cinthia O. A. Freitas[2]([⊠]) [iD],
and Mauren L. Sguario[1] [iD]

[1] Universidade Tecnológica Federal do Paraná, Ponta Grossa, Paraná, Brazil
{sbkaminski,mlsguario}@utfpr.edu.br
[2] Pontifícia Universidade Católica do Paraná, Curitiba, Paraná, Brazil
cinthia.freitas@pucpr.br

Abstract. The paper contributes to the CBIR systems applied to trademark retrieval. The proposed method seeks to find dynamically the best feature extractor that represents the trademark queried. In the experiments are applied four feature extractors: Concavity/Convexity deficiencies (CC), Freeman Chain (FC), Scale Invariant Feature Transform (SIFT) and Hu Invariant Moments (Hu). These extractors represent a set of classes of features extractors, which are submitted to a classification process using two different classifiers: ANN (Artificial Neural Networks) and SVM (Support Vector Machines). The selecting the best feature extractor is important to processing the next levels in search of similar trademarks (i.e. applying zoning mechanisms or combining the best feature extractors), because it is possible restrict the number of operations in large databases. We carried out experiments using UK Patent Office database, with 10,151 images. Our results are in the same basis of the literature and the average in the best case for the normalized recall (R_n) is equal to 0.91. Experiments show that dynamic selection of extractors can contribute to improve the trademarks retrieval.

Keywords: Dynamic selection · Feature extractor · ANN · SVM

1 Introduction

Trademarks are used as marketing tools, communicating a certain assurance of quality and innovation which the companies seek to promote and maintain [7]. They are important reputational assets, and your violation can have serious consequences [11].

In order to maintain the integrity and visibility of their trademarks, companies constantly search in the web and the media (magazines, newspapers, videos, among others) existence of trademarks similar to yours. For Abe *et al.* [4], the labor and cost associated with this effort increase every year. However, these concerns should be considered from the beginning of the process of registering a new trademark [11], verify conflicting with a trademark already registered, as well as avoid infringement of copyright. Trademark retrieval is highly complex task, due to the diversity of form and abstract elements that a trademark has. Thus, recognition systems need to have mechanisms to ensure efficiency at retrieval task.

© Springer Nature Switzerland AG 2018
I. Batyrshin et al. (Eds.): MICAI 2018, LNAI 11288, pp. 219–231, 2018.
https://doi.org/10.1007/978-3-030-04491-6_17

A pattern recognition problem can involve a number of patterns with each class consisting of various features. For Aires *et al.* [9] is very hard to a single feature extractor solves the complexity and reaches the best solution. Each feature extractor looks at the image for different types of information, such as: topological, geometric, texture, edge, and others. However, each represents the image differently, and often the representation of this image by a particular extractor achieves more accurate results than other features extractors. Thus, using a lot of extractors allow the system to determine an extractor that highlights features that best represent the various information in the image and can differentiate it from the other images in the database.

This work show the CBIR (Content-based Image Retrieval) systems applied to trademark retrieval, seeking to a classifier can contribute by finding the feature extractor that best represents the elements of the trademark required.

The paper is organized as follows. Section 2 presents the concepts and related works to CBIR Systems. Section 3 presents the feature extractors, classifiers and the baseline system. The experiments and results are summarized in the Sect. 4. Finally, the Sect. 5 discusses the experimental results and present the conclusion and future works.

2   Related Works

Because of economical relevance of trademarks the company's request intelligent image analysis systems [8]. In this way, such intelligent systems start the process by registering a new trademark. However, all of these applications need to handle great amount of images. Thus, this kind of system is a challenge for many researchers.

The work presented by Kumar *et al.* [12] considered the three most important systems for trademark retrieval: Artisan, Star, and. Trademark Systems. Different methodologies and approaches have been applied by these systems. The Artisan System proposed an approach taking into account principles of human perception, which consists of some abstract geometric designs [1]. Star System applied CBIR techniques, including the Fourier descriptors, grey level projection, and moment invariants [14]. The Trademark System applies graphical feature vectors (GF-vector) to describe the image and calculate the similarity based on human perception [4]. For Anuar *et al.* [7] these three systems are very important and significant researches on trademarks.

The study presented by Shaaban *et al.* [13] presented the approach for retrieving trademark based on integrating multiple classifiers; the idea is speed up the retrieving process and to improve retrieving accuracy. The system applied three feature extractors: Invariant Moments; Decomposition in Singular Values (SVD - Singular Value Decomposition) and Discrete Cosine Transform (DCT - 2D Discrete Cosine Transform). For [13] feature extraction is the most important step in these kind of systems; obtaining good results in the selected features to utilize them in the classification stage.

Haitao *et al.* [5] extracted the features based on the contour of the image using Fourier Moments. In the experimentation was applied SVM model to solve the problems of poor generalization performance, local minimum and over fitting. In addition, kernel function was applied in SVM maps data set linear inseparable to a higher dimensional space where the training set is separable. For [5] this is the reason for the SVM classifiers are widely used in pattern recognition.

Consider the popularization and increasing use of the deep learning methods, well-known Convolutional Neural Network (CNN), Aker *et al.* [15] applied these models to the trademark retrieval problem. Models were tested using a large scale trademark dataset. Some solutions were presented, such as, fine-tuning, distance metric learning, using CNN features locally, and making them invariant to aspect ratio of the trademark.

As presented before, several authors [12, 13, 5, 15] have investigated the similarities between trademarks. However, the development of retrieval systems is challenged because the high degree of difficulty to find features extractors that may represent the trademark queried in a way that distinguishing the others trademarks in large database (with a high degree of dissimilarity).

In this paper, our efforts were to determine which feature extractor best represent a trademark. In the experiments carried out we have class of features extractors and for classification we test four ANNs Class Modular and a SVM.

3   Baseline System

The experimental protocol uses as input a 256 grey-level image. Then, a preprocessing step is applied, which is composed to binarization (OTSU) and bounding box definition. The feature set is based on four methods of extraction. Two methods contour-based: Freeman Chain Code (FCC), Concavity/Convexity deficiencies (CC), and two methods region-based: Scale Invariant Features Transform (SIFT) [6] and Hu Invariant Moments (Hu). Matching of the similarities was calculated by Euclidian distance. The best results of Top-1% were compared with [1] and [3].

3.1   Database

We use the UK database Patent Office that belongs to the Intellectual Property Office (IPO) from United Kingdom (http://www.ipo.gov.uk) to perform our experiments. This database contains 10,745 images of trademarks, all in gray-levels. The experiments

Table 1. Truth set

No.	TM	Similar	No.	TM	Similar	No.	TM	Similar	No.	TM	Similar
01		25	06		17	11		09	16		11
02		15	07		10	12		15	17		20
03		11	08		19	13		12	18		09
04		09	09		24	14		12	19		22
05		09	10		10	15		16	20		12

			Total	287
		Total Overall (287+ 20 test)		307

were carried out using a set of 20 image queries selected by experienced trademark examiners from UK Patent Office. The 20 image queries are shown in Table 1.

3.2   Features Extractors

In this section we will briefly present the feature extractors applied in this work. More details can be found in [9].

Concavity/Convexity Deficiencies (CC)

The Concavity/Convexity deficiencies feature set puts on evidence the topological and geometrical properties of the shape to be recognized and is computed by labeling the background pixels of the input images. The idea of concavity/convexity deficiencies is check for each background pixel in the image; we search in four-directions: North, South, East, and West. When black pixels are attain in all directions, we verify at four auxiliary directions in order to confirm if the current white pixel is really inside a closed contour. The entire and definitive symbols were adapted to trademark retrieval, and then we have 24 different symbols [9]. Figure 1a presents the labeling process of background pixels from a trademark query.

Freeman Chain Code (FC)

Chain codes are used to represent borders of objects, through a sequence of straight line segments of specified length and direction [9]. Figure 1b presents Freeman Chain Code from a trademark query.

Scale Invariant Feature Transform (SIFT)

Lowe [2] presents a framework to extracting distinctive invariant features from images and shows that it can be used to perform matching between different views of an object or scene. The features are invariant to image scale and rotation, and provide robust matching across a substantial range of affine distortion, noise, and illumination changing's. Lowe [2] presents four important stages to generate the set of image features: (1) scale-space extreme detection; (2) key-point localization; (3) orientation assignment; (4) key-point description.

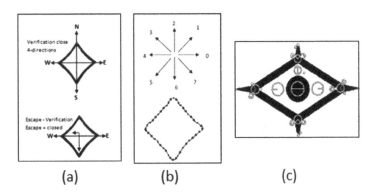

Fig. 1. (a) Concavity/Convexity deficiencies (b) Freeman Chain Code (c) SIFT in trademark database UK

This approach is widely applied in many researches [8, 11, 15] for retrieval objects in images databases. For this reason, we tested SIFT in our experiments. Figure 1c shows SIFT features extracted from a trademark contained in the UK/IPO database.

Hu Invariant Moments (Hu)
The seven moments proposed by Hu are widely used in image processing because of the robustness to image translation, scale and rotation transformation. These moments are represented by seven equations nominated as Hu invariant moments and Hu moments. Moment is a robust technique for decomposing an image into a finite set of invariant features. In practical terms, the use of Moments for image recognition requires the selection of a subset of moment values that contains enough information to characterize each image only [9].

3.3 Dynamic Selection

It was necessary to define a strategy that would be able to dynamically select the best extractor. For this task, experiments were performed using two classifiers: Artificial Neural Networks (ANN) and Support Vector Machines (SVM). Details on the construction of these classifiers are contained in the following sections. Both classifiers were trained and tested based on "truth set" contained in the UK Patent Office database [17], presented in Table 1.

Artificial Neural Network
Artificial Neural Networks based on supervised learning have a set of input variables and an expected output set. ANN compares the output value to the desired value, making corrections to the model so that it encounters an acceptable error. After the training step, a new input set unknown can be presented to ANN and its task is to correctly classify this new class.

Based on the individual results presented in Table 2 of Sect. 4.1, the ANN training and validation sets were constructed. Neural network training was performed using four (4) MLP (Multi-Layer Perceptron) networks with hidden layers, considering that the number of neurons in the hidden layer is the half number of neurons in the input layer. The number of training epochs is variable according to the extractor used in each ANN, such values were obtained observing the learning curve from JNNS during the training. The learning algorithm used was Back Standard Propagation, with learning parameter 0.2. The weights were randomly initialized with values between −1 and 1.

The Fig. 2 presents an overview of the building of ANN under the premise of making Multiple Classifiers. Thus, each trained ANN is specialized in recognizing a class of extractor. Each ANN has as output values between 1 and 0; '1' represents "recognized" image and '0' represents "unrecognized" image for the evaluated extractor class. Given a query trademark, information is extracted applying all four extractors described in Sect. 3.2. Then the features vectors are normalized and sent to the respective ANN (Fig. 2).

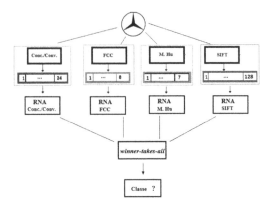

Fig. 2. ANNs multiple classifiers.

SVM

The SVM classifier used in this work was built by LIBSVM tool widely used and available in [16]. For building the SVM classifier, it is necessary to define training and test sets. These sets were created using information from the 307 trademark images contained in the "truth set" (Table 1). The SVM training set, which has 287 images, was constructed based on the individual results presented in Table 2 of Sect. 4.1, such as the methodology used for the construction of the ANN sets. Unlike the ANNs, only one SVM classifier was built for the classification problem. The composition of the characteristics vector is accomplished by the union of the feature vectors of the four extractors and the class to which each trademark belongs, as shown in Fig. 3.

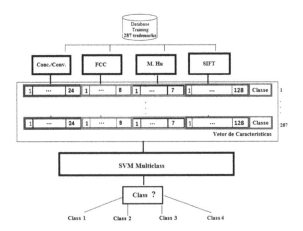

Fig. 3. SVM classifier.

The Fig. 3 shows the composition of the SVM Multiclass training set used in the experiments, being trained and tested for the four classes representing the four extractors applied to the Model: class 1 - Concavity/Convexity Deficiencies - CC (dimension vector 24), class 2 - Freeman Chains - FCC (vector of dimension 8), class 3 - Invariant Moments of Hu (vector of size 7), and class 4 - SIFT extractor (dimension vector 128). The feature vector has a total size equal to 167. The 20 query images were inserted only in the test database. As a final result, SVM indicates the best extractor to be applied to the trademark query.

3.4   Matching and Measure of Retrieval

The similarity calculation between the trademarks is performed through the Euclidean distance of the feature vectors. To evaluate the results the trademarks retrieval, we used two measures usually applied in any CBIR system that generates output in ranked order: Normalized Recall R_n and Normalized Precision P_n [1]. The system retrieval performance, in return to a query, is 0 (worst case) to 1 (best case). Normalized Recall and Precision are defined by following Eq. (1) and (2).

$$R_n = 1 - \frac{\sum_{i=1}^{n} R_i - \sum_{i=1}^{n} i}{n(N-n)} \tag{1}$$

$$P_n = 1 - \frac{\sum_{i=1}^{n} (\log R_i - \sum_{i=1}^{n} (\log i)}{\log\left(\frac{N!}{(N-n)!n!}\right)} \tag{2}$$

Where R_i is the rank at which relevant trademark, i is actually retrieved, n is the total number of relevant trademarks, and N is the size of the whole trademark database.

4   Experimental Results

The proposed method to the dynamic classification of the feature extractor is presented in Figs. 2 and 3. The matching among the query image and the images in the database is performed to calculate the similarity using Euclidian Distance, and an overview of the complete method is presented in Fig. 4. At the end, a ranking of the Top-100 of the images most similar to the query image is presented as result.

4.1   General Results of Features Extractors

In this section a comparison is made between the results of all extractors used in the experimentations. The best results are presented in Table 2, highlighting the best R_n rate for each trademark contained in the true set (Table 1). Table 2 is important because the individual results of each feature extractor were used to define training and validation sets to be used by the classifiers (as discussed in Sect. 3.3).

We observe in Table 2 that by selecting the extractor that best represents the trademark, it is possible to improve the recovery rate. Also, we did not combine feature extractors to increase the R_n rate. Considering that the features extractor is an important component in CBIR systems to obtain good results, this knowledge allows to design

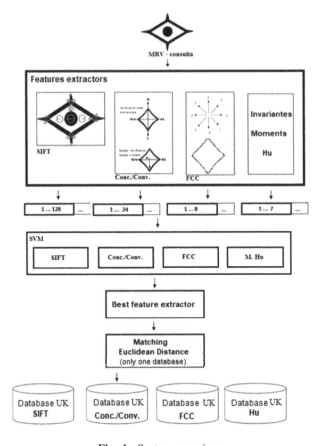

Fig. 4. System overview.

strategies to improve its performance in the recovery of similar trademarks (for example, applying zoning mechanisms or combining the best feature extractors).

4.2   Results ANNs

Four ANNs were constructed, each one specialized in one of the four features extractors. The goal is to obtain a 'vote' or score from each one ANN for each trademark queried. By means of the Majority Rule, the class of extractor that trademark belongs to is determined, that is, the best extractor to be used for the trademark in question was determined. This task is important to maximize the obtained results. The Table 3 presents the votes of each ANN for the 20 trademarks queried; the 'Best' extractor is presented in Table 2.

We can observe at Table 3 that 14 trademarks were able to be classified correctly in their extractor classes. Therefore, 6 images did not obtain the expected results and, for 4 images, the second largest vote ranks correctly (, , ,). This demonstrates that the Majority Rule strategy could be replaced so as to also consider the

Table 2. Best feature extractor - R_n

TM	CC	FCC	SIFT	Hu	Best	TM	CC	FCC	SIFT	Hu	Best
⬤	0,91	0,79	0,89	0,91	0,91	▽	0,68	0,47	0,67	0,87	0,87
🌀	0,90	0,79	0,93	0,81	0,93	//	0,68	0,77	0,94	0,57	0,94
⊛	0,74	0,79	1,00	0,64	1,00	⛩	0,65	0,86	0,98	0,81	0,98
◈	0,96	0,80	0,89	0,77	0,96	◎	0,82	0,79	0,96	0,58	0,96
▣▣	0,85	0,89	0,99	0,50	0,99	◯	0,66	0,85	0,65	0,48	0,85
◆	0,85	0,39	0,76	0,70	0,85	◕	0,48	0,62	0,87	0,44	0,87
▣	0,68	0,89	0,60	0,59	0,89	ᗐ	0,76	0,45	0,58	0,59	0,76
🌿	0,95	0,58	0,93	0,63	0,95	▲	0,83	0,89	0,45	0,59	0,89
⫽	0,75	0,73	0,81	0,30	0,81	▲	0,75	0,97	0,61	0,52	0,97
❖	0,79	0,90	0,42	0,74	0,90	▲	0,89	0,85	0,96	0,82	0,96
						Average	0,78	0,75	0,79	0,64	0,91
						SD	0,12	0,17	0,19	0,16	0,06

second largest vote. The trademark (▽) that was supposed to be classified belonging to the class of extractor Moments de Hu, was not successful. This fact is due to the amount of examples of this trademark contained in the truth set (only nine). It is important to note that only this trademark obtained better results with Hu Moments, resulting in a small set of trademarks for the training, validation and testing set, this amount was insufficient to solve this task.

The results obtained by the ANNs were not enough to solve the classification to determine the best extractor to be used by queried trademark. In order to obtain better results, experiments were performed applying SVM.

4.3 Results SVM

The Support Vector Machines (SVM) was developed with the purpose of performing classification tasks, being successfully used in pattern recognition applications [5, 6, 10].

In order to evaluate the performance of SVM in the classification of extractors, comparing with the results obtained by ANNs presented in Sect. 4.2, a SVM Multiclass was constructed. The features vectors of each extractor were combined (juxtaposed) in a single feature vector. This vector has a dimension equal to 167, that is: 128 features obtained by the average of the SIFT key points, 24 features of Concavity and Convexity, 8 features of Freeman Chains Code and 7 features of Hu Moments. Two sets were constructed: training and testing. Table 4 shows the results obtained from the trademarks contained in the test set.

Table 3. Vote 4 (four) ANNs

Trademark	1o.	2o.	Best	Trademark	1o.	2o.	Best
	CC	SIFT	CC		CC	SIFT	HU
	SIFT	M. Hu	SIFT		SIFT	M. Hu	SIFT
	SIFT	CC	SIFT		SIFT	FCC	SIFT
	CC	M. Hu	CC		SIFT	M. Hu	SIFT
	SIFT	FCC	SIFT		FCC	M. Hu	FCC
	CC	SIFT	CC		SIFT	CC	SIFT
	SIFT	FCC	FCC		CC	FCC	CC
	FCC	CC	CC		SIFT	FCC	FCC
	SIFT	M. Hu	SIFT		SIFT	FCC	FCC
	CC	M. Hu	FCC		SIFT	M. Hu	SIFT

Table 4. SVM multiclass

TM	Predict	Table 2	R_n Best	TM	Predict	Table 2	R_n Best
	CC	CC	0,91		HU	HU	0,87
	SIFT	SIFT	0,93		SIFT	SIFT	0,94
	SIFT	SIFT	1		SIFT	SIFT	0,98
	CC	CC	0,96		SIFT	SIFT	0,96
	SIFT	SIFT	0,99		FCC	FCC	0,85
	CC	CC	0,85		SIFT	SIFT	0,87
	FCC	FCC	0,89		CC	CC	0,76
	CC	CC	0,95		FCC	FCC	0,89
	SIFT	SIFT	0,81		FCC	FCC	0,97
	CC	FCC	0,79		SIFT	SIFT	0,96
						Average	0,91
						SD	0,06

In the results presented in Table 4 we can observe that the SVM was able to classify the trademarks better in relation to the results obtained by the ANNs. Of the 20 trademarks in the test set, 19 trademarks were classified correctly. The trademark (❖)

did not obtain the expected result, its class should be FCC, but was classified as Concavity/Convexity. This fact also occurred for ANNs.

Based on 20 trademarks from the test set, only one trademark presented confusion (Table 4). However, this result does not affect the General Average of the best extractor for R_n, since the difference between the results obtained by the extractors is 0.1, Concavity/Convexity obtained $R_n = 0.79$ and FCC obtained $R_n = 0.90$. These results confirm that SVMs are successful in pattern recognition systems. According to [6], the SVMs are efficient in relation to speed and complexity. This method equates the minimum search of a convex function, that is, without local minimums. Thus, many problems that occur in ANNs and decision trees are eliminated. This observation may explain the good results obtained by the SVM in relation to the ANNs.

Table 5. Comparative results average overall

Authors	R_n	P_n
Our method – best extractor	0.91	0.75
Eakins et al. [1]	0.89	0.67
Cerri et al. [3]	0,81	0,56

5 Discussion and Conclusion

Select an extractor features that best represents the trademark reduces the search in databases. The features extracted are concatenated in a single vector only when sent to SVM (167 features). When the SVM selects which extractor use, the search in the database is restricted to number of features of best extractor, it reduces the cost, because the system has fewer values to compute between the queried trademark and the 10,151 trademarks of the database.

We can observe in Fig. 4 a trademark query, a feature extraction by four proposed features extractors, the features being sent to SVM, the matching with the other images of the database is restricted to database referring to the extractor indicated by SVM. The trademark (⬥) obtained the best results with the Concavity/Convexity extractor, so the matching will only be performed in the database of the images with the Concavity/Convexity features, only 24 features are used to perform the matching.

Additionally, zoning mechanisms can be applied considering only the features extractor defined by the SVM. Zoning mechanisms were applied in trademark retrieval in reference [11]. More, interesting experiments can be performed combining the best features extractors.

The Table 5 compares our results with the literature. The references [1] and [3] also used in their experiments the database of UK Patent Office. However, [1] and [3] do not use any classifiers in their experiments. It is important to observe that in this work and in [1] and [3] the feature extractors used are different. However, the Matching (Euclidian Distance) and the Measure of Retrieval R_n and P_n (presented in Sect. 3.4) are the same. The rates compared in Table 5 relate to the recovery rate of the trademarks contained in the true set (Table 1).

We understand that using SVM to classify the feature extractors made it possible to improve trademark retrieval rates (Table 5) and additionally reduces the cost (making fewer comparisons) during the retrieval task in large databases.

So, our results are better to those found in the literature and it convinced us to carry on the research, including experiments using Deep learning [15] and increasing the number of query images. These observations are opening the way for new interesting directions of research.

Acknowledgments. The authors wish to thanks the CAPES, Federal Technological University of Parana (UTFPR-PG, Brazil) and Pontifical Catholic University of Parana (PUCPR, Brazil), which have supported this work.

References

1. Eakins, J.P., Boardman, J.M., Graham, M.E.: Similarity retrieval of trademarks image. IEEE Multimedia **5**, 53–63 (1998)
2. Lowe, D.: Distinctive image features from scale-invariant keypoints. Int. J. Comput. Vis. **60** (2), 91–110 (2004)
3. Cerri, A., Ferri, M., Gioirgi, D.: Retrieval of trademarks images by means of size functions. Graph. Models **68**(5–6), 451–471 (2006)
4. Abe, K., Iguchi, H., Tian, H., Roy, D.: Recognition of plural grouping patterns in trademarks for CBIR according to the Gestalt psychology. Inst. Eletron. Inf. Commun. Eng. **89**(6), 1798–1805 (2006)
5. Haitao, R., Yeli, L., Likun, L.: Single closed contour trademark classification based on support vector machine. In: 3rd International Congress on Image and Signal Processing (CISP), pp. 1942–1946 (2010)
6. Rufino, H.L.P.: Algoritmo de Aprendizado Supervisionado - Baseado em Máquina de Vetores de Suporte - Uma contribuição Para o Reconhecimento de Dados Desbalanceados. Universidade Federal de Uberlândia. Tese de Doutorado (2011)
7. Anuar, F.M., Setchi, R., Yu-Kun, L.: Trademark image retrieval using an integrated shape descriptor. Expert Sys. Appl. **40**(1), 105–121 (2013)
8. Sahbi, H., Ballan, L., Serra, G., Bimbo, A.D.: Context-dependent logo matching and recognition. IEEE Trans. Image Process. **22**(3), 1018–1030 (2013)
9. Aires, S.B.K., Freitas, C.O.A., Oliveira, L.S.: Feature Analysis for Content-based Trademark Retrieval. In: 27th International Conference on Computer Applications in Industry and Engineering, New Orleans, USA, pp. 245–249 (2014)
10. Liu, Y.: Sistema de recomendação dos amigos na rede social online baseado em Máquinas de Vetores Suporte. Dissertação de Mestrado. UnB, Brasília (2014)
11. Aires, S.B.K., Freitas, C.O.A., Soares, L.E.O.: SIFT applied to perceptual zoning for trademark retrieval. In: IEEE International Conference on Systems, Man, and Cybernetics (SMC 2015), v. 1 (2015)
12. Kumar, R., Tripathi, R.C., Tiwari, M.D.: A comprehensive study on content based trademark retrieval system. Int. J. Comput. Appl. **13**, 18–22 (2011)
13. Shaaban, Z.: Trademark image retrieval system using neural network. Int. J. Comput. Sci. Netw. (IJCSN) **3**, 73–82 (2014)
14. Wu, J.K., Lam, C.P., Mehtre, B.M., Gao, Y.J., Narasimhalu, A.: Content based retrieval for trademark registration. Multimedia Tools Appl. **3**(3), 245–267 (1996)

15. Aker, C., Tursun, O., Kalkan, S. :Analyzing deep features for trademark retrieval. In: 25th Signal Processing and Communications Applications Conference (SIU) (2017)
16. Chang, C.C., Lin, C.J.: LIBSVM - a library for support vector machines. Disponível em http://www.csie.ntu.edu.tw/~cjlin/libsvm. Acessado em 07 de maio de 2018
17. INTELLECTUAL Property Office. Disponível em http://www.ipo.gov.uk/. Acesso em 20 de junho de 2018

Bayesian Chain Classifier with Feature Selection for Multi-label Classification

Ricardo Benítez Jiménez$^{(\boxtimes)}$, Eduardo F. Morales, and Hugo Jair Escalante

Instituto Nacional de Astrofísica, Óptica y Electrónica (INAOE),
Sta. María Tonantzintla, 72840 Puebla, Mexico
{ricardo.benitez,emorales,hugojair}@inaoep.mx

Abstract. Multi-label classification task has many applications in Text Categorization, Multimedia, Biology, Chemical data analysis and Social Network Mining, among others. Different approaches have been developed: Binary Relevance (BR), Label Power Set (LPS), Random k label sets (RAkEL), some of them consider the interaction between labels in a chain (Chain Classifier) and other alternatives around this method are derived, for instance, Probabilistic Chain Classifier, Monte Carlo Chain Classifier and Bayesian Chain Classifier (BCC). All previous approaches have in common and focus on is in considering different orders or combinations of the way the labels have to be predicted. Given that feature selection has proved to be important in classification tasks, reducing the dimensionality of the problem and even improving classification model's accuracy. In this work a feature selection technique is tested in BCC algorithm with two searching methods, one using Best First (BF-FS-BCC) and another with GreedyStepwise (GS-FS-BCC), these methods are compared, the winner is also compared with BCC, both tests are compared through Wilcoxon Signed Rank test, in addition it is compared with others Chain Classifier and finally it is compared with others approaches (BR, RAkEL, LPS).

Keywords: Multi-label classification · Chain classifier
BCC · Feature selection

1 Introduction

Classical classification task consists of determining a label or class for an instance, which is characterized by n features, on the other hand, multi-label classification consists in predicting a subset $l \subseteq L$ of binary labels (0, 1). The task of multi-label classification has many applications in Text Categorization, Multimedia, Biology, Chemical data analysis and Social Network Mining, among others. This task has to deal with problems like high computational cost, derived from considering dependencies between labels and the complexity of the resulting model.

There are two well-known approaches for multi-label classification task: Binary Relevance (BR) and Label Power Set (LPS), the BR approach creates

© Springer Nature Switzerland AG 2018
I. Batyrshin et al. (Eds.): MICAI 2018, LNAI 11288, pp. 232–243, 2018.
https://doi.org/10.1007/978-3-030-04491-6_18

n binary classifiers, one per label and determines the subset of labels according to each binary classifier [17]. In contrast, LPS produces n classifiers according to all the possible combinations of binary labels existing on the training data [12], which in the worst case yields 2^n classifiers. These methods have some disadvantages, for instance, BR does not consider the relationship between labels and LPS might suffer from a high computational complexity depending on the number of n different subset of labels.

With the intention to deal with the complexity of the models and the interaction between labels, other approaches have been developed. One strategy is Random k label sets (RAkEL), which by following the LPS logic creates n classifiers but considering only a subset of labels with a size of k [11]. On the other hand, the Chain Classifier (CC) has been proposed by considering the interaction between labels, it uses a classifier for each label, considers the previous prediction l_i to classify the next label l_j [6] and others alternatives around this method are derived, for instance, Probabilistic Chain Classifier (PCC) which estimates the optimal chain but at a high computational cost, therefore, it is recommended to use it in datasets with no more than 15 labels [1], Monte Carlo Chain Classifier (MCC) estimates the optimal chain through a Monte Carlo approach [5]. Furthermore, there is also the Bayesian Chain Classifier (BCC) which builds a Bayesian Network and then a undirected tree using Chow and Liu's algorithm, then it chooses a node as the root and starting off from there it creates a directed tree [15].

Nevertheless, one thing all previous approaches have in common and focus on is in considering different orders or combinations of the way the labels have to be predicted. In this work, a feature selection technique is applied in the middle of the BCC algorithm, the idea is to improve its performance by improving separately the individual internal classifiers with the feature selection.

2   Related Work

Feature selection has proved to be important in classification tasks, reducing the dimensionality of the problem and even improving classification model's accuracy.

Recently, strategies been developed to reduce the number of features used in the multi-label classification task. In [8] they used feature selection with two methods, RefielF and Information Gain, in a BR strategy, in both cases using a filter approach for each label in two different scenarios, one with C4.5 decision trees and the other one with Support Vector Machine (SVM) as internal classifier, in [9] they extend the work with the LPS approach using lazy k Nearest Neighbor algorithm as internal classifier.

Fast Correlation-Based Filter is also explored in [4] by creating a Maximum Expanding Tree between each label and each of the features, again implemented along with a BR strategy. Other measures have been explored in [14], in order to determine relevant features, including a fast calculation method for chi-square statistics is developed in an attempt to reduce the time this schema requires on BR.

Furthermore, in [16] they compare Principal Component Analysis (PCA) and Genetic Algorithms (GA) as a wrapper style. Naive Bayes is assigned as internal classifier and using transformed algorithms for multi-label classification like Adaboost and Rank-SVM. Although the base approach, Multi-Label Naive Bayes (MLNB) maintains better results than MLNB-PCA and MLNB-GA, however, these are competitive with a smaller amounts of features.

Hence, this work proposes a basic approach to select features for the BCC algorithm using a search of the subset of features in two different ways, Best First (BF) and GreedyStepwise (GS), while using CfsSubsetEval as the evaluator of the subset, these techniques are implemented within the BCC algorithm. BF, GS and Correlation-based Feature Selection [3] (CfsSubsetEval) are implemented in Weka [13].

3   Methodology and Development

Two searching methods are applied to find the subset of features to be used in the BCC algorithm (FS-BCC), the first one: with a BF technique (BF-FS-BCC). The second one: GreedyStepwise Technique (GS-FS-BCC), this approaches are tested with the Naive Bayes as the internal classifier.

3.1   Feature Selection in BCC

BCC algorithm [10] is modified with the intention to introduce a feature selection technique in this case BF or GS resulting on FS-BCC.

Given the structure and the chain sequence of the BCC algorithm, for each label a subset of features X is selected and a classifier is built. This can be summarize as following:

1. Obtain the undirected tree of the classes using Chow and Liu's algorithm.
2. Create a directed tree selecting randomly one class as a root node.
3. For each class C_j in the chain do:
   (a) Do until the worth of the subset doest not improve:
       i. Search with BF or GS the subset of features using CfsSubsetEval as heuristic.
       ii. Eval the worth of a subset with respect to accuracy.
   (b) Build a Navie Bayes classifiers for the class C_j with the selected features and its parent $Pa(C_j)$.
4. To determinate the subset of classes that correspond a new instance, the output of the each internal classifier is concatenated.

Figure 1 illustrates this approach using a Naive Bayes as internal classifier. On the left, original BCC algorithm building a classifier with all the features and its parent for each class $(C_{j1}, C_2..., C_5)$, on the right, FS-BCC build a classifier with a subset of the features for each class, hence each internal classifier can contain different features. It is important to mention that this approach continues using the parent in the tree as an additional feature.

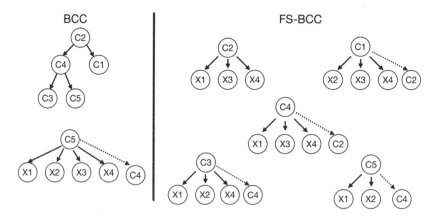

Fig. 1. Original BCC in the left and BCC with Feature Selection (FS-BCC) in the right.

FS-BCC has been implemented in Meka [7], a software for multi-label classification that includes algorithms implemented in Weka [13] like BF and GS, furthermore include CfsSubsetEval evaluator. BF and GS are used with the forward variant, this variant begins with a subset of one feature and continues adding other new feature until the worth of the subset stops improving[1].

3.2   Datasets

In Table 1 is possible to see number of labels (classes), instances, features and the domain of each dataset. It can be observed that the domains are Music, Image, Text, and Biology, the number of labels and features goes from 6 to 159 and 71 to 1836 respectively, on the other hand, the number of instances goes from 593 to 120,919. These datasets are available in http://mulan.sourceforge.net/datasets-mlc.html and http://waikato.github.io/meka/datasets/.

3.3   Evaluation Measures

Four evaluation measures are considered in this work, Multi-label Accuracy also called Accuracy, Macro Average (by example), Hamming Score (Mean Accuracy) and elapsed time, the last one with the intention to have more information when choosing BF-FS-BCC or GS-FS-BCC, these measures can be found in Meka [7].

Accuracy: Known as multi-label accuracy is given by the Eq. 1, where c_j is the real value of the label in the vector of subset of labels and c'_j is the predicted value for the classification model.

$$Accuracy = \frac{1}{N} \sum_{i=i}^{N} \frac{|c_i \wedge c'_i|}{|c_i \vee c'_i|} \tag{1}$$

---

[1] The implementation code is available in: https://github.com/R-Benitez-J/FS-BCC.

Table 1. Description of each dataset.

Num.	Dataset	Labels	Instances	Features	Domain
1	Music	6	593	71	Music
2	Scene	6	2,407	294	Image
3	Slashdot	22	3,782	1,079	Text
4	Yeast	14	2,417	1,03	Biology
5	20NG	20	19,300	1,006	Text
6	Enron	53	1,702	1,001	Text
7	LangLog	75	1,460	1,004	Text
8	Medical	45	978	1,449	Text
9	Ohsumed	23	13,929	1,002	Text
10	IMDB	28	120,919	1,001	Text
11	Bibtex	159	7,395	1,836	Text

Macro-Average (by example): Corresponds to a half of F1-Score but in this case evaluated by example.

$$MacroAverage = \frac{1}{2} \sum_{1}^{N} (2 * Precision * Recall/(Precision + Recall)) \quad (2)$$

Hamming Score (Mean Accuracy): Refers to the average of the accuracy for each label, where, c'_{ij} corresponds to the prediction given by the model and c_{ij} to the real value (Eq. 3).

$$HammingScore = \frac{1}{d} \sum_{j=1}^{d} \frac{1}{N} \sum_{i=1}^{N} \delta \left(c'_{ij}, c_{ij} \right) \quad (3)$$

Time: The time is presented in seconds and represents the average elapsed time per fold, in this case, all the experiments were tested with 10-Cross-Fold Validation.

Percentage of Selected Features: Indicates the percentage of the features that an internal classifier is built, in Eq. 4, N_{sf} corresponds to the selected features through FS-BCC and N_f to the original number of attributes.

$$\%SelectedFeatures = \frac{N_{sf} \times 100}{N_f} \quad (4)$$

4   Experiments and Results

In the same way BCC algorithm was proposed [10, 15], Naive Bayes is used as the internal classifier.

Table 2. Hamming Scores (Ham. Score) and Macro Average (Ma. Avg.) of BF-FS-BCC and GS-FS-BCC.

Num.	Dataset	BF-FS-BCC		GS-FS-BCC		Abs. Diff. (Rank)	
		Ham. Score	Ma. Avg.	Ham. Score	Ma. Avg.	Ham. Score	Ma. Avg.
1	Music	0.7473	**0.601**	**0.7482**	0.597	0.0009 (9)	0.004 (6)
2	Scene	0.8142	0.614	0.8142	0.614	0 (n/a)	0 (n/a)
3	Slashdot	0.944	**0.363**	**0.9443**	0.361	0.0003 (6)	0.002(3)
4	Yeast	**0.7473**	**0.572**	0.7472	0.571	0.0001 (1.5)	0.001 (1.5)
5	20NG	0.9471	0.494	**0.9474**	0.494	0.0003 (5)	0 (n/a)
6	Enron	0.9264	0.471	**0.9268**	0.471	0.0004 (7)	0 (n/a)
7	LangLog	0.9609	0.176	**0.961**	**0.179**	0.0001 (1.5)	0.003 (4.5)
8	Medical	0.9871	**0.772**	0.9871	0.767	0 (n/a)	0.005 (7)
9	Ohsumed	**0.9270**	0.478	0.9269	0.478	0.0001 (3)	0 (n/a)
10	IMDB	0.9181	**0.077**	**0.9183**	0.074	0.0002 (4)	0.003 (4.5)
11	Bibtex	0.9423	0.417	**0.9429**	**0.418**	0.0006 (8)	0.001(1.5)
	Avg.	0.8965	**0.4577**	**0.8968**	0.4567		

Table 3. Accuracy and Elapsed Time of BF-FS-BCC and GS-FS-BCC

Num.	Dataset	BF-FS-BCC		GS-FS-BCC		Abs. Diff. (Rank)	
		Accuracy	Time	Accuracy	Time	Accuracy	Time
1	Music	**0.508**	0.186	0.505	**0.137**	0.003 (6.5)	0.049 (1)
2	Scene	0.51	**19.444**	0.51	21.726	0 (n/a)	2.282 (3)
3	Slashdot	**0.324**	294.425	0.323	**68.729**	0.001 (2.5)	225.696(9)
4	Yeast	0.453	**1.657**	0.453	1.751	0 (n/a)	0.094(2)
5	20NG	0.44	**893.808**	0.441	961.932	0.001 (2.5)	68.124(7)
6	Enron	0.357	146.746	**0.358**	**133.658**	0.001 (2.5)	13.087(4)
7	LangLog	0.272	280.189	**0.275**	**256.435**	0.003 (6.5)	23.757(6)
8	Medical	**0.725**	41.987	0.721	**27.99**	0.004 (8)	13.997(5)
9	Ohsumed	0.393	661.65	0.393	**517.866**	0 (n/a)	143.784(8)
10	IMDB	**0.06**	5779.67	0.058	**5137.666**	0.002 (5)	642.004(10)
11	Bibtex	0.323	**16866.793**	0.324	18375.316	0.001 (2.5)	1508.523(11)
	Avg.	**0.3968**	**2271.505**	0.3965	2318.4731		

The experiments were development in four phases, in the first phase, BF-FS-BCC and GS-FS-BCC are compared looking for significance difference between them. In the second phase the winner of the previous phase was compared with the original BCC algorithm, in both cases (first and second phase) the method to compared the differences was Wilcoxon signed rank [2] with a significance level of 0.05. In the third phase FS-BCC and BCC are compared against Chain Classifiers approaches like CC, MCC and PCC (limited to 15 labels). At the end, in the fourth phase BCC and FS-BCC is compared against other approaches like LPS and RAkEL with k = 3 also used in [11].

Table 4. Average (Avg.) and standard deviation (STD) of Selected Features expressed in number and percentage.

Num.	Dataset	Avg. and STD of Selected Features	Avg. and STD. of Selected Features in %
1	Music	19.48 (±5.69)	3.29% (±0.96%)
2	Scene	69.43 (±22.45)	23.62% (±7.63%)
3	Slashdot	21.37 (±12.97)	1.98% (±1.2%)
4	Yeast	18.68 (±9.91)	18.13% (±0.51%)
5	20NG	21.76 (±0.89)	2.17% (±0.89%)
6	Enron	12.39 (±8.48)	1.24% (±0.85%)
7	LangLog	18.19 (±14.09)	1.81% (±1.4%)
8	Medical	8.94 (±6.43)	0.62% (±0.44%)
9	Ohsumed	20.98 (±3.05)	2.09% (±0.3%)
10	IMDB	13.127 (±5.223)	1.311% (±0.522%)
	Avg.	22.43 (±8.918)	5.626% (±1.47%)

Table 5. Hamming Score, Macro Average and Wilcoxon Test of BCC and BF-FS-BCC

Num.	Dataset	BCC		BF-FS-BCC		Abs. Diff. (Rank)	
		Ham. Score	Ma. Avg.	Ham. Score	Ma. Avg.	Ham. Score	Ma. Avg.
1	Music	**0.7513**	**0.636**	0.7473	0.601	0.004 (1)	0.035 (2)
2	Scene	0.7597	0.566	**0.8142**	**0.614**	0.0545 (7)	0.048 (4)
3	Slashdot	0.9306	**0.448**	**0.9442**	0.363	0.0134 (2)	0.085 (6)
4	Yeast	0.6982	0.54	**0.7473**	**0.572**	0.0491 (6)	0.032 (1)
5	20NG	0.8988	0.457	**0.9471**	**0.494**	0.0483 (5)	0.037 (3)
6	Enron	0.8038	0.34	**0.9264**	**0.471**	0.1226 (10)	0.131 (10)
7	LangLog	0.7252	0.078	**0.9609**	**0.176**	0.2357 (11)	0.098 (8)
8	Medical	0.967	0.636	**0.9871**	**0.772**	0.0201 (3)	0.136 (11)
9	Ohsumed	0.8575	0.399	**0.927**	**0.478**	0.0695 (8)	0.079 (5)
10	IMDB	0.8757	**0.194**	**0.9181**	0.077	0.0424 (4)	0.117 (9)
11	Bibtex	0.8597	0.319	**0.9423**	**0.417**	0.0826 (9)	0.098 (7)
	Avg.	0.8298	0.4194	**0.8965**	**0.4577**		

4.1 BF-FS-BCC Compared to GS-FS-BCC

In this phase, four Wilcoxon Signed Rank tests were developed, one for each evaluation measure and also the time, in Table 2, it is possible to observe the Wilcoxon Signed Rank test for Hamming Score and Macro Average, the first two columns correspond to BF-FS-BCC, the next two columns to GS-FS-BCC and the last two correspond to the absolute difference and rank. Table 3 presents Accuracy and Elapsed Time in the same form.

Table 6. Accuracy and Wilcoxon Test of BCC and BF-FS-BCC

Num.	Dataset	BCC	BF-FS-BCC	Abs. Diff. (Rank)
1	Music	**0.534**	0.508	0.026 (1)
2	Scene	0.453	**0.51**	0.057 (3)
3	Slashdot	**0.389**	0.324	0.065 (4)
4	Yeast	0.421	**0.453**	0.032 (2)
5	20NG	0.37	**0.44**	0.07 (5)
6	Enron	0.227	**0.357**	0.13 (10)
7	LangLog	0.183	**0.272**	0.089 (7)
8	Medical	0.537	**0.725**	0.188 (11)
9	Ohsumed	0.292	**0.393**	0.101 (8)
10	IMDB	**0.144**	0.06	0.084 (6)
11	Bibtex	0.21	**0.323**	0.113 (9)
	Avg.	0.3418	**0.3968**	0.055

Table 7. Hamming Score of BF-FS-BCC compared to other chain classifiers

Dataset	BCC	BF-FS-BCC	CC	MCC	PCC
Music	0.7513	0.7473	**0.7522**	0.7512	0.751
Scene	0.7597	**0.8142**	0.7630	0.7630	0.763
Slashdot	0.9306	**0.9440**	0.9357	0.9343	–
Yeast	0.6982	**0.7473**	0.6952	0.6944	–
20NG	0.8988	**0.9471**	0.9135	0.9134	–
Enron	0.8038	**0.9264**	0.7962	0.7960	–
LangLog	0.7252	**0.9609**	0.6723	0.6724	–
Medical	0.9670	**0.9871**	0.9680	0.9679	–
Ohusumed	0.8575	**0.9270**	0.8627	0.8624	–
Avg.	0.8213	**0.889**	0.8176	0.8172	–

Wilcoxon Signed Rank test indicates that there are no significant differences in Macro Average, Accuracy and elapsed time between GS-FS-BCC and BF-FS-BCC, only in Hamming Score GS-FS-BCC is significantly better than BF-FS-BCC, however in average BF-FS-BCC is better than GS-FS-BCC, hence BF-FS-BCC is chosen to be compared with BCC algorithm and other approaches.

Selected Features: Table 4 indicates the average (Avg.) of number and percentage of selected features for each dataset, including the corresponding standard deviation (STD). These values correspond to the BS-FS-BCC method. In summary, this method uses only the 5.63% of the total number of features with a STD of 1.47%.

Table 8. Accuracy of BF-FS-BCC compared to other chain classifiers

Dataset	BCC	BF-FS-BCC	CC	MCC	PCC
Music	**0.534**	0.508	0.532	0.532	0.531
Scene	0.453	**0.51**	0.458	0.458	0.458
Slashdot	0.389	0.324	0.422	**0.434**	–
Yeast	0.421	**0.453**	0.418	0.42	–
20NG	0.37	**0.44**	0.399	0.4	–
Enron	0.227	**0.357**	0.228	0.228	–
LangLog	0.183	**0.272**	0.18	0.18	–
Medical	0.537	**0.725**	0.549	0.547	–
Ohsumed	0.292	**0.393**	0.294	0.294	–
Avg.	0.3784	**0.4424**	0.3866	0.3881	–

Table 9. Macro Average of BF-FS-BCC compared to other chain classifiers

Dataset	BCC	BF-FS-BCC	CC	MCC	PCC
Music	**0.636**	0.601	0.634	0.633	0.633
Scene	0.566	**0.614**	0.571	0.571	0.571
Slashdot	0.448	0.363	0.48	**0.494**	–
Yeast	0.54	**0.572**	0.532	0.534	–
20NG	0.475	**0.494**	0.482	0.483	–
Enron	0.34	**0.471**	0.34	0.34	–
LangLog	0.078	**0.176**	0.071	0.071	–
Medical	0.636	**0.772**	0.643	0.642	–
Ohsumed	0.399	0.077	**0.401**	**0.401**	–
Avg.	0.4575	0.46	0.4615	**0.4632**	–

4.2 BCC Compared to BF-FS-BCC

Table 5, shows the Hamming Score and Macro Average of BCC and BF-FS-BCC, the Wilcoxon Signed rank indicates a significant difference in Hamming Score where BF-FS-BCC is the winner, on the other hand there is no difference in Macro Average and Accuracy (Table 6). Nevertheless, BF-FS-BCC is better than BCC at least on average.

4.3 BF-FS-BCC Compared to Others Chain Classifiers

This section presents the results of the comparative of BF-FS-BCC with CC, MCC and PCC multi-label classifiers, in these results only the nine first datasets are used given the required time to test the IMDB and Bibtex datasets.

Table 10. Hamming Score of BF-FS-BCC, BCC and other approaches

Num.	Dataset	BCC	BF-FS-BCC	BR	RAkEL	LPS
1	Music	0.7513	0.7473	0.7463	0.7025	**0.766**
2	Scene	0.7597	0.8142	0.7582	0.7908	**0.863**
3	Slashdot	0.9306	0.944	0.9318	0.93	**0.95**
4	Yeast	0.6982	0.7473	0.6983	0.6774	**0.759**
5	20NG	0.8988	0.9471	0.8971	0.9162	**0.964**
6	Enron	0.8038	**0.9264**	0.8037	0.8932	–
7	LangLog	0.7252	**0.9604**	0.7246	0.9251	–
8	Medical	0.967	**0.9871**	0.9671	0.9697	–
9	Ohsumed	0.8575	**0.927**	0.8577	0.8764	–
	Avg.	0.8213	**0.8889**	0.8205	0.8534	

Table 11. Accuracy of BF-FS-BCC, BCC and other approaches

Num.	Dataset	BCC	BF-FS-BCC	BR	RAkEL	LPS
1	Music	**0.534**	0.508	0.526	0.503	0.507
2	Scene	0.453	0.51	0.452	0.507	**0.615**
3	Slashdot	0.389	0.324	0.377	0.247	**0.51**
4	Yeast	0.421	0.453	0.422	0.426	**0.473**
5	20NG	0.37	0.44	0.367	0.357	**0.548**
6	Enron	0.227	**0.357**	0.227	0.046	–
7	LangLog	0.183	**0.272**	0.183	0.172	–
8	Medical	0.537	**0.725**	0.541	0.36	–
9	Ohsumed	0.292	**0.393**	0.293	0.208	–
	Avg.	0.3784	**0.4424**	0.3764	0.314	

In Table 7, Hamming Score is presented, Accuracy in Table 8 and Macro Average in Table 9, BF-FS-BCC is better on average in at least two measures, Accuracy and Hamming Score, in Macro Average, MCC is better but BF-FS-BCC is competitive.

4.4 BF-FS-BCC Compared to Others Approaches

BCC algorithm is highly competitive compared to other approaches (BR, RAkEL and LPS), hence given that BF-FS-BCC is better than BCC algorithm at least on average, this method is compared with these other approaches.

From Tables 10, 11 and 9 we can determine, BF-FS-BCC is highly competitive with other approaches, even outperforms some of them (BR and RAkEL) in some cases. LPS was not tested in all the dataset given the number of labels in the dataset (Table 12).

Table 12. Macro accuracy of BF-FS-BCC, BCC and other approaches

Num.	Dataset	BCC	BF-FS-BCC	BR	RAkEL	LPS
1	Music	**0.636**	0.601	0.629	0.626	0.595
2	Scene	0.566	0.614	0.567	0.621	**0.641**
3	Slashdot	0.448	0.363	0.435	0.288	**0.537**
4	Yeast	0.54	**0.572**	0.54	0.553	0.568
5	20NG	0.457	0.494	0.454	0.424	**0.652**
6	Enron	0.34	**0.471**	0.34	0.074	–
7	LangLog	0.078	**0.176**	0.078	0.051	–
8	Medical	0.636	**0.772**	0.638	0.429	–
9	Ohsumed	0.399	**0.478**	0.4	0.281	–
	Avg.	0.4555	**0.5045**	0.4534	0.3718	

5   Conclusions and Future Work

BF-FS-BCC was tested using 10-Cross-Fold validation and through these results, it is possible to determine that BF-FS-BCC is highly competitive compared to other Chain Classifier and even better in some cases than BR, RAkEl, using only in average 5.6% of the features. That is an important reduction of features, nevertheless, as a disadvantage, the time required to find the subset of features needs to be considered, although this elapsed time affects only in the model building phase.

As future work in this same approach, another internal classifier can be considered for more test. Also to compare BF-FS-BCC with transformed algorithms for multi-label classification like Adaboost and Rank-SVM, among others.

Other techniques of feature selection can apply with the intention to reduce the time of building model and improve the FS-BCC approach.

References

1. Dembczynski, K., Cheng, W., Hüllermeier, E.: Bayes optimal multilabel classification via probabilistic classifier chains. In: ICML, vol. 10, pp. 279–286 (2010)
2. Demšar, J.: Statistical comparisons of classifiers over multiple data sets. J. Mach. Learn. Res. **7**(Jan), 1–30 (2006)
3. Hall, M.A.: Correlation-based feature selection for machine learning (1999)
4. Lastra, G., Luaces, O., Quevedo, J.R., Bahamonde, A.: Graphical feature selection for multilabel classification tasks. In: Gama, J., Bradley, E., Hollmén, J. (eds.) IDA 2011. LNCS, vol. 7014, pp. 246–257. Springer, Heidelberg (2011). https://doi.org/10.1007/978-3-642-24800-9_24
5. Read, J., Martino, L., Luengo, D.: Efficient Monte Carlo optimization for multi-label classifier chains. In: 2013 IEEE International Conference on Acoustics, Speech and Signal Processing (ICASSP), pp. 3457–3461. IEEE (2013)

6. Read, J., Pfahringer, B., Holmes, G., Frank, E.: Classifier chains for multi-label classification. Mach. Learn. **85**(3), 333 (2011)
7. Read, J., Reutemann, P., Pfahringer, B., Holmes, G.: MEKA: a multi-label/multi-target extension to WEKA. J. Mach. Learn. Res. **17**(21), 1–5 (2016). http://jmlr.org/papers/v17/12-164.html
8. Spolaôr, N., Cherman, E.A., Monard, M.C., Lee, H.D.: Filter approach feature selection methods to support multi-label learning based on relieff and information gain. In: Barros, L.N., Finger, M., Pozo, A.T., Gimenénez-Lugo, G.A., Castilho, M. (eds.) SBIA 2012. LNCS (LNAI), pp. 72–81. Springer, Heidelberg (2012). https://doi.org/10.1007/978-3-642-34459-6_8
9. SpolaôR, N., Cherman, E.A., Monard, M.C., Lee, H.D.: A comparison of multi-label feature selection methods using the problem transformation approach. Electron. Notes Theor. Comput. Sci. **292**, 135–151 (2013)
10. Sucar, L.E., Bielza, C., Morales, E.F., Hernandez-Leal, P., Zaragoza, J.H., Larrañaga, P.: Multi-label classification with Bayesian network-based chain classifiers. Pattern Recogn. Lett. **41**, 14–22 (2014)
11. Tsoumakas, G., Katakis, I., Vlahavas, I.: Random k-labelsets for multilabel classification. IEEE Trans. Knowl. Data Eng. **23**(7), 1079–1089 (2011)
12. Tsoumakas, G., Katakis, I., et al.: Multi-label classification: an overview. Int. J. Data Warehous. Min. (IJDWM) **3**(3), 1–13 (2007)
13. Witten, I.H., Frank, E., Hall, M.A., Pal, C.J.: Data Mining: Practical Machine Learning Tools and Techniques. Morgan Kaufmann, Burlington (2016)
14. Xu, H., Xu, L.: Multi-label feature selection algorithm based on label pairwise ranking comparison transformation. In: 2017 International Joint Conference on Neural Networks (IJCNN), pp. 1210–1217. IEEE (2017)
15. Zaragoza, J.H., Sucar, L.E., Morales, E.F., Bielza, C., Larranaga, P.: Bayesian chain classifiers for multidimensional classification. IJCAI, vol. 11, pp. 2192–2197 (2011)
16. Zhang, M.L., Peña, J.M., Robles, V.: Feature selection for multi-label naive Bayes classification. Inf. Sci. **179**(19), 3218–3229 (2009)
17. Zhang, M.L., Zhou, Z.H.: Ml-KNN: a lazy learning approach to multi-label learning. Pattern Recogn. **40**(7), 2038–2048 (2007)

A Time Complexity Analysis to the ParDTLT Parallel Algorithm for Decision Tree Induction

Joel Suárez-Cansino[✉], Anilú Franco-Árcega, Linda Gladiola Flores-Flores, Virgilio López-Morales, and Ruslan Gabbasov

Intelligent Computing Research Group, Information and Systems Technologies Research Center, Engineering and Basic Sciences Institute, Autonomous University of the State of Hidalgo, 42184 Mineral de la Reforma, Hidalgo, Mexico
`jsuarez@uaeh.edu.mx`

Abstract. In addition to the usual tests for analyzing the performance of a decision tree in a classification process, the analysis of the amount of time and the space resource required are also useful during the supervised decision tree induction. The parallel algorithm called "Parallel Decision Tree for Large Datasets" (or ParDTLT for short) has proved to perform very well when large datasets become part of the training and classification process. The training phase processes in parallel the expansion of a node, considering only a subset of the whole set of training objects. The time complexity analysis proves a linear dependency on the cardinality of the complete set of training objects, and that the dependence is asymptotic and log–linear on the cardinality of the selected subset of training objects when categoric and numeric data are applied, respectively.

Keywords: Entropy · ParDTLT algorithm · Time complexity Synthetic data · Real data

1 Introduction

The decision–making process uses historical information to improve the performance of an institution, a business or an enterprise, which usually define places where the proper management of the corresponding data is invaluable. Data are classified accordingly to the values of their attributes, which can be numerical or categorical, in order of making predictions or discovering new knowledge. Prediction and discovering are two tasks, among many others that are also important and interesting, that institutions can do with their information. Data mining is related to the search for hidden information in a database and is the area that provides some techniques to help in the decision–making process [1]. Currently,

This research project has been supported by the Mexican Public Education Bureau (PRODEP).

I. Batyrshin et al. (Eds.): MICAI 2018, LNAI 11288, pp. 244–256, 2018.
https://doi.org/10.1007/978-3-030-04491-6_19

the decision tree induction is one of the most used techniques for solving this kind of problems.

In particular, given an object as an input to be classified, a decision tree is a graph with a series of pathways that the classifier follows to make a decision. Traditionally the graph is defined by internal and external nodes (leaf nodes), and every internal node connects to a subset of nodes and so on until a leaf node appears. Every node coming from a previous node is called a child node. Each internal node has associated a test attribute, which determines the path to be followed by the object along the induction process.

In addition to the existing common tests to analyze the performance of the decision tree in the classification process, tests that are usually oriented to describe the percentage of positive–false and false–positive classification results, it is also required to analyze the amount of time and space resources needed during the tree induction. In other words, during the training phase, just before the tree becomes a classifier. Test attributes support the training phase, and the training objects inside the node to expand are evaluated using an information–based method, such as the gain ratio criterion [2].

An experimental analysis of the quality of classification of the ParDTLT algorithm, along with the space requirements during the stage of the tree induction, has been made in a previous work [3], and this paper is aimed to discuss the time complexity of the same parallel algorithm during the construction of the decision tree. The algorithm ParDTLT uses parallel computing techniques, specifically in the node splitting process.

The use of relatively small subsets of instances of training objects, to make the node expansion during the training phase, and the distribution of the attributes among the available processors are the main characteristics of the algorithm. Additionally, an experimental setting supports the complexity analysis of the classical and deterministic decision tree training, making the possibility of comparing the results with some theoretic measures of complexity in the future [7]. Furthermore, the reported results do not consider categoric and numerical data [8,9] as this work does.

1.1   Paper Structure

This paper makes an experimental and time complexity analysis of the different phases involved in the parallel induction of a decision tree when the ParDTLT algorithm is applied. This analysis suggests the bounds or an estimate of the time complexity for the parallel algorithm. The incremental processing of the training objects is the main idea of the algorithm, and this procedure helps to reduce the use of the primary memory. At the same time, this idea decreases the time complexity of the algorithm.

The rest of the paper organizes the sections in the following way: Sect. 2 describes the complete scenario of the formal description to use in the analysis. On the other hand, Sect. 3 describes the analysis for obtaining the time complexity related to the ParDTLT algorithm. Finally, Sect. 4 presents the results of some experiments that have been conducted to compare them with the obtained analysis, along with the conclusions of this work.

2   Algorithm ParDTLT

A set $\mathcal{T}$ of N_t training objects defines the context of the problem. The objects in the set $\mathcal{T}$ initially follow an order, in such a way that two consecutive of them do not belong to the same class, with the possible exception of those located at the very final end of the set. A set $\mathcal{A}$ of N_a attributes characterizes to all the objects in $\mathcal{T}$, and a given tuple defines an object. Furthermore, the attribute A_i, $A_i \in \mathcal{A}$, can have N_i different values, which can be categorical or numeric.

At the same time, every object in the list belongs to one of N_c classes previously specified. Therefore, the required ingredients to carry on the supervised induction are defined; namely, the set $\mathcal{T}$ of training patterns and the desired classification given by the set $\mathcal{C}$ of N_c classes.

Parallel computing techniques support the decision tree construction, where a number N_p of processors are available. Given a node of the decision tree, the processing of the N_a attributes is distributed inside the N_p processors. In other words, the technique, to be explained later on, sequentially processes every node of the decision tree, and the N_p processors share the required computation on the whole set of attributes.

If the attribute value is categorical, then the criterion of node expansion is supported by the discrete property of this kind of attribute; namely, the number of different values is not difficult to compute. This number is then used to calculate the entropy for every particular value of the attribute and its contribution to the total entropy.

Things are a little bit different when the attribute to process is a numeric one, with a dense set of different values. Because of their continuous nature, it is impossible to compute the total amount of different values for this kind of attribute, but there does exist a very well known methodology to overcome this problem [2]. The method described by the authors in the cited reference applies the concept of cut point, which is used to compute an entropy, for every particular value of the cut points, and their contributions to the total entropy.

The knowledge of the entropic contributions for every attribute value to the total entropy makes possible to decide the next division of the current node, because of this division is based on the best information gain provided by the entropic contributions.

Some of the authors of this paper give a detailed explanation of the working of the algorithm in a previous publication [3]; however, Fig. 1 briefly describes the expansion process containing the parallel processing when numerical data are involved.

The method previously described can be applied to the original large dataset of training objects $\mathcal{T}$, but the situation requires the existence of sufficient amount of primary memory and an unacceptable time of processing because of the frequent access to secondary memory. To overcome both difficulties, the proposed

Some available processor analyses an attribute.

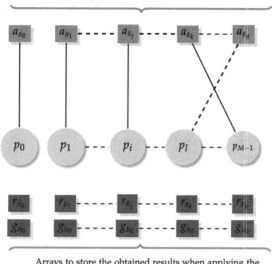

Arrays to store the obtained results when applying the
gain ratio criterion to each attribute.

Fig. 1. Paralelization of the numerical attribute's analysis.

method in this paper assumes that every node in the tree accepts, at most, a number N of training objects during the processing of node's division [3].

3   Time Complexity Analysis

The Algorithm 1 briefly describes the complete process for the tree induction. Three steps in the algorithm contribute substantially to the time complexity; namely, the step related to nodes creation (line number 2) and the steps to computing the entropy (lines number 6 and 9).

The first case decides the depth of the tree, while the second and third cases depend on the type of attribute, which can be numerical or categorical. The computation of the total entropy considers the training objects with the complete set of attributes, $\mathcal{A}$, and the result measures how much order the given classes introduced in the object's distribution.

Similarly, the calculation of the partial entropy uses subsets of the N training objects, where the different values of the attribute assigned to the processor are used to construct these subsets. The object's distribution is constrained to the attribute selection. However, the algorithm to compute the entropy is almost the same as the algorithm for the computation of the total entropy.

The Algorithm 2 shows how calculations are done. The algorithm requires a set of training objects, in addition to the classes where the objects in this set belong. In the case of the total entropy, the N training objects stored in

Algorithm 1: Brief description of the ParDTLT algorithm.
Input: An ordered set of training objects, $\mathcal{T}$, a set $\mathcal{C}$ of classes where the training objects belong to, maximum number of objects to be processed per current node, N.
Output: A decision tree, D, which is trained in a supervised form based on the set of training objects, $\mathcal{T}$, and the set of classes, $\mathcal{C}$.

ParDTLT$(\mathcal{T}, \mathcal{C}, N)$

```
(1)  repeat
(2)       Create nodes of the decision tree
(3)       repeat
(4)            Feed training objects to nodes already created
(5)       until one node has N training objects (current node)
(6)       Compute total entropy of N objects in current node
(7)       Distribute tasks among the available processors based on the attributes
(8)       in parallel for each available processor
(9)            Compute partial entropy per attribute or cut point value
(10)           Store gain ratio per attribute
(11)           Select attribute with best gain ratio
(12)           Node division based on the the values of the selected attribute
(13) until all the training objects in T have been fed
(14) return D
```

Algorithm 2: Entropy computation, to calculate gain ratio later on.
$\mathcal{O}(Entropy) = |C||S|$.
Input: A set S of training objects, the classes where every training object belongs to, C.
Output: The entropy, E, of the training set, S.

Entropy(S, C)

```
(1)  E ← 0
(2)  for i = 1 to |C|
(3)       counter ← 0
(4)       for j = 1 to |S|
(5)            if C_j = i − 1
(6)                 counter ← counter + 1
(7)            endif
(8)       endfor
(9)       p ← counter/|S|
(10)      if 0 < counter
(11)           E ← E − p · log_2(p)
(12)      endif
(13) endfor
(14) return E
```

the current node and the set of classes defined by $\mathcal{C}$ represent the input to the algorithm. Therefore, the equation $\mathcal{O}(N, N_c) = N_c N$ gives the big $\mathcal{O}$ order in time complexity for this case.

The case of the calculation of the partial entropy must consider one of two possibilities of the attribute's values, which can be categorical or numerical. Let assume that finite enumerable sets, with numerical or alphanumerical elements, define the categorical attributes, and that non–enumerable sets, with numeric elements, define numerical attributes.

If the attribute's values are categorical, then the computation of the partial entropy is direct. The contribution of every possible attribute's value defines the partial entropy. Let assume that A_i is an attribute with values that are categorical and that N_i represents the number of different values for this attribute. Furthermore, let assume that $N_{i,1}$ training objects are related to the first value, $N_{i,2}$ training objects with the second value, and so on until N_{i,N_i} training objects with the N_ith value. If so, the following equation must be satisfied in the current node

$$N = \sum_{m=1}^{N_i} N_{i,m}$$

Next, the $N_{i,m}$ training objects with the complete set of classes, $\mathcal{C}$, are sent to the entropy function, which produces a big $\mathcal{O}$ time complexity order equal to $\mathcal{O}(N_{i,m}, N_c) = N_c N_{i,m}$. However, the search for the set of N_i different values implies the addition of a time complexity $\mathcal{O}(N) = N$, which means that the equation

$$\mathcal{O}(N_c, N_{i,m}) = N + N_c N_{i,m}$$

gives the time complexity for the computation of the partial entropy of just one of the N_i different values of the attribute A_i (the search of a single subset included).

This process is repeated N_i times and the final big $\mathcal{O}$ contribution is given by the following equation

$$\sum_{m=1}^{N_i} (N + N_c N_{i,m}) = N N_i + N_c N$$
$$= \mathcal{O}(N, N_c, N_i)$$

Hence, the final time complexity for the computation of the partial entropy related with the attribute A_i is given by the equation

$$\mathcal{O}(N_c, N, N_i) = N_i N + N_c N$$

Since every categorical attribute contributes with a given number of different values, the total parallel contribution to the partial entropy of the whole set of categorical attributes is given as follows,

$$\mathcal{O}(N_c, N, N_i) = N_{max} N + N_c N$$

where $N_{max} = max\{N_1, N_2, \ldots\}$, and N_i is the number of different values of the categorical attribute A_i.

Let assume that the attribute A_i has N_i different numeric values. From the total number of different numeric values, only N_{i1} are equal in value, N_{i2} have also another equal value, and so on until N_{iN_i} objects with another different value.

In this case, the training considers the attribute's values provided by the N_{ij} and $N_{i,j+1}$ objects (two different numeric values, in fact) to produce a cut point. This process is repeated so many times as the total amount of different values of the corresponding numeric attribute.

Since there are N_i different numeric values, then the expression $N_i - 1$ provides the number of different cut points. Next, the two sets of numerical values, above and below the corresponding point, are used to compute the contribution to the partial entropy. Assume that the jth cut point is analyzed, then the expression

$$\sum_{m=1}^{j} N_{i,m}$$

provides the total number of different objects above, while the expression

$$\sum_{m=j+1}^{N_i} N_{i,m}$$

gives the total number of different objects below.

The contribution to the time complexity of the partial entropy computation in this case is just given by the following expression

$$N_c \sum_{m=1}^{j} N_{i,m} + N_c \sum_{m=j+1}^{N_i} N_{i,m} = N_c N$$

Table 1. Task distribution, total entropy computation, partial entropy computation, sorting and tree induction are the main big $\mathcal{O}$ time complexity contribution in the algorithm ParDTLT

Comment	Data type	
	Categoric	Numeric
Task distribution	$\frac{N_a}{N_p}$	$\frac{N_a}{N_p}$
Total entropy	$N_c N$	$N_c N$
Partial entropy	$N_{max} N + N_c N$	$2(N_{max} - 1)N + (N_{max} - 1)N_c N$
Sort		$N \ln N + N$
Depth	$\frac{N_t}{N}$	$\frac{N_t}{N}$
	Contribution to time complexity	

This procedure needs to be made for every cut point and, since there are $N_i - 1$ cut points, the total contribution is then given by the expression $(N_i - 1)N_cN$. On the other hand, since every subset above and below the cut point requires to compute the corresponding subset of training objects, then the total contribution to the partial entropy is given by the equation

$$\mathcal{O}(N_i, N_c, N) = 2(N_i - 1)N + (N_i - 1)N_cN$$

The Table 1 shows the main contributions to the time complexity in the algorithm ParDTLT. The computation of the partial entropy depends on the type of data and, when numerical data are involved, then this calculation requires to sort the corresponding set of objects.

3.1   Complexity of the Algorithm

Table 1 shows that the time complexity of the algorithm ParDTLT depends on several important variables, which include the total number of training objects, N_t. If the rest of the variables are defined by some convenient constant values, then the time complexity linearly depends on the variable N_t. This statement is true even for mixed categorical and numeric data.

On the other hand, the denominator N in the expression coming from the contribution of the depth of the tree assumes the worst case, where every node in the tree has only one child node. In fact, one node in the tree can have no children at all, one child or more than one child, which means that the depth of the tree can be less than $\frac{N_t}{N}$.

The depth of the tree should be expected to depend inversely on N, while for N sufficiently large and approximately equal to N_t, the contribution of the depth becomes a constant equal to the unity $\left(\frac{N_t}{N} \approx 1\right)$. For a given and fixed N_t, the expression $\frac{aN+b}{cN+d}N_t$ models this behavior, where the variable N in the total and partial entropy is included into the numerator.

This expression converges to a fraction of the total number of training objects, N_t, and this fraction plays the role of a lower asymptote, since $1 < \frac{N_t}{N}$. Furthermore, the value of the asymptote should be expected to be in a direct proportion with N_t. This analysis corresponds to categorical data only, although the numerical data analysis is almost similar, with the exception of including the sorting contribution to the time complexity, as well.

On the other hand, the maximum number, N_{max}, of numerical values in a given numerical attribute should be expected to be directly proportional to N, which means that the expression $a \ln N + bN + c$ models the time complexity contribution quite well, where the variables N_a, N_p, N_c, N_t are assumed to be defined by some convenient constant values.

4   Results and Conclusions

Several experimental runs were realized to test the previous time complexity analysis. On the basis of the proper selection of large datasets containing an acceptable number of categorical and numeric attributes, the model's prediction compares to the experimental result. The Table 2 shows the main characteristics of the selected datasets.

Table 2. Datasets used in the experiments

Dataset	Agrawal	GalStar	Numeric	Categoric
Dataset type	Synthetic	Real	Synthetic	Synthetic
No. of categorical attributes	3	0	0	50
No. of numerical attributes	6	30	50	0
No. of classes	2	2	2	2
Size	6,000,000	4,000,000	3,500,000	3,500,000

In particular, the experimentation considers the synthetic datasets named Agrawal [4], Numeric and Categoric, which the Weka system created [5]. Additionally, the experiment also includes the real dataset GalStar that is given by the SDSS. The dataset contains several attributes oriented to the classification of stars and galaxies [6].

4.1   Time Complexity Behavior

First of all, the attention is focused into the behavior of the time complexity as a function of the size N_t of the whole set of training objects, $\mathcal{T}$. Agrawal provides the training objects, and they are a mixture of categorical and numeric attributes.

The variables N_a, N_p, N_c, N_{max} and N are fixed during the experimental data acquisition (execution time and size of $\mathcal{T}$). The experimental results show that this behavior is linear in N_t and the Fig. 2 graphically describes how a linear function fits into the experimental data.

The same figure shows the 95% confidence interval for the computed fitting parameters, and three different normality tests (one sample t-test statistic, z-test statistic and the chi-square goodness of fit test statistics) do not reject the null hypothesis, which states that the "residual data come from a normal distribution with mean zero and standard deviation approximated by the data". The reliability being at the 5% level (p-value). The difference between the experimental and theoretical data (as given by the fitting function) defines the residual data. Thus, the time complexity model acceptably predicts the time complexity behavior of the algorithm applied to experimental data.

If numeric attributes participate, then similar results appear. In this case, GalStar provides the training objects. The variables N_a, N_p, N_c, N_{max} and N are

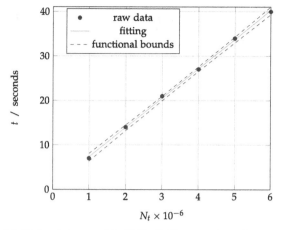

(a) Dataset Agrawal in Table 2. Six experiments were made with the variable N_t running from 1,000,000 training objects to 6,000,000.

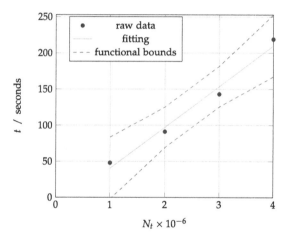

(b) Dataset GalStar in Table 2. Four experiments were made with the variable N_t running from 1,000,000 training objects to 4,000,000.

Fig. 2. Time versus size, N_t, of training set of objects $\mathcal{T}$.

fixed during the development of the experimentation. The experimental results show that the behavior is linear in N_t and the Fig. 2b graphically describes how a linear function fits into the experimental data.

The same figure shows the 95% confidence interval for the computed fitting parameters and the same three different normality tests do not reject the same null hypothesis at the 5% level (p–value). Hence, the experimental results confirm the prediction of the behavior of the time complexity as given by the previous analysis.

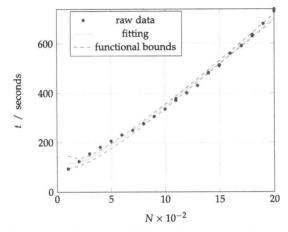

(a) Dataset Numeric in Table 2. Twenty experiments were made with the variable N (the number of training objects in the current node) running from 100 training objects to 2,000.

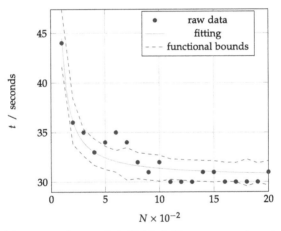

(b) Dataset Categoric in Table 2. Twenty experiments were made with the variable N (the number of training objects in the current node) running from 100 training objects to 2,000.

Fig. 3. Time versus size, N, of training set of objects in current node.

The third scenario considers numeric attributes as well, although the experiment looks for confirming the relationship between the time complexity and the number of objects to be processed in the current node, as given by the variable N. The process assigns constant values to the other variables. Figure 3 describes how a log–linear function fits into the experimental data. The dataset is Numeric in Table 2.

The same figure shows the 95% confidence interval for the computed fitting parameters and the same three different normality tests do not reject the same null hypothesis at the 5% level (p–value). Hence, the experimental results confirm the prediction of the behavior of the time complexity as given by the previous analysis.

Finally, the fourth experimental scenario considers only categorical attributes from the dataset Categoric in Table 2. The experiment looks for validating the prediction of the time complexity analysis given in a previous section, as a function of the variable N. The process assigns constant values to the rest of the variables, as well.

In this case, a rational function fits quite well into the experimental data and Fig. 3b shows the 95% confidence interval for the values of the fitting parameters. A normality test shows that the residuals follow, at the 5% level as well, a normal distribution with zero mean and with a standard distribution approximated by the experimental data.

The behavior of the algorithm when the number of processors increases needs an experimental analysis, as well. Thus, the experimental procedure for this analysis considers the dataset Agrawal and GalStar, being both the largest ones of the set of datasets used in this paper. Figure 4 shows the experimental results that say that the algorithm's execution time decreases as the number of processors increases. Furthermore, there is not an appreciable change in the execution time above some number of processors.

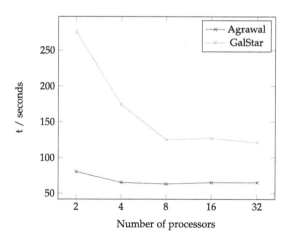

Fig. 4. Experimental scalability analysis of the algorithm ParDTLT.

4.2   Conclusions

The algorithm ParDTLT efficiently induces a decision tree and does not require the existence of extraordinary resources of primary and secondary memories.

Furthermore, the time complexity of the algorithm is at most linear in the number of training objects, N, that the current node in fact processes.

The time complexity behavior for categorical and numeric data is almost the same, with the exception being numerical data involving a log–linear component coming from the requirement of a previous sorting of training objects in the current node, thus implying, when compared with the processing of categorical data, a much bigger time execution.

The execution time can be drastically reduced as a function of the inverse of the number of processors, N_p, although for a fixed and sufficiently large N_t, the time complexity can be also reduced, above some lower limit that depends on N_t, with an increasing N, as shown in the Fig. 3b. The impossibility of accurately predicting the real depth of the final decision tree, which directly affects the time complexity, explains the unpredictable variation in the residuals.

Therefore, in general terms, the experimental results confirm the conclusions of the previous time complexity analysis, making a strong emphasis on the fact that numerical data requires much more execution time than categorical data.

References

1. Dunham, M.H.: Data Mining, Introductory and Advanced Topics. 1st edn. Alan R. Apt (2003)
2. Ross Quinlan, J.: C4.5: Programs for Machine Learning. Morgan Kaufmann, Burlington (1993)
3. Franco-Árcega, A., Suárez-Cansino, J., Flores-Flores, L.: A parallel algorithm to induce decision trees for large datasets. In: XXIV International Conference on Information, Communication and Automation Technologies (ICAT 2013), Sarajevo, Bosnia and Herzegovina, 30 October–01 November 2013. IEEE Xplore, Digital Library (2013)
4. Agrawal, R., Imielinski, T., Swami, A.: Database mining: a performance perspective. IEEE Trans. Knowl. Data Eng. **5**, 914–925 (1993)
5. Hall, M., Frank, E., Holmes, G., Pfahringer, B., Reutemann, P., Witten, I.H.: The WEKA data mining software: an update. SIGKDD Explor. Newsl. **11**(1), 10–18 (2009)
6. Adelman-McCarthy, J., Agueros, M.A., Allam, S.S.: The sixth data release of the sloan digital sky survey. ApJS **175**(2), 297 (2008)
7. Buhrman, H., de Wolf, R.: Complexity measures and decision tree complexity: a survey. Theor. Comput. Sci. **288**, 21–43 (2002)
8. Kent Martin, J., Hirschberg, D.S.: On the complexity of learning decision trees. In: Proceedings of the 4th International Symposium on Artificial Intelligence and Mathematics (AI/MATH96), pp. 112–115 (1996)
9. Moshkov, M.J.: Time complexity of decision trees. In: Peters, J.F., Skowron, A. (eds.) Transactions on Rough Sets III. LNCS, vol. 3400, pp. 244–459. Springer, Heidelberg (2005). https://doi.org/10.1007/11427834_12

Infrequent Item-to-Item Recommendation via Invariant Random Fields

Bálint Daróczy[1]([envelope]), Frederick Ayala-Gómez[2], and András Benczúr[1]

[1] Institute for Computer Science and Control,
Hungarian Academy of Sciences (MTA SZTAKI), Budapest 1111, Hungary
daroczyb@ilab.sztaki.hu, benczur@sztaki.mta.hu
[2] Faculty of Informatics, Eötvös Loránd University,
Pázmóny P. sny. 1/C., Budapest 1117, Hungary
fayala@caesar.elte.hu

Abstract. Web recommendation services bear great importance in e-commerce and social media, as they aid the user in navigating through the items that are most relevant to her needs. In a typical web site, long history of previous activities or purchases by the user is rarely available. Hence in most cases, recommenders propose items that are similar to the most recent ones viewed in the current user session. The corresponding task is called session based item-to-item recommendation. Generating item-to-item recommendations by "people who viewed this, also viewed" lists works fine for popular items. These recommender systems rely on item-to-item similarities and item-to-item transitions for building next-item recommendations. However, the performance of these methods deteriorates for rare (i.e., infrequent) items with short transaction history. Another difficulty is the cold-start problem, items that recently appeared and had no time yet to accumulate a sufficient number of transactions. In this paper, we describe a probabilistic similarity model based on Random Fields to approximate item-to-item transition probabilities. We give a generative model for the item interactions based on arbitrary distance measures over the items including explicit, implicit ratings and external metadata. We reach significant gains in particular for recommending items that follow rare items. Our experiments on various publicly available data sets show that our new model outperforms both simple similarity baseline methods and recent item-to-item recommenders, under several different performance metrics.

Keywords: Recommender systems · Fisher information
Markov random fields

1 Introduction

Recommender systems [23] have become common in a variety of areas including movies, music, videos, news, books, and products in general. They produce a list of recommended items by either collaborative or content based filtering.

© Springer Nature Switzerland AG 2018
I. Batyrshin et al. (Eds.): MICAI 2018, LNAI 11288, pp. 257–275, 2018.
https://doi.org/10.1007/978-3-030-04491-6_20

Collaborative filtering methods [17,24,28] build models of the past user-item interactions, while content based filtering [18] typically generates lists of similar items based on item properties. To assess the attitude towards the items viewed by the user, recommender systems rely on users explicit feedback (e.g., ratings, like/dislike) or implicit feedback (e.g., clicks, plays, views).

The Netflix Prize Challenge [2,15] revolutionized our knowledge of recommender systems but biased research towards the case where user profiles and item ratings (1–5 stars) are known. However, for most Web applications, users are reluctant to create logins and prefer to browse anonymously. Or, we purchase certain types of goods (e.g., expensive electronics) so rarely that our previous purchases will be insufficient to create a meaningful user profile. Several practitioners [14] argue that most of the recommendation tasks they face are implicit feedback and without sufficient user history. In [20] the authors claim that 99% of the recommendations systems they built for industrial application tasks are implicit, and most of them are item-to-item. For these cases, recommender systems rely on the recent items viewed by the user in the actual shopping session.

In this paper, we consider the task of recommending relevant items to a user based on items seen during the current session [17,24] rather than on user profiles. Best known example of this task is the Amazon list of books related to the last visited one [17]. An intuitive approach to building the list of relevant items to recommend is to consider item pair frequencies. However, for rare items, it is necessary to use global similarity data to avoid recommendations based on low support. In addition, we have to devise techniques that handle new items well. In the so-called cold start case [25], the new items have yet insufficient number of interactions to reliably model their relation to the users.

Our key idea is to utilize the known, recent or popular items for item-to-item recommendation via multiple representations. The starting point of our method is the idea of [14], to utilize the entire training data and not just the item-item conditional probabilities. Our item-to-item model is able to use single or combined similarity measures such as Jaccard or cosine based on collaborative, content, multimedia and metadata information.

We evaluate the top-n recommendation [5] performance of our models by Recall and DCG. We present our proposed approach in Sects. 3, 4, and 6. The experimental results are presented in Sect. 7.

2   Related Work

Recommender systems are surveyed in [23]. Several recommender systems consider a setting similar to the Netflix Prize Competition [2], where users and their explicit feedback (1–5 stars) are given, and the task is to predict unseen ratings. In this paper, we consider cases where users do not give explicit ratings, and we have to infer their preferences from their implicit feedback [14]. And, we assume that a rich user history is not available, so we rely on the present items of the user's session.

The first item-to-item recommender methods [17,24] used similarity information to find nearest neighbor transactions [6]. Another solution is to extract association rules [4]. Both classes of these methods deteriorate if the last item of the session has a low item transition support (e.g., rare or recent items). Nearest neighbor methods were criticized for two reasons. First, the similarity metrics typically have no mathematical justification. Second, the confidence of the similarity values is often not involved when finding the nearest neighbor, which leads to overfitting in sparse cases. In [16], a method is given that learns similarity weights for users, however the method gives global and not session based user recommendation.

Rendle et al. [21] proposed a session-based recommender system that models the users by factorizing personal Markov chains. Their method is orthogonal to ours in that they provide more accurate user based models if more data is available, while we concentrate on extracting actionable knowledge from the entire data for the sparse transactions in a session.

The item-to-item recommendation can be considered a particular context-aware recommendation problem. In [9] sequentiality as a context is handled by using pairwise associations as features in an alternating least squares model. They mention that they face the sparsity problem in setting minimum support, confidence, and lift of the associations, and they used the category of last purchased item as a fallback. In a follow-up result [10], they use the same context-aware ALS algorithm. However, they only consider seasonality as a context in that paper.

Closest to our work is the *Euclidean Item Recommender* (EIR) [14] by Koenigstein and Koren. They model item-to-item transitions using item latent factors where the Euclidean distance between two vectors approximates the known transition probabilities in the training dataset. Our model differs in that we do not need to optimize a vector space to learn the transition probabilities in a lower dimensional space. Instead, we start from an arbitrary similarity definition, and we may extend similarity for all items, by using all training data, in a mathematically justified way. We use Fisher information, that is applied for DNA splice site classification [11] and computer vision [19], but we are the first to apply it in recommender systems. We applied – to the most extent reproducible – the experimental settings of EIR.

3 Similarity Graph

The starting point of our item-to-item recommender model is a set of arbitrary item pair similarity measures, which may be based on implicit or explicit user feedback, user independent metadata such as text description, linkage or even multimedia content. By the pairwise similarity values and potentially other model parameters θ, we model item i as a random variable $p(i|\theta)$. From $p(i|\theta)$, we will infer the distance and the conditional probability of pairs of items i and j by using all information in θ.

Formally, let us consider a certain sample of items $S = \{i_1, i_2, \ldots, i_N\}$ (e.g., most popular or recent items), and assume that we can compute the distance of any item i from each of $i_n \in S$. We will consider our current item i along with its distance from each $i_n \in S$ as a random variable generated by a Markov Random Field (MRF). Random fields are a set of (dependent) random variables. In case of MRF the connection between the elements is described by an undirected graph satisfying the Markov property [3]. For example, the simplest Markov Random Field can be obtained by using a graph with edges between item i and items $i_n \in S$, as shown in Fig. 1.

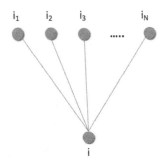

Fig. 1. Similarity graph of item i with sample items $S = \{i_1, i_2, ..., i_N\}$ of distances $\text{dist}(i, i_n)$ from i.

Let us assume that we are given a Markov Random Field generative model for $p(i|\theta)$. By the Hammersley-Clifford theorem [8], the distribution of $p(i|\theta)$ is a Gibbs distribution, which can be factorized over the maximal cliques and expressed by a potential function U over the maximal cliques as follows:

$$p(i \mid \theta) = e^{-U(i|\theta)}/Z(\theta), \tag{1}$$

where $U(i \mid \theta)$ is the energy function and

$$Z(\theta) = \sum_i e^{-U(i|\theta)}$$

is the sum of the exponent of the energy function over our generative model, a normalization term called the partition function. If the model parameters are previously determined, then $Z(\theta)$ is a constant.

Given a Markov Random Field defined by a certain graph such as the one in Fig. 1 (or some more complex graph defined later), a wide variety of proper energy functions can be used to define a Gibbs distribution. The weak but necessary restrictions are that the energy function has to be positive real valued, additive over the maximal cliques of the graph, and more probable parameter configurations have to have lower energy.

Given a finite sample set $S = \{i_1, .., i_N\}$, we define the simplest similarity graph as seen in Fig. 1 by describing the energy function for (1) as

$$U(i \mid \theta = \{\alpha_1, .., \alpha_N\}) := \sum_{n=1}^{N} \alpha_n \text{dist}(i, i_n), \tag{2}$$

where dist is an arbitrary distance or divergence function of item pairs and the hyperparameter set θ is the weight of the elements in the sample set.

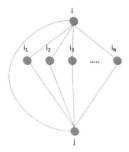

Fig. 2. Pairwise similarity graph with sample set $S = \{i_1, i_2, ..., i_N\}$ for a pair of items i and j.

In a more complex model, we capture the connection between pairs of items by extending the generative graph model with an additional node for the previous item as shown in Fig. 2. In the pairwise similarity graph, the maximal clique size increases to three. To capture the joint energy with parameters $\theta = \{\beta_n\}$, we can use a heuristic approximation similar to the pseudo-likelihood method [3]: we approximate the joint distribution of each size three clique as the sum of the individual edges by

$$U(i, j \mid \theta) := \sum_{n=1}^{N} \beta_n(\mathrm{dist}(i, i_n) + \mathrm{dist}(j, i_n) + \mathrm{dist}(i, j)), \qquad (3)$$

At first glance, the additive approximation seems to oversimplify the clique potential and falls back to the form of Eq. (2). However, the effect of the clique is apparently captured by the common clique hyperparameter β_n, as also confirmed by our experiments.

4   Fisher Information

Items with low support usually cannot be captured by traditional similarity models. To handle the similarity of rare items, in this section we introduce the Fisher information to estimate distinguishing properties by using the similarity graphs.

Let us consider a general parametric class of probability models $p(i|\theta)$, where $\theta \in \Theta \subseteq \mathbb{R}^\ell$. The collection of models with parameters from a general hyperparameter space Θ can then be viewed as a (statistical) manifold M_Θ, provided that the dependence of the potential on Θ is sufficiently smooth. By [13], M_Θ can be turned into a Riemann manifold by giving an inner product (kernel) at the tangent space of each point $p(i|\theta) \in M_\Theta$, where the inner product varies smoothly with p.

The notion of the inner product over $p(i|\theta)$ allows us to define the so-called Fisher metric on M. The fundamental result of Čencov [26] states that the Fisher metric exhibits a unique invariance property under some maps which are quite natural in the context of probability. Thus, one can view the use of Fisher kernel as an attempt to introduce a natural comparison of the items on the basis of the generative model [11].

We start defining the Fisher kernel over the manifold M_Θ of probabilities $p(i|\theta)$ as in Eq. (1) by considering the tangent space. The tangent vector

$$G_i = \nabla_\theta \log p(i|\theta) = \left(\frac{\partial}{\partial \theta_1} \log p(i|\theta), \dots, \frac{\partial}{\partial \theta_l} \log p(i|\theta) \right) \tag{4}$$

is called the *Fisher score* of item i. The *Fisher information matrix* is a positive semidefinite matrix defined as

$$F(\theta) := \mathbf{E}_\theta (\nabla_\theta \log p(i|\theta) \nabla_\theta \log p(i|\theta)^T), \tag{5}$$

where the expectation is taken over $p(i|\theta)$. In particular, the nm-th entry of $F(\theta)$ is

$$F_{nm} = \sum_i p(i|\theta) \left(\frac{\partial}{\partial \theta_n} \log p(i|\theta) \right) \left(\frac{\partial}{\partial \theta_m} \log p(i|\theta) \right).$$

Thus, to capture the generative process, the gradient space of M_Θ is used to derive the Fisher vector, a mathematically grounded feature representation of item i. The corresponding kernel function

$$K(i,j) := G_i^T F^{-1} G_j \tag{6}$$

is called the *Fisher kernel*. An intuitive interpretation is that G_i gives the direction where the parameter vector θ should be changed to fit item i the best [19]. In addition, we prove a theorem for our kernels on a crucial reparametrization invariance property that typically holds for Fisher kernels [22].

Theorem 1. *For all $\theta = \rho(\mu)$ for a continuously differentiable function ρ, K_θ is identical.*

Proof. The Fisher score is

$$G_i(\mu) = G_i(\rho(\mu)) \left(\frac{\partial \rho}{\partial \mu} \right)$$

and therefore

$$K_\mu(i,j) = G_i(\mu) F_\mu^{-1} G_j(\mu)$$

$$= G_i(\rho(\mu)) \left(\frac{\partial \rho}{\partial \mu} \right) \left(F_{\rho(\mu)} \left(\frac{\partial \rho}{\partial \mu} \right)^2 \right)^{-1} G_j(\rho(\mu)) \left(\frac{\partial \rho}{\partial \mu} \right)$$

$$= G_i(\rho(\mu)) F_{\rho(\mu)}^{-1} G_j(\rho(\mu)) = K_\rho(i,j). \qquad \square$$

Essentially, the theorem states that the kernel will not depend on the hyperparameters θ.

5   Item Similarity by Fisher Information

Based on the similarity graphs introduced in Sect. 3 and by taking advantage of the invariance properties of the Fisher metric, we propose two ranking methods for item-item transitions.

5.1   Item-Item Fisher Conditional Score (FC)

Our first item-to-item recommender method will involve similarity information in the item-item transition conditional probability computation by using Fisher scores as in Eq. (4). By the Bayes theorem,

$$G_{j|i} = \nabla_\theta \log p(j \mid i; \theta) = \nabla_\theta \log \frac{p(i, j \mid \theta)}{p(i \mid \theta)}$$

$$= \nabla_\theta \log p(i, j \mid \theta) - \nabla_\theta \log p(i \mid \theta), \tag{7}$$

thus we need to determine the joint and the marginal distributions for a particular item pair.

First, let us calculate the Fisher score of (4) with $p(i|\theta)$ of the single item generative model defined by (2),

$$G_i^k(\theta) = \nabla_{\theta_k} \log p(i|\theta)$$

$$= \frac{1}{Z(\theta)} \sum_i e^{-U(i|\theta)} \frac{\partial U(i \mid \theta)}{\partial \theta_k} - \frac{\partial U(i \mid \theta)}{\partial \theta_k}$$

$$= \sum_i \frac{e^{-U(i|\theta)}}{Z(\theta)} \frac{\partial U(i \mid \theta)}{\partial \theta_k} - \frac{\partial U(i \mid \theta)}{\partial \theta_k}.$$

By (1), our formula can be simplified as

$$G_i^k(\theta) = \sum_i p(i \mid \theta) \frac{\partial U(i \mid \theta)}{\partial \theta_k} - \frac{\partial U(i \mid \theta)}{\partial \theta_k}$$

$$= \mathbf{E}_\theta \big[\frac{\partial (U(i|\theta))}{\partial \theta_k} \big] - \frac{\partial U(i|\theta)}{\partial \theta_k}. \tag{8}$$

For an energy function as in Eq. (2), the Fisher score of i has a simple form,

$$G_i^k(\theta) = \mathbf{E}_\theta[\text{dist}(i, i_k)] - \text{dist}(i, i_k), \tag{9}$$

and similarly for Eq. (3),

$$G_{ij}^k(\theta) = \mathbf{E}_\theta[\text{dist}(i, i_k) + \text{dist}(j, i_k) + \text{dist}(i, j)]$$

$$- (\text{dist}(i, i_k) + \text{dist}(j, i_k) + \text{dist}(i, j)). \tag{10}$$

Now, if we put (9) and (10) into (7), several terms cancel out and the Fisher score becomes

$$G_{j|i}^k = \mathbf{E}_\theta[\text{dist}(j, i_k) + \text{dist}(i, j)] - (\text{dist}(j, i_k) + \text{dist}(i, j)).$$

The above formula involves the distance values on the right side, which are readily available, and the expected values on the left side, which may be estimated by using the training data. We note that here we make a heuristic approximation: instead of computing the expected values (e.g., by simulation), we substitute the mean of the distances from the training data.

As we discussed previously, the Fisher score resembles how well the model can fit the data, thus we can recommend the best fitting next item j^* based on the norm of the Fisher score,

$$j^* = \arg\min_{j \neq i} \|G_{j|i}(\theta)\|,$$

where we will use ℓ_2 for norm in our experiments.

5.2 Item-Item Fisher Distance (FD)

In our second model, we rank the next item by its distance from the last one, based on the Fisher metric. With the Fisher kernel $K(i, j)$, the *Fisher distance* can be formulated as

$$\text{dist}_F(i, j) = \sqrt{K(i, i) - 2K(i, j) + K(j, j)}, \tag{11}$$

thus we need to compute the Fisher kernel over our generative model as in (6). The computational complexity of the Fisher information matrix estimated on the training set is $\mathcal{O}(T|\theta|^2)$, where T is the size of the training set. To reduce the complexity to $\mathcal{O}(T|\theta|)$, we can approximate the Fisher information matrix with the diagonal as suggested in [11,19]. Hence we will only use the diagonal of the Fisher information matrix,

$$F_{k,k} = \mathbf{E}_\theta[\nabla_{\theta_k} \log p(i|\theta)^T \nabla_{\theta_k} \log p(i|\theta)]$$

$$= \mathbf{E}_\theta[(\mathbf{E}_\theta[\frac{\partial U(i \mid \theta)}{\partial \theta_k}] - \frac{\partial (U(i \mid \theta)}{\partial \theta_k})^2].$$

For the energy functions of Eqs. (2) and (3), the diagonal of the Fisher kernel is the standard deviation of the distances from the samples. We give the Fisher vector of i for (2):

$$\mathcal{G}_i^k = F^{-\frac{1}{2}} G_i^k \approx F_{kk}^{-\frac{1}{2}} G_i^k$$

$$= \frac{\mathbf{E}_\theta[\text{dist}(i, i_k)] - \text{dist}(i, i_k)}{\mathbf{E}_\theta^{\frac{1}{2}}[(\mathbf{E}_\theta[\text{dist}(i, i_k)] - \text{dist}(i, i_k))^2]}.$$

The final kernel function is

$$K(i, j) = G_i^T F^{-1} G_j \approx G_i^T F_{diag}^{-1} G_j$$

$$= G_i^T F_{diag}^{-\frac{1}{2}} F_{diag}^{-\frac{1}{2}} G_j = \sum_k \mathcal{G}_i^k \mathcal{G}_j^k.$$

By substituting into (11), the recommended next item after item i will be

$$j^* = \arg\min_{j \neq i} \text{dist}_F(i, j).$$

5.3   Multimodal Fisher Score and Distance

So far we considered only a single distance or divergence measure over the items. We may expand the model with additional distances with a simple modification to the graph of Fig. 1. We expand the points of the original graph into new points $R_i = \{r_{i,1}, .., r_{i,|R|}\}$ corresponding to R representatives for each item i_n in Fig. 3. There will be an edge between two item representations $r_{i,\ell}$ and $r_{j,k}$ if they are the same type of representation ($\ell = k$) and the two item was connected in the original graph. This transformation does not affect the maximal clique size and therefore the energy function is a simple addition, as

$$U(i \mid \theta) = \sum_{n=1}^{N} \sum_{r=1}^{|R|} \alpha_{nr} \mathrm{dist}_r(i_r, i_{nr}), \tag{12}$$

and if we expand the joint similarity graph to a multimodal graph, the energy function will be

$$U(i, j \mid \theta) = \sum_{n=1}^{N} \sum_{r=1}^{|R|} \beta_{nr} (\mathrm{dist}_r(i_r, i_{nr})$$
$$+ \mathrm{dist}_r(j_r, i_{nr}) + \mathrm{dist}_r(i_r, j_r)). \tag{13}$$

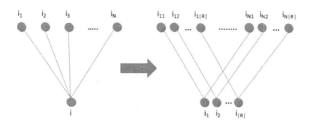

Fig. 3. The single and multimodal similarity graph with sample set $S = \{i_1, i_2, ..., i_N\}$ and $|R|$ modalities.

Now, let the Fisher score for any distance measure $r \in R$ be G_{ir}, than the Fisher score for the multimodal graph is concatenation of the unimodal Fisher scores as

$$G_i^{multi} = \{G_{i1}, .., G_{i|R|}\},$$

and therefore the norm of the multimodal Fisher score is a simple sum over the norms:

$$\|G_i^{multi}\| = \sum_{r=1}^{|R|} \|G_{ir}\|. \tag{14}$$

The calculation is similar for the Fisher kernel of Eq. (12), thus the multimodal kernel can be expressed as

$$K_{multi}(i,j) = \sum_{r=1}^{|R|} K_r(i,j). \tag{15}$$

6 Similarity Measures

Next we enumerate distance and divergence measures that can be used in the energy functions (2) and (3). Without using the Fisher information machinery, these measures yield the natural baseline methods for item-to-item recommendation. We list both implicit feedback collaborative filtering and content based measures.

6.1 Feedback Similarity

For user implicit feedback on item pairs, various joint and conditional distribution measures can be defined based on the frequency f_i and f_{ij} of items i and item pairs i, j, as follows.

1. Cosine similarity (Cos):
$$cos(i,j) = \frac{f_{ij}}{\sqrt{f_i f_j}}.$$

2. Jaccard similarity (JC):
$$JC(i,j) = \frac{f_{ij}}{f_i + f_j - f_{ij}}.$$

3. Empirical Conditional Probability (ECP): estimates the item transition probability:
$$ECP(j|i) = \frac{f_{ij}}{f_i + 1},$$
   where the value 1 is a smoothing constant.

Additionally, in [14] the authors suggested a model, the Euclidean Item Recommender (EIR) to approximate the transition probabilities with the following conditional probability

$$p(j|i) = \frac{\exp^{-||x_i - x_j||^2 + b_j}}{\sum \exp^{-||x_i - x_k||^2 + b_k}},$$

where they learn the item latent vector x_i and bias b_i.

All of the above measures can be used in the energy function as the distance measure after small modifications.

Now, let us assume that our similarity graph (Fig. 1) has only one sample element i and the conditional item is also i. The Fisher kernel will be,

$$K(i,j) = \frac{1}{\sigma_i^2}(\mu_i - \text{dist}(i,i))(\mu_i - \text{dist}(i,j))$$

$$= \frac{\mu_i^2}{\sigma_i^2} - \frac{\mu_i}{\sigma_i^2}\text{dist}(i,j))$$

$$= C_1 - C_2 * \text{dist}(i,j),$$

where μ_i and σ_i are the expected value and variance of distance from item i. Therefore if we fix θ, C_1 and C_2 are positive constants and the minimum of the Fisher distance will be

$$\min_{j \neq i} \text{dist}_F(i,j) = \min_{j \neq i} \sqrt{K(i,i) - 2K(i,j) + K(j,j)}$$

$$= \min_{j \neq i} \sqrt{2C_2 * \text{dist}(i,j)} = \min_{j \neq i} \text{dist}(i,j).$$

Hence if we measure the distance over the latent factors of EIR, the recommended items will be the same as defined by EIR, see Eq. (10) in [14].

6.2 Content Similarity

Besides item transitions, one can measure the similarity of the items based on their content (e.g., metadata, text, title). The content similarity between two items is usually measured by the cosine, Jaccard, tf-idf, or the Jensen-Shannon divergence of the "bag of words".

7 Experiments

We performed experiments on four publicly available data sets. As baseline methods, we computed four item-item similarity measures: Empirical Conditional Probability (ECP), Cosine (Cos), Jaccard (JC) as defined in Sect. 6, and we also implemented the Euclidean Item Recommender of [14]. As content similarity, we mapped the movies in the MovieLens dataset to DBpedia [1] [1]. DBpedia represents Wikipedia as a graph. For instance, a movie in DBPedia is represented as a node connected by labeled edges to other nodes such as directors, actors or genre. We compute the Jaccard similarity between two items using the nodes connected to the movies as a "bag of words". For evaluation, we use Recall, and Discounted Cumulative Gain (DCG) [12].

We conducted experiments by adding 200 sampled items to the testing item to evaluate recommendations. That is, given the current item in a session i and a known co-occurrence j we add randomly 200 items and rank them based on the score of the models. The best models should preserve j on the top of the sorted list.

---

[1] http://wiki.dbpedia.org.

Table 1. Co-occurrence quartiles

Dataset	25%	50%	75%	Max
Books	1	1	2	1,931
Yahoo! Music	4	9	23	160,514
MovieLens	29	107	300	2,941
Netflix	56	217	1,241	144,817

7.1 Data Sets and Experimental Settings

We carried out experiments over four data sets: Netflix [2], MovieLens[2], Ziegler's Books [27] and Yahoo! Music [7].

To generate item transitions by creating pairs from the items consumed by the users in the data set. For example, if a user consumed items a, b and c we create three co-occurrence pairs. That is, $[(a, b), (b, c), (c, a)]$. We do this for all the users and then we calculate the frequency of each pair. Figure 4 and Table 1 shows that most of the co-occurrence in the datasets are infrequent. 75% of the pairs have low item support. Since our research is focused on infrequent items, we filtered out the items with high support. The maximum co-occurrence frequency that we considered for the data sets in our experiments are 2 for Books, 23 for Yahoo! Music, 300 for MovieLens and 1241 for Netflix.

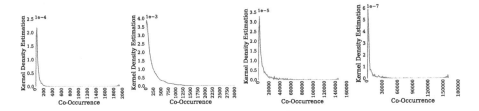

Fig. 4. The Kernel Density Estimation of the item co-occurrence saturates in items that are infrequent. From left to right and top to bottom: Books, MovieLens, Netflix, Yahoo! Music.

To generate the training and testing set we place most of the users in a training set and the rest of the users in the testing set. Then, we generate the pairs as described before. The number of training and testing pairs and the properties of the data sets can be seen in Table 2. During testing the evaluation was performed over a sampled set of 200 items as in [14] for all three metrics and solved ties arbitrary.

In our experiments, all algorithms use the item frequencies of the training period as input parameters. However, it could be possible to keep the cur-

---

[2] http://grouplens.org/datasets/movielens/.

Table 2. Data sets used in the experiments.

Data set	Items	Users	Training pairs	Testing pairs
Netflix	17,749	478,488	7,082,109	127,756
MovieLens	3,683	6,040	670,220	15,425
Yahoo! Music	433,903	497,881	27,629,731	351,344
Books	340,536	103,723	1,017,118	37,403

rent frequencies up to date and recalculate the prediction of each algorithm on the fly.

7.2 Experimental Results

This section presents different experiments related to the size of the sample set, the modalities used (e.g., implicit, content), the performance on infrequent items and finally the overall performance. As acronyms FC stands for Fisher conditional score from Sect. 5.1 followed by similarity, FD for Fisher distance from Sect. 5.2 followed by similarity. In case of multimodal the model use both content and collaborative similarity values.

Sample Set. The similarity graphs are defined via the set of items used as samples (Figs. 1, 2 and 3). To smooth the Fisher vector representation of sparse items we choose the most popular items in the training set as elements for the sample set. As we can see in Figs. 5 and 6, recommendation quality saturates at a certain

Table 3. Experiments with combination of collaborative filtering for the least frequent (25%) conditional items of the MovieLens data.

	Recall@20	DCG@20
Cosine	0.0988	0.0553
Jaccard	0.0988	0.0547
ECP	0.0940	0.0601
EIR	0.1291	0.0344
FC Cosine	0.1020	0.0505
FD Cosine	0.1578	0.0860
FC Jaccard	0.1770	**0.1031**
FD Jaccard	**0.1866**	0.1010
FC ECP	0.0940	0.0444
FD ECP	0.1626	0.0856
FC EIR	0.0861	0.0434
FD EIR	0.1068	0.0560

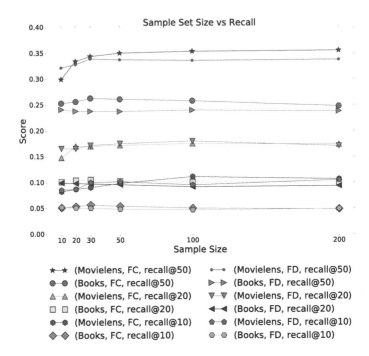

Fig. 5. The sample set size improves Recall until it saturates at 50. In this example we use Jaccard similarity.

Table 4. Experiments on MovieLens with DBPedia content, all methods using Jaccard similarity.

	Recall@20	DCG@20
Collaborative baseline	0.139	0.057
Content baseline	0.131	0.056
FC content	0.239	0.108
FD content	0.214	0.093
FC multimodal	0.275	0.123

sample set size. Therefore we set the size of the sample set to 20 for the remaining experiments.

Performance of Similarity Functions. Another relevant parameter of the similarity graphs is the choice of the similarity functions. Table 3 presents the performance of the different similarity functions. Overall, Jaccard similarity is the best performing and we used it for the rest of our experiments.

Performance on Infrequent Items. One of the main challenges in the field of recommendation systems is the "cold start" problem, therefore we examine the per-

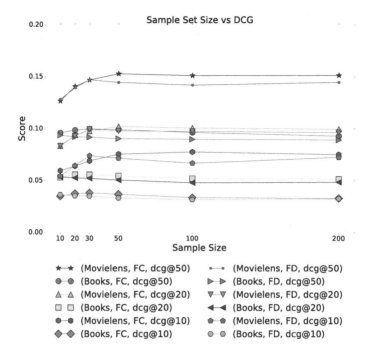

Fig. 6. The sample set size improves DCG until it saturates at 50. In this example we use Jaccard similarity.

formance in case of low item support. Figure 7 shows the advantage of the Fisher methods for infrequent items. As support increases, best results are reached by blending based on item support. If the current session ends with an item of high support, we may take a robust baseline recommender. And if the support is lower, less than around 100, Fisher models can be used to compile the recommendation.

Modalities: Implicit Feedback and Content. In Table 4 we show our experiments with DBPedia content as a modality on MovieLens. For simplicity, we set the size of the sample set for both Fisher models to 10. The overall best performing model is the multimodal Fisher with Jaccard similarity, while every unimodal Fisher method outperform the baselines. By using Eq. (15), we could blend different modalities such as content and feedback without the need of setting external parameters or applying learning for blending.

Summary of Performance vs Baselines. Tables 3, 4 and 5 present our implicit feedback results. The choice of the distance function strongly affects the performance of the Fisher models. As seen in Table 3, the overall best performing distance measure is Jaccard for both types of Fisher models. The results in Table 5 show that the linear combination of the standard normalized scores of the Fisher

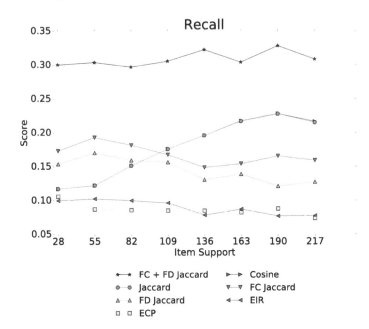

Fig. 7. Recall@20 as the function of item support for the Netflix data set.

methods outperforms the best unimodal methods (Fisher with Jaccard) for Netflix and Books, while for MovieLens and Yahoo! Music, Fisher distance with Jaccard performs best.

8   Discussion and Future Work

Recommending infrequent item-to-item transitions without personalized user history is a challenging problem. We consider our results for simple, non-personalized item-to-item recommendation as the first step towards demonstrating the power of the method. As a key feature, the model can fuse different modalities including collaborative filtering, content, and side information, without the need for learning weight parameters or using wrapper methods. In the near future, we plan to extend our methods to personalized recommendation settings and refine the underlying similarity measures with complex models (e.g., neural networks [28]). The publicly available datasets we used limited our experiments. Datasets containing a session id, item, and timestamp are scarce. Because of this, future work could be to experiment with real sessions, especially within a short period (e.g., news recommendation). Also, we constrained our similarity graphs for simple item-to-item transitions, defining the next item in the "random walk" depending only on the last seen item. To find out the limitation of this hypothesis we intend to expand the generative model to utilize the previous items in a session.

Table 5. Summary of experiments results for the four datasets. The *Max freq* are defined in the frequency quartile (Table 1). For most methods, there are (up to rounding errors) two best baseline and two best Fisher models, except for Recall where a third method Cosine appears in the cell marked by a star (*). The best methods are usually FD Jaccard and FC + FD.

	Best baseline & new method	Max freq	MovieLens	Books	Yahoo! Music	Netflix
Recall@20	Jaccard	25%				0.13
		50%				0.18
		75%	0.12*			0.20
	EIR	25%	0.12	0.10	0.13	
		50%	0.11	0.10	0.11	
		75%		0.10	0.12	
	FD Jaccard	25%	**0.18**		**0.23**	
		50%	**0.19**		**0.23**	
		75%	**0.14**		**0.20**	
	FC + FD	25%		**0.14**		**0.30**
		50%		**0.14**		**0.30**
		75%		**0.13**		**0.31**
DCG@20	ECP	25%	0.05			
		50%	0.05			
		75%	0.05			
	EIR	25%		0.06	0.05	0.12
		50%		0.06	0.05	0.12
		75%		0.06	0.05	0.12
	FD Jaccard	25%	**0.10**		**0.11**	
		50%	**0.11**		**0.11**	
		75%	**0.08**		**0.10**	
	FC + FD	25%		**0.08**		**0.17**
		50%		**0.08**		**0.17**
		75%		**0.08**		**0.17**

9 Conclusions

In this paper, we considered the session based item-to-item recommendation task, in which the recommender system has no personalized knowledge of the user beyond the last items visited in the current user session. We proposed Fisher information based global item-item similarity models for this task. We reached significant improvement over existing methods in case of infrequent item-to-item transitions by experimenting with a variety of data sets as well as evaluation metrics.

Acknowledgments. The publication was supported by the Hungarian Government project 2018-1.2.1-NKP-00008: Exploring the Mathematical Foundations of Artificial Intelligence and by the Momentum Grant of the Hungarian Academy of Sciences. F.A. was supported by the Mexican Postgraduate Scholarship of the Mexican National Council for Science and Technology (CONACYT). B.D. was supported by 2018-1.2.1-NKP-00008: Exploring the Mathematical Foundations of Artificial Intelligence.

References

1. Auer, S., Bizer, C., Kobilarov, G., Lehmann, J., Cyganiak, R., Ives, Z.: DBpedia: a nucleus for a web of open data. In: Aberer, K., et al. (eds.) ASWC/ISWC -2007. LNCS, vol. 4825, pp. 722–735. Springer, Heidelberg (2007). https://doi.org/10.1007/978-3-540-76298-0_52
2. Bennett, J., Lanning, S.: The netflix prize. In: Proceedings of KDD Cup and Workshop (2007)
3. Besag, J.: Statistical analysis of non-lattice data. Statistician **24**(3), 179–195 (1975)
4. Davidson, J., et al.: The YouTube video recommendation system. In: Proceedings of the Fourth ACM RecSys, pp. 293–296 (2010)
5. Deshpande, M., Karypis, G.: Item-based top-n recommendation algorithms. ACM Trans. Inf. Syst. (TOIS) **22**(1), 143–177 (2004)
6. Desrosiers, C., Karypis, G.: A comprehensive survey of neighborhood-based recommendation methods. In: Ricci, F., Rokach, L., Shapira, B., Kantor, P.B. (eds.) Recommender Systems Handbook, pp. 107–144. Springer, Boston, MA (2011). https://doi.org/10.1007/978-0-387-85820-3_4
7. Dror, G., Koenigstein, N., Koren, Y., Weimer, M.: The Yahoo! music dataset and KDD-Cup'11. In: KDD Cup, pp. 8–18 (2012)
8. Hammersley, J.M., Clifford, P.: Markov fields on finite graphs and lattices. Seminar (1971, unpublished)
9. Hidasi, B., Tikk, D.: Fast ALS-based tensor factorization for context-aware recommendation from implicit feedback. In: Flach, P.A., De Bie, T., Cristianini, N. (eds.) ECML PKDD 2012. LNCS (LNAI), vol. 7524, pp. 67–82. Springer, Heidelberg (2012). https://doi.org/10.1007/978-3-642-33486-3_5
10. Hidasi, B., Tikk, D.: Context-aware item-to-item recommendation within the factorization framework. In: Proceedings of the 3rd Workshop on Context-awareness in Retrieval and Recommendation, pp. 19–25. ACM (2013)
11. Jaakkola, T.S., Haussler, D.: Exploiting generative models in discriminative classifiers. In: Advances in Neural Information Processing Systems, pp. 487–493 (1999)
12. Järvelin, K., Kekäläinen, J.: Cumulated gain-based evaluation of IR techniques. ACM Trans. Inf. Syst. (TOIS) **20**(4), 422–446 (2002)
13. Jost, J.: Riemannian Geometry and Geometric Analysis. Springer, Heidelberg (2011). https://doi.org/10.1007/978-3-642-21298-7
14. Koenigstein, N., Koren, Y.: Towards scalable and accurate item-oriented recommendations. In: Proceedings of the 7th ACM RecSys, pp. 419–422. ACM (2013)
15. Koren, Y.: The bellkor solution to the netflix grand prize. Netflix Prize Documentation **81**, 1–10 (2009)
16. Koren, Y.: Factor in the neighbors: scalable and accurate collaborative filtering. ACM Trans. Knowl. Disc. Data (TKDD) **4**(1), 1 (2010)
17. Linden, G., Smith, B., York, J.: Amazon.com recommendations: item-to-item collaborative filtering. IEEE Internet Comput. **7**(1), 76–80 (2003)

18. Lops, P., de Gemmis, M., Semeraro, G.: Content-based recommender systems: state of the art and trends. In: Ricci, F., Rokach, L., Shapira, B., Kantor, P.B. (eds.) Recommender Systems Handbook, pp. 73–105. Springer, Boston, MA (2011). https://doi.org/10.1007/978-0-387-85820-3_3
19. Perronnin, F., Dance, C.: Fisher kernels on visual vocabularies for image categorization. In: IEEE CVPR 2007 (2007)
20. Pilászy, I., Serény, A., Dózsa, G., Hidasi, B., Sári, A., Gub, J.: Neighbor methods vs. matrix factorization - case studies of real-life recommendations. In: ACM RecSys 2015 LSRS (2015)
21. Rendle, S., Freudenthaler, C., Schmidt-Thieme, L.: Factorizing personalized Markov chains for next-basket recommendation. In: Proceedings of the 19th International Conference on WWW, pp. 811–820. ACM (2010)
22. Janke, W., Johnston, D., Kenna, R.: Information geometry and phase transitions. Physica A: Stat. Mech. Appl. **336**(1), 181–186 (2004)
23. Ricci, F., Rokach, L., Shapira, B.: Introduction to Recommender Systems Handbook. Springer, Boston (2011). https://doi.org/10.1007/978-0-387-85820-3_1
24. Sarwar, B., Karypis, G., Konstan, J., Reidl, J.: Item-based collaborative filtering recommendation algorithms. In: Proceedings of the 10th International Conference on WWW, pp. 285–295 (2001)
25. Schein, A.I., Popescul, A. Ungar, L.H. Pennock, D.M.: Methods and metrics for cold-start recommendations. In: Proceedings of the 25th ACM SIGIR, pp. 253–260. ACM (2002)
26. Čencov, N.N.: Statistical Decision Rules and Optimal Inference, vol. 53. American Mathematical Society (1982)
27. Ziegler, C.-N., McNee, S.M., Konstan, J.A., Lausen, G.: Improving recommendation lists through topic diversification. In: Proceedings of the 14th International Conference on WWW, pp. 22–32. ACM (2005)
28. Wang, H., Yeung, D.-Y.: Towards bayesian deep learning: a framework and some existing methods. IEEE Trans. Knowl. Data Eng. **28**(12), 3395–3408 (2016)

An Approach Based on Contrast Patterns for Bot Detection on Web Log Files

Octavio Loyola-González[1(✉)], Raúl Monroy[2], Miguel Angel Medina-Pérez[2], Bárbara Cervantes[2], and José Ernesto Grimaldo-Tijerina[3]

[1] School of Science and Engineering, Tecnologico de Monterrey,
Vía Atlixcáyotl No. 2301, Reserva Territorial Atlixcáyotl,
72453 Puebla, Mexico
octavioloyola@itesm.mx
[2] School of Science and Engineering, Tecnologico de Monterrey,
Carretera al Lago de Guadalupe Km. 3.5,
52926 Atizapán, Estado de México, Mexico
{raulm,migue,bcervantesg}@itesm.mx
[3] Network Information Center Mexico, Tecnologico de Monterrey,
Avenida Eugenio Garza Sada 427 L4-6,
64840 Monterrey, Nuevo León, Mexico
jgrimaldo@nic.mx

Abstract. Nowadays, companies invest resources in detecting non-human accesses on their web traffics. Usually, non-human accesses are a few compared with the human accesses, which is considered as a class imbalance problem, and as a consequence, classifiers bias their classification results toward the human accesses obviating, in this way, the non-human accesses. In some classification problems, such as the non-human traffic detection, high accuracy is not only the desired quality, the model provided by the classifier should be understood by experts. For that, in this paper, we study the use of contrast pattern-based classifiers for building an understandable and accurate model for detecting non-human traffic on web log files. Our experiments over five databases show that the contrast pattern-based approach obtains significantly better AUC results than other state-of-the-art classifiers.

Keywords: Bot detection · Contrast pattern
Supervised classification

1 Introduction

Nowadays, there is a large number of free Internet services, such as web search engines, social networks, and webmails. Nevertheless, maintaining an Internet service is usually very expensive, because it must be available 24/7. Also, Internet services subsist due to the revenue generated through Internet advertising. Online ads services is an industry that generates over $22 billion USD a year [16].

© Springer Nature Switzerland AG 2018
I. Batyrshin et al. (Eds.): MICAI 2018, LNAI 11288, pp. 276–285, 2018.
https://doi.org/10.1007/978-3-030-04491-6_21

Companies use internet ads in order to obtain a positive outcome (getting additional sales, positioning brand image, etc.). Internet ads services often provide a payment model, where a customer bill is proportional to the volume of internet traffic driven to the customer web site. For example, the pay-per-click model is a function of the number of clicks operated over a banner or a link, associated to a customer account. Payment models of this type can be easily misused, for an adversary may use an automated tool to artificially increase the click rate on an ad, with the intend of either increasing the profit of the associated Internet ad service provider, or draining the advertising budget of a target victim (possibly a competitor). This situation has given rise to the bot detection problem, where the chief task is to timely detect whether an internet activity originates from a human, or an automated tool [15].

Usually, bot detection has to deal with the class imbalance problem, where objects (in our case, instances of web visitor behavior) are not equally distributed among classes (in our case, bot or human). This yields a bias in the classification results, which tend to favor the majority class. Also, the most interesting class (the bot class), is frequently the one that contains significantly fewer objects [13].

There are several classifiers designed to deal with class imbalance [13]. However, most classification mechanisms produce a model that cannot be understood by experts in the application domain [18]. Developing understandable classifiers is a desired property [1,12,14]. Indeed, for some applications, this is even mandatory. For example, the Equal Credit Opportunity Act of the US considers illegal the use of indefinite or vague reasons in the denial of a credit application; hence, some institutions should only use classifiers with a explanatory model [14]

Contrast pattern-based classifiers are an important family of both understandable and accurate classifiers. They have been used on a number of real-world applications, such as characterization for subtypes of leukemia, prediction of a heart disease, gene transfer and microarray concordance analysis, classification of spatial and image data, and gene expression profiles [13]. Contrast pattern-based classification play a major role in bot detection, for there is a need to explicitly characterize bot behavior, as well as for explaining actions taken against an offending account. This, however, has been hitherto gone unnoticed.

In this paper we present a study of the use of contrast pattern-based classification for bot detection on web log files. In our experiments, we have used five datasets, three of which we have borrowed from the literature, and the rest we have collected from a real e-commerce site. We shall show that a pattern-based classification approach is up to obtain a significantly larger classification results. The main contribution of this paper is, hence, the use of a contrast pattern-based classifier to bot detection; this classifier is able to create an accurate and understandable model that help experts approach web server bot access.

The rest of the paper has the following structure. Section 2 provides a brief review about bot detection, creating emphasis on click-fraud detection. Section 3 contains a brief introduction to the contrast pattern-based approach and a review of the most prominent contrast pattern-based algorithms. Section 4 presents our study about bot detection using supervised classifiers based and not based on patterns over five databases, including the experimental setup and a brief interpretation, issued by both marketing and TI experts, of the obtained pattern-based model. Finally, Sect. 5 provides conclusions and future work.

2   Bot Detection

According to the Interactive Advertising Bureau trade group about 36% of all web traffic is considered fake [10]. In fact, a report issued by the digital security firm White Ops and the Association of National Advertisers estimates that advertisers could have lost $6.3 billion USD in 2015 due to click-fraud automated tools (a.k.a bots), which artificially increase the number of visits to a website [10]. Hence, bot traffic detection has become an important task for industry and academia.

One approach to bot detection consists of analyzing every web log file registered by a server after a web access [15–17]. Usually, after a user clicks on an ad or a link of a website, the server, using a background process, records relevant information about the user, such as user IP, visit time, visited URL, number of bytes consumed in that access, and so on. Commonly, servers use the Extended Log File Format provided by the World Wide Web Consortium [8], which includes relevant features regarding web access.

Other approach for bot detection combines web log files and some features captured during the session of a website visitor, such as the number of clicks and keystrokes [10,15,17]. Nonetheless, this approach is difficult to deploy because people are reluctant to be monitored by script codes that are a potential threat to their privacy. These kinds of tools, in particular key loggers, have been used for collecting and then communicating sensitive information from a computer.

There exists a number of bot detection mechanisms reported on in the literature; however, decision-tree based bot detectors have proven to output the best identification performance (at least, in terms of accuracy) [10,15,17].

In [15], authors proposed a learning-based ensemble approach for click-fraud detection, especially designed to mobile advertisement. For testing their click-fraud detection mechanism, authors used a dataset, divided into two parts. The first dataset contains the information of 3,081 publishers, where each object is captured via four features, namely: publisher ID, account No, address, and status. The second data set contains 3,173,834 click details on advertisements gathered from all of the above publishers. The latter dataset includes nine features and three distinct classes. Authors show how applying resampling methods over each dataset does not improve the results regarding the use of the dataset without applying resampling methods. The bot detection mechanism proposed in [15] attained an accuracy of 59.39%. The main drawbacks of this model are: (i) computational complexity, the ensemble makes use of six different classifiers, and (ii) fair comparison, authors did not compare their method against others reported on in the literature.

In [17], authors also approached click-fraud detection, using the databases of [15]. However, the work of [17] centered around the extraction new features.

To this aim, authors computed statistical measures out from the features proposed in the original database. Statistical measures include maximum value, average value, skewness, variance, and ratio. The extended database contains 21 new features as well as seven of the original features. An important contribution of [17] is that authors reduced the number of objects from the original database in two orders of magnitude, making the new databases more suitable for testing different classification mechanisms. Authors showed how the combination of recursive feature elimination and a decision tree algorithm (using Hellinger distance for assessing a split candidate) is up to obtain 64.07% of accuracy, hence improving on detection performance.

In [10], authors developed a dataset out from recorded activity of 25 different websites. By inspection, authors identified that, out of 7,708 (HTTP) ad requests, 809 were generated by click-bots. Having selected six classifiers for their study, authors found out that RandomForest obtained the highest average accuracy (99.57%) and precision (97.43%). Also, Random Forrest attained the lowest false positive rate (0.30%). Although it achieved a good average accuracy, average precision, an false positive rate, an SVM classifier attained a recall value of only 69.96%. The main limitations of the method of [10] are: (i) it is unable to detect complex JavaScript and encrypted requests that is almost impossible to identify at the operating system level without the help of a browser and (ii) it has been tested on other datasets, as proposed in [17].

Concluding, bot detection has become an important task for both industry and academia; some proposed mechanisms have obtained good classification performance. Nevertheless, web activity should be understood by experts in the application domain with the aim of creating rules in the server to prevent new bot intrusion or for understanding human behavior. For that, contrast pattern-based classifiers provide patterns in a language that is easy for a human to understand, and they have demonstrated to be more accurate than other popular classification models. Also, as far as we know, there is not any study about applying the contrast pattern-based approach for bot detection on web log files.

3   Contrast Patterns

Over the past years, several classifiers have been proposed in the literature but nowadays obtaining a high accuracy is not the only desired characteristic for a classifier; experts in the application domain should understand the results associated to each class of the problem [13,18]. Contrast pattern-based classifiers are an important family of both understandable and accurate classifiers. They provide a model that is easy for a human to understand, and they have demonstrated to be more accurate than other popular classification models [13,18].

A *classifier based on contrast patterns* uses a collection of contrast patterns to create a classifier that predicts a query object class [18].

A *pattern* is represented by a conjunction of relational statements, each with the form: $[f_i \# v_j]$, where v_j is a value in the domain of feature f_i, and $\#$ is a relational operator from the set $\{=, \neq, \leq, >\}$ [13,18]. For example,

$[Time\_Hour \in [0,5]] \wedge [Bytes \leq 200] \wedge [Country = "Tasmania"]$ is a pattern describing a collection of bot accesses. Let p be a pattern and T be a dataset; then, the support of p is a fraction resulting from dividing the number of objects in T described (covered) by p by the total number of objects in T. Now, a *contrast pattern* (CP) for a class c is a pattern whereby the support of CP for c is significantly higher than any support of CP for every class other than c [1,2,13].

CP-based classifiers are used in various real-world applications, in which they have reported effective classification results [13]. Nevertheless, to our knowledge, CP-based approach has not been used for the bot detection problem.

For building a contrast pattern-based classifier, there are two main phases: *Mining* and *Classification* [11,13]. Mining is dedicated to finding a set of candidate patterns by an exploratory analysis using a search-space; using a set of inductive constraints provided by the user. Classification is responsible for searching the best strategy for combining the information provided by a subset of patterns and so builds an accurate model based on patterns.

Pattern mining based on decision trees has two advantages: (i) the local discretization performed by decision trees with numeric features avoids doing an "a priori" global discretization, which might cause information loss; (ii) decision trees provide a significant reduction of the search space of potential patterns [5].

There are several algorithm for mining contrast patterns based on decision trees but those following the diversity approach has shown better results than other approaches. In [5], shown an experimental comparison about diversity generation procedures. Authors have shown that Bagging, Random Forest, and LCMine [6] are the CP mining algorithms that allow obtaining a set of patterns which produce better classification results than other pattern-based solutions.

LCMine [6] extracts a collection of patterns from a set of decision trees. LCMine selects the best k splits in first levels and the best split in lower levels. This way, for $k = \{5, 4, 3, 2\}$, LCMine creates $(5*4*3*2) = 120$ different trees. Patterns are extracted from the paths from the root node to the leaves. Finally, for each pattern, the class with highest support determines the pattern's class.

In [5], the authors have shown how Bagging miner creates diversity by generating each tree with a bootstrap replicate of the training set. Since small changes in the training sample lead to significant changes in the model of a decision tree, Bagging is an excellent way to obtain a diverse collection. Also, they explain that Random Forest Miner (RFMiner) creates diverse trees by selecting a random subset of features at each node. The success of RFMiner can also be explained because injecting randomness at the level of nodes tends to produce higher accuracy models. In [5], the authors claim that RFMiner and Bagging miner are the best for mining patterns because they obtain high-quality patterns.

On the other hand, there are several CP-based classifiers [3,12,13,19] but PBC4cip [13] has reported good classification results; including class imbalance problems. PBC4cip weights the sum of supports in each class, for all patterns covering a query object, taking into account the class imbalance level. Another prominent CP-based classifier is CAEP [3], which aggregates the supports of the patterns matching the query object per class. After, CAEP normalizes all the votes with the average votes per class for assigning the class with higher vote to the query object. CAEP has shown good classification results in several contexts but in a recent paper [13] PBC4cip improves the results obtained by CAEP.

4   Bot Detection Through Contrast Pattern-Based Approach

This section presents a study on using contrast pattern-based classifiers for bot detection in web log files, which is a class imbalance problem. Section 4.1 presents our experimental setup, and Sect. 4.2 presents the experimental results.

4.1   Experimental Setup

For our experiments, we used five databases from which two were collected by us jointly with the NIC[1] Mexico (Network Information Center Mexico) company. The first database contains 55,538 objects (3,338 bot and 52,200 human) and the second database contains 154,790 (3,499 bot and 151,291 human). Both databases contain 11 mixed features from the web log files hosted in a web server belonging to the NIC company. Each object corresponding to web accesses records by the web server, and each object was labeled as bot or human traffic. Those objects labeled as human were validated by the NIC's webmaster taking into account all accesses from login users; while those objects labeled as bot were traffic generated by us using the following automated software: Traffic Spirit[2], Jingling Traffic Bot[3], and w3af[4]. Also, we used three state-of-the-art databases proposed by [17], which were stated in Sect. 2.

All databases were partitioned using 5-fold-cross-validation and Distribution Optimally Balanced Stratified Cross-Validation (DOB-SCV) for avoiding problems into data distribution on class imbalance problems [13].

As we stated in Sect. 3, LCMine, Bagging, and Random Forest miners allow obtaining a set of patterns which produce better classification results than other pattern-based solutions reported in the literature. For this reason, we selected these miners in our experimental setup. Also, we selected PBC4cip [13] as the pattern-based classifier because it has reported better classification results than other state-of-the-art classifiers.

We compare the classification results obtained by the different tested pattern-based classifiers against other nine state-of-the-art classifiers, such as k-Nearest Neighbor with k = 3 (**3NN**), Bagging using C4.5 (**Bag_DT**), Bayesian network (**BN**), C4.5 (**J48**), Logistic Regression (**Log**), Multilayer Perceptron (**MLP**), Naïve Bayes (**NaiveBayes**), Random Forest (**RF**), and Support Vector Machine (**SVM**); all these were executed using the WEKA Data Mining tool [7], using the parameter values recommended by their authors.

---

[1] www.nic.mx.

[2] www.ipts.com.

[3] ipjingling.blogspot.com.

[4] w3af.org.

To evaluate the performance of supervised classifiers, we used Area Under the receiver operating characteristic Curve (AUC) [9], which evaluates the true positive rate (TP_{rate}) versus the false positive rate (FP_{rate}). Finally, we used the Friedman's test and the Shaffer's dynamic post-hoc procedure to statistically compare our results. Both test and procedure are those recommended by [4] for this experimental setup; which assume non-normality in the classification results. Post-hoc results will be shown by using CD (critical distance) diagrams [13]. In a CD diagram, the rightmost classifier is the best classifier. The position of the classifier within the segment represents its rank value, and if two or more classifiers share a thick line it means that they have statistically similar behavior.

4.2 Experiment Resutls

Table 1 shows the average of AUC for all compared classifiers on the tested databases as well as the average (Avg) and the standard deviation (SD) obtained by each classifier. From Table 1, we can see that all classifiers attained an AUC higher than 0.93 for our collected databases (NIC_DB1 and NIC_DB2). Notice that the average AUC for all classification results in these databases was higher than 0.98. On the other hand, for the remaining three databases those pattern-based classifiers obtained the best classification results with an average AUC higher than 0.91. Regarding these results, we want to emphasize that Random Forest (RF) obtained the best average for all databases (0.9561); nevertheless, the average obtained by Bagging miner was 0.9192 and it obtained a lower SD compared to the SD obtained by RF.

Table 1. Results of average AUC for all compared classifiers on the tested databases. The best results appear boldface and those results not available because of a lot of delay during execution time appear as N/A.

Classifiers/Databases	NIC_DB1	NIC_DB2	T_08mar12	T_09feb12	T_23feb12	Avg	SD
3NN	0.9988	0.9988	0.8294	0.7752	0.7687	0.8742	0.1039
Bag_DT	0.9864	N/A	0.5000	0.5000	0.5000	0.6216	0.2106
BN	0.9999	0.9999	0.8827	0.8633	0.8634	0.9218	0.0641
J48	0.9989	0.9955	0.8499	0.8144	0.8399	0.8997	0.0804
Log	N/A	N/A	0.9185	0.8888	0.8894	0.8989	**0.0138**
MLP	N/A	N/A	0.5696	0.7175	0.6821	0.6564	0.0631
NaiveBayes	0.9997	0.9999	0.8741	0.8605	0.8499	0.9168	0.0682
RF	**1.0000**	**1.0000**	0.9393	0.8240	0.9174	**0.9561**	0.0365
SVM	N/A	N/A	0.8763	0.8831	0.8690	0.8762	0.0057
RF Miner	0.9773	0.9851	**0.9382**	0.8443	0.8166	0.9123	0.0693
Bagging Miner	0.9366	0.9588	0.8680	**0.9186**	**0.9141**	0.9192	0.0301
LCMine Miner	0.9624	0.9652	0.9210	0.8459	0.7756	0.8940	0.0732
Average	0.9844	0.9879	0.8306	0.8196	0.8072	-	-
SD	0.0210	0.0157	0.1370	0.1114	0.1128	-	-

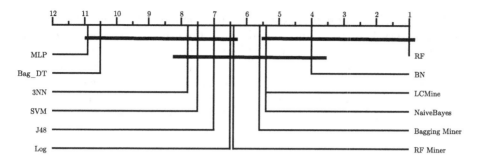

Fig. 1. CD diagram with a statistical comparison (using $\alpha = 0.05$) of the Friedman's ranking results for all compared classifiers on the tested databases.

Figure 1 shows a CD diagram with the classification ranking for each tested classifier using all the databases described in Sect. 4.1. From this figure, we can see that RF, BN, LCMine, and NaiveBayes are the most accurate classifiers for the bot detection problem. Note that although RF is the best into the Friedman's ranking, LCMine has no statistical difference with RF. Also, the results provided by LCMine can be interpreted through the patterns associated with each class, which RF does not offer in an easy way to the experts.

Table 2 shows some of the contrast patterns extracted from the T_08mar12 database using LCMine miner. This table shows the pattern's class (*Class*), the items describing the pattern (*Pattern*), and the support (*Supp*) of each pattern regarding its class. From this table, we can see how more than 50% of the bot traffic can be described by patterns, which contain between four and five items; whereas, patterns that contain two items can describe more than 60% of the human traffic. As we can see, patterns are suitable for introducing them as rules into a firewall software.

Table 2. Some of the contrast patterns extracted from the T_08mar12 database using LCMine miner.

Class	Pattern	Supp
Bot	$[clicks > 108.00] \wedge [agent\_click \leq 0.09] \wedge [cntry \leq 105.50] \wedge [cid\_click > 0.02]$	0.53
	$[AvgClickPerMin > 0.03] \wedge [agent\_click \leq 0.09] \wedge [cntry \leq 105.50] \wedge [cid\_click > 0.02]$	0.53
	$[clicks > 108.00] \wedge [agent\_click \leq 0.08] \wedge [refer \leq 4692.50] \wedge [cntry \leq 96.50]$	0.51
Human	$[clicks \leq 108.00] \wedge [ip\_click\_ratio > 0.60]$	0.65
	$[AvgClickPerMin \leq 0.03] \wedge [ip\_click\_ratio > 0.60]$	0.65
	$[clicks \leq 108.00] \wedge [refer\_click > 0.20]$	0.63

The opinion of the NIC's marketing expert was that the explanatory model proposed by the contrast pattern-based approach is very helpful because it obtains high classification results and provides a model understandable model, which can be converted into rules. These rules can be introduced, in an easy way, into a firewall used to manage the web accesses in the company. Although Bayesian network obtains better classification results than LCMine jointly with

PBC4cip, the NIC's TI experts comment that the model provided by Bayesian network is very hard to be introduced into a firewall, and it does not provide an explanatory model to detect when there is a bot access. On the other hand, Random Forest provides a model that could be converted into rules, but it contains significantly more rules than those extracted from the model provide by LCMine. Therefore, the NIC's marketing and TI experts have endorsed the CP-based approach as the best solution for bot detection in their web server.

5   Conclusions

The main contribution of this paper is a study of the use of contrast pattern-based classifiers for bot detection on web log files. Another contribution of this paper is the databases collected by us, which can be used in future research for validating new algorithms for bot detection.

From our study, we can conclude that LCMine jointly with PBC4cip obtains the accurate classification results for the bot detection problem. Statistical tests prove that the differences among the contrast pattern-based approach and many other classifiers, not based on patterns, are statistically significant. Also, through our study, we find useful patterns which were analyzed by both marketing and TI experts and these patterns were introduced into a system to forewarn about bot traffic in the web servers of the NIC company.

Finally, as future work, following the same approach presented in this paper, we will extend our study for identifying bots through a one-class classifier based on contrast patterns. Also, we will increase our databases with millions of verified human traffic and also we will apply visualization techniques to help the marketing experts to obtain a visual model of the web traffic. These studies would help to improve the accuracy of the bot detection problem.

Acknowledgment. This research was partly supported by Google incorporation under the APRU project "AI for Everyone". Authors are thankful to Robinson Mas del Risco and Fernando Gómez Herrera for providing bot software, and for helping on bot execution throughout our experimentations, respectively.

References

1. Dong, G.: Preliminaries. In: Dong, G., Bailey, J. (eds.) Contrast Data Mining: Concepts, Algorithms, and Applications. Data Mining and Knowledge Discovery Series, chap. 1, pp. 3–12. Chapman & Hall/CRC (2012)
2. Dong, G., Li, J.: Efficient mining of emerging patterns: discovering trends and differences. In: Proceedings of the fifth ACM SIGKDD International Conference on Knowledge Discovery and Data Mining, KDD 1999, pp. 43–52. ACM, New York (1999)
3. Dong, G., Zhang, X., Wong, L., Li, J.: CAEP: classification by aggregating emerging patterns. In: Arikawa, S., Furukawa, K. (eds.) DS 1999. LNCS (LNAI), vol. 1721, pp. 30–42. Springer, Heidelberg (1999). https://doi.org/10.1007/3-540-46846-3_4

4. García, S., Fernández, A., Luengo, J., Herrera, F.: Advanced nonparametric tests for multiple comparisons in the design of experiments in computational intelligence and data mining: experimental analysis of power. Inf. Sci. **180**(10), 2044–2064 (2010)
5. García-Borroto, M., Martínez-Trinidad, J.F., Carrasco-Ochoa, J.A.: Finding the best diversity generation procedures for mining contrast patterns. Expert Syst. Appl. **42**(11), 4859–4866 (2015)
6. García-Borroto, M., Martínez-Trinidad, J.F., Carrasco-Ochoa, J.A., Medina-Pérez, M.A., Ruiz-Shulcloper, J.: LCMine: an efficient algorithm for mining discriminative regularities and its application in supervised classification. Pattern Recogn. **43**(9), 3025–3034 (2010)
7. Hall, M., Frank, E., Holmes, G., Pfahringer, B., Reutemann, P., Witten, I.H.: The WEKA data mining software: an update. SIGKDD Explor. **11**(1), 10–18 (2009)
8. Hallam-Baker, P.M., Behlendorf, B.: W3C - Extended Log File Format. www.w3. org, https://www.w3.org/TR/WD-logfile.html
9. Huang, J., Ling, C.X.: Using AUC and accuracy in evaluating learning algorithms. IEEE Trans. Knowl. Data Eng. **17**(3), 299–310 (2005)
10. Iqbal, M.S., Zulkernine, M., Jaafar, F., Gu, Y.: FCFraud: fighting click-fraud from the user side. In: 17th International Symposium on High Assurance Systems Engineering (HASE), pp. 157–164, January 2016
11. Knobbe, A., Crémilleux, B., Fürnkranz, J., Scholz, M.: From local patterns to global models: the LeGo approach to data mining. In: International Workshop from Local Patterns to Global Models (ECML 2008), pp. 1–16. LeGo (2008)
12. Loyola-González, O., Martínez-Trinidad, J.F., Carrasco-Ochoa, J.A., García-Borroto, M.: Study of the impact of resampling methods for contrast pattern based classifiers in imbalanced databases. Neurocomputing **175**(Part B), 935–947 (2016)
13. Loyola-González, O., Medina-Pérez, M.A., Martínez-Trinidad, J.F., Carrasco-Ochoa, J.A., Monroy, R., García-Borroto, M.: PBC4cip: a new contrast pattern-based classifier for class imbalance problems. Knowl.-Based Syst. **115**, 100–109 (2017)
14. Martens, D., Baesens, B., Gestel, T.V., Vanthienen, J.: Comprehensible credit scoring models using rule extraction from support vector machines. Eur. J. Oper. Res. **183**(3), 1466–1476 (2007)
15. Perera, K.S., Neupane, B., Faisal, M.A., Aung, Z., Woon, W.L.: A novel ensemble learning-based approach for click fraud detection in mobile advertising. In: Prasath, R., Kathirvalavakumar, T. (eds.) MIKE 2013. LNCS (LNAI), vol. 8284, pp. 370–382. Springer, Cham (2013). https://doi.org/10.1007/978-3-319-03844-5_38
16. Soldo, F., Metwally, A.: Traffic anomaly detection based on the IP size distribution. In: International Conference on Computer Communications, pp. 2005–2013 (2012)
17. Taneja, M., Garg, K., Purwar, A., Sharma, S.: Prediction of click frauds in mobile advertising. In: Eighth International Conference on Contemporary Computing (IC3), pp. 162–166 (2015). https://doi.org/10.1109/IC3.2015.7346672
18. Zhang, X., Dong, G.: Overview and analysis of contrast pattern based classification. In: Dong, G., Bailey, J. (eds.) Contrast Data Mining: Concepts, Algorithms, and Applications. Data Mining and Knowledge Discovery Series, chap. 11, pp. 151–170. Chapman & Hall/CRC (2012)
19. Zhang, X., Dong, G., Ramamohanarao, K.: Information-based classification by aggregating emerging patterns. In: Leung, K.S., Chan, L.-W., Meng, H. (eds.) IDEAL 2000. LNCS, vol. 1983, pp. 48–53. Springer, Heidelberg (2000). https://doi.org/10.1007/3-540-44491-2_8

User Recommendation in Low Degree Networks with a Learning-Based Approach

Marcelo G. Armentano[1(✉)], Ariel Monteserin[1], Franco Berdun[1],
Emilio Bongiorno[2], and Luis María Coussirat[2]

[1] ISISTAN Research Institute (CONICET-UNICEN), Tandil, Argentina
marcelo.armentano@isistan.unicen.edu.ar
[2] Facultad de Ciencias Exactas, UNICEN,
Campus Universitario, Paraje Arroyo Seco, Tandil, Argentina

Abstract. User recommendation plays an important role in microblogging systems since users connect to these networks to share and consume content. Finding relevant users to follow is then a hot topic in the study of social networks. Microblogging networks are characterized by having a large number of users, but each of them connects with a limited number of other users, making the graph of followers to have a low degree. One of the main problems of approaching user recommendation with a learning-based approach in low-degree networks is the problem of extreme class imbalance. In this article, we propose a balancing scheme to face this problem, and we evaluate different classification algorithms using as features classical metrics for link prediction. We found that the learning-based approach outperformed individual metrics for the problem of user recommendation in the evaluated dataset. We also found that the proposed balancing approach lead to better results, enabling a better identification of existing connections between users.

Keywords: User recommendation · Online social networks · Link prediction

1 Introduction

In last years, the popularity of the social networks has significantly increased, mainly due to the wide variety of user-generated content and to the availability of a variety of devices to access to such networks. The user-generated content published in these networks plays an important role in the dynamic of the networks, since users usually relate to other users who provide relevant content to them. In this context, the recommendation of interesting users is not a trivial task because of the large number of users in certain social networks.

For example, Twitter is one of the most popular microblogging service that manages hundreds of millions of users since of July 2009. Twitter users are able to "tweet" about any topic in no more than 140 characters; additionally they can follow other users to receive their tweets or news. Recently, these functionalities were enhanced by the addition of pictures, short-videos, livestream and moments. Sina Weibo is another example of a popular microblogging system with more than three hundred million active

I. Batyrshin et al. (Eds.): MICAI 2018, LNAI 11288, pp. 286–298, 2018.
https://doi.org/10.1007/978-3-030-04491-6_22

users that produce more than one hundred million posts each day. Finding users that post relevant and interesting information in these networks is then a challenging task.

Traditionally, recommender systems follow two main approaches: content-based and collaborative based filtering. Both approaches consider information about which items a user is interested in to generate recommendations. However, in the context of user recommendation in social networks, we only count with the information that users share in the network, and with the social interactions among members that define the structure of the social network. Then, the information about which users other users can be interested in has to be derived from these two sources. However, the social graph structure has a low degree meaning that the number of connections per user is low with respect to the total number of users (Al Hasan and Zaki 2011). This fact makes it difficult to apply learning-based approaches to predict new connections among users, due to the extreme class imbalance (connected versus not connected nodes).

In this work, we present a balancing approach to evaluate classification models for predicting connections among users in microblogging systems. A classifier is a model that describes and distinguishes data classes or concepts, in order to use the model to predict the class of objects whose class label is unknown (Han et al. 2011). We propose as the inputs of the classifier a set of topological and content-based metrics. Topological metrics compute different measures between pairs of nodes taking into account the graph structure of the social networks, such as common neighbors or paths. On the other hand, content-based metrics assign to each pair of nodes a similarity score that is computed using the attributes of the nodes, such as the users profiles (Bhattacharyya et al. 2011) or user-generated content (Armentano et al. 2013). Particularly, these metrics reduce their performance when the social network has a low degree. For this reason, we claim that the combination of these metrics in a classification model aid us to overcome this problem.

To evaluate our approach, we ran several experiments in order to determine the effects of the proposed balancing approach and to compare the performance of the learning-based approach and the individual metrics. The results clearly showed that (a) the balancing approach improved the performance of the learning-based approach, and (b) the learning-based approach outperformed the individual metrics.

This article is organized as follows. Section 2 introduces some concepts related to user recommendations and states the research questions that guided our research. Section 3 presents the supervised learning approach to user recommendation and the balancing approach. Sections 4 shows the results obtained from the experiments. Finally, in Sect. 5, we discuss the results and present future works.

2 Literature Review

Social networks can be represented with a graph structure in which each node corresponds to a user and each edge corresponds to a relationship among users. These relationships can be interpreted in different ways, for example, friendship relationships, messages sent from one user to another, mentions to other users, etc.

One of the advantages of seeing a social network as a graph is that we can apply concepts taken from the graph theory and interpret them in the context of social networks. The graph theory enable the study of how the topology of the networks might affect individual users, how each user might affect other users, the patterns that might arise by the communication among users, among many other applications.

A well-known problem in the context of social networks is the study of the evolution of the social relationships among members. This problem is known as link prediction, and consists in predicting the establishment of future relationships among users that are not observed in a given snapshot of the network (Liben-Nowell and Kleinberg 2007). In the context of microblogging systems, link prediction techniques can be applied to recommend users that a target user might be interested in following. This implies the understanding on how users relate to each other, which is a problem that strongly depends on the structure and objective of the social network.

Recommending users in a social network is a complex task due to the dynamics of the networks in which new users appear and disappear all the time. However, link prediction techniques assume that the users of the network remain static and that the network only evolves by adding new relationships among users. From a recommendation point of view, this is still useful, since when new users enter the network, a recommender system can suggest users to follow according the new user's interests and the first relationships that he/she establishes with the members of the network.

Link prediction is usually studied by two main approaches: similarity-based approaches and learning-based approaches (Wang et al. 2015). To recommend new users to a target user, similarity-based approaches measure the proximity of the target user with every other user in the network. Then, those nodes with higher values of proximity are recommended assuming that a high proximity value implies a higher probability of the nodes being connected. The proximity between two nodes is measured with different metrics, which can be roughly grouped into local and global metrics. Local metrics measure information that is local to the node and its direct neighbors, assuming that users tend to create relationships with users who are similar to them (regarding for example, location, interests, religion, etc.). Global metrics, instead, measure information that extends beyond the local neighborhood of the target user and consider path lengths between nodes and random walks. The main disadvantage of global metrics is that there are inefficient to be computed, and are impractical to be computed in real time for large networks.

Several metrics have been proposed to measure the similarities between users in a social network. Liben-Nowell and Kleinberg (Liben-Nowell and Kleinberg 2007) provided a good review of the most used similarities metrics and compare their performance for link prediction in the context of social collaboration networks. Authors conclude that there is not a "magic" metric that perform better in all circumstances and, depending on each particular network, some metrics behave better than others do. Armentano et al. (2012) proposed an algorithm for recommending relevant users that explores the topology of the network considering different factors to identify users that can be considered good information sources. In a posterior study, Armentano et al. (2013) proposed a user recommendation approach based on both the exploration of the topology of the network and the users' interests. They found that user-generated content is a rich source of information for profiling users and finding like-minded people. Chen et al. (2017) took the similarity metrics proposed in (Armentano et al. 2012) and combine them with a linear equation. The weights of this linear combination were computed with a simulated annealing algorithm and concluded that the resulting combined metric outperformed the individual metrics.

On the other hand, learning based methods approach the link prediction problem as a classification problem. Each pair of nodes is described by a set of features (usually taken or derived from similarity-based approaches) with the class being the existence of absence of a link in the network. One of the main challenges in learning classification models for link prediction is the extreme class imbalance: the number of possible links in a graph is quadratic in the number of nodes; however, the number of actual links is only a very small fraction of this number (Al Hasan and Zaki 2011). For example, Rattigan and Jensen (2005) showed that for a uniformly sampled dataset taken from DBLP in the year 2000 with one million training instances, only 20 positive instances (real links) are expected. This problem has been addressed in previous work with two main approaches: by altering the training sample with up-sampling or down-sampling (Chawla et al. 2002), and by altering the learning method (Ertekin et al. 2007) (Karakoulas and Shawe-Taylor 1998).

In this direction, Han and Xu (2016) proposed a link prediction system that combines multiple features to learn different classifiers (SVM, Naive Bayes, Random Forest and Logistic Regression). They evaluate the approach on a dataset collected from Sina microblogging system concluding that the combination of multiple features achieve better classification performance. The evaluation procedure followed in this experiment was as follows: (1) 500 active users were selected from the dataset, (2) all positive links were selected for each active user, (3) an equal number of negative links were randomly selected from the remaining connections, and (4) a set of 100.000 samples were randomly selected from the combination of positive and negative examples. In our approach, we follow a different procedure for balancing the dataset, as described in Sect. 3.2. Similarly, Ahmed et al. (2016) also proposed a supervised learning approach to link prediction in Twitter. In this case, authors considered only topological features, disregarding any content information. The problem of the class imbalance was also solved in this article by randomly selecting a number of negative links equals to the number of available positive links.

Considering the literature review, we observe that the performance of a learning-based approach to user recommendation in microblogging systems has not been analyzed for low degree networks. For this reason, we based our research on the following research questions:

RQ1 Does the balancing of the dataset affect the classification performance?

RQ2 Does a supervised learning approach outperform individual metrics for link prediction in low degree microblogging systems?

3 Supervised Learning Approach to User Recommendation

Supervised learning aims at obtaining a function or model from training examples. Each example consists in a pre-computed set of features and the class to which it belongs. The set of features is used to abstract the characteristics that define the examples of each class. The learnt model is then used to determine the class of new examples for which the set of features is known, but not the class.

In order to approach the user recommendation problem in social networks as a classification problem, we first need to define the features that will describe the examples and the class to which it example belongs.

Given a graph G(V,E), where V is the set of nodes and E the set of edges between nodes, a label function $L_{x,y}$ is defined for two nodes $x,y \in V$ as follows:

$$L_{x,y} = \begin{cases} + \; if \, (x,y) \in E \\ -if \, (x,y) \notin E \end{cases}$$

With this labeling function, we can see the link prediction problem as a classic binary classification problem in which positive instances correspond to existent links and negative instances correspond to pairs of unconnected nodes.

Then, as illustrated in Fig. 1, we first compute different local topology similarities along with content-based similarities between each pairs of nodes in the network. This set of values constitutes our feature set that describe each training example (Sect. 3.1). Then, we split the dataset into positive instances (real links that exist in the social graph) and negative examples (not observed links). In order to improve the classification performance, we built a set of N balanced datasets, each of which contains all the positive examples, and an equal number of negative examples, randomly selected from the set of negative examples (Sect. 3.2). We test the performance of each dataset with a 10-fold cross validation approach with different well-known classification models: Naïve Bayes, Decision Trees, Random Forest and Support Vector Machines (Sect. 3.3).

3.1   Feature Set

As in any supervised learning problem, a key aspect to get a good classification model is to choose a good set of features that describes the training examples. Once the set of features is selected, different classification algorithms should be compared in order to find the one that is able to better model the training examples.

Since online microblogging networks, such as Twitter and Sina Weibo has millions of users we chose to use only local topological metrics to measure similarities among users. Local topological metrics can be easily computed since they depend only on the direct neighborhood of each node. Most of these metrics are based in the number of common neighbors of two users, following the assumption that the more friends two users have in common, the higher the probability that those two users will became friends in the future.

We describe next the feature used in this work. In all cases, given a user x, we denote the set of x's neighbors with $\Gamma(x)$ and the total number of x's neighbors with $|\Gamma(x)|$.

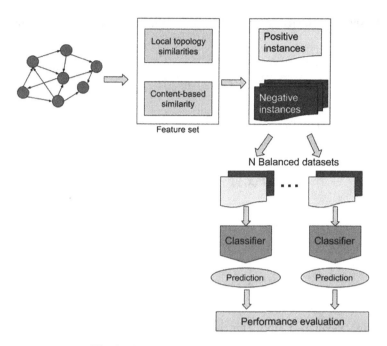

Fig. 1. Scheme of the proposed approach

Common neighbors (CN) Represents the number of common neighbors between two users x and y

$$CN(x,y) = |\Gamma(x) \cap \Gamma(y)|$$

Jaccard coefficient (JC) Normalizes the CN metric with the total number of different neighbors that both users have. This metrics penalize users that have a high number of friends with respect to the number of common friends.

$$JC(x,y) = \frac{|\Gamma(x) \cap \Gamma(y)|}{|\Gamma(x) \cup \Gamma(y)|}$$

Sørensen Index (SI) Also normalizes CN, but with the total number of neighbors that both user have. Then, nodes with few neighbors would have higher link probability with this metric.

$$SI(x,y) = \frac{|\Gamma(x) \cap \Gamma(y)|}{|\Gamma(x)| + |\Gamma(y)|}$$

Salton Cosine Similarity (SC) is defined as the inner product of the links of x and y.

$$SC(x,y) = \frac{|\Gamma(x) \cap \Gamma(y)|}{\sqrt{|\Gamma(x)| * |\Gamma(y)|}}$$

Hub Promoted (HP) Define the topological overlap of x and y. The number of common neighbors is normalized by the lower out-degree of x and y.

$$HP(x,y) = \frac{|\Gamma(x) \cap \Gamma(y)|}{min(|\Gamma(x)|, |\Gamma(y)|)}$$

Hub Depressed (HD) Similar to HP, but the number of common neighbors is normalized but the higher out-degree of x and y.

$$HD(x,y) = \frac{|\Gamma(x) \cap \Gamma(y)|}{max(|\Gamma(x)|, |\Gamma(y)|)}$$

Licht-Holme-Nerman (LHN) This metric is similar to SC, but do not apply a square root in the denominator. Then, it assigns more weight to those pairs of nodes with many common neighbors compared to the expected number of neighbors.

$$LHN(x,y) = \frac{|\Gamma(x) \cap \Gamma(y)|}{|\Gamma(x)| * |\Gamma(y)|}$$

Parameter-Dependent (PD) PD introduces a free parameter λ to improve accuracy for predicting both popular and unpopular links. When $\lambda = 0$, we get CN, when $\lambda = 0.5$ we get SC, and if $\lambda = 1$ we obtain the same similarity than that given by LHN.

$$PD(x,y) = \frac{|\Gamma(x) \cap \Gamma(y)|}{(|\Gamma(x)| * |\Gamma(y)|)^{\lambda}}$$

Adamic-Adar coefficient (AA) This metric was originally proposed for computing similarity between two web pages. AA assigns more weight to those common neighbors with low number of connections.

$$AA(x,y) = \sum_{z \epsilon \Gamma(x) \cap \Gamma(y)} \frac{1}{\log(|\Gamma(z)|)}$$

Preferential attachment (PA) Follows the idea that there is more probability to create a link from nodes with many links. Then, it defines the similarity of two nodes as the product of the number of neighbors of each node.

$$PA(x, y) = |\Gamma(x)| * |\Gamma(y)|$$

Resource allocation (RA) This metric is very similar to AA, but penalizes more heavily those neighbors with high outdegree. In networks with low average degrees both AA and RA perform similarly. However, in networks with high average degrees, RA usually works better.

$$RA(x, y) = \sum_{z \in \Gamma(x) \cap \Gamma(y)} \frac{1}{|\Gamma(z)|}$$

Content-based similarity (CB) This metric measures the similarity between the content published by two users in the microblogging system. All the textual information in the tweets published by the users is aggregated and represented in the Vector Space Model (Salton and Mcgill 1986). This metric works as follows: we take the list of words used in all the tweets published by each user; we remove stopwords and run the Porter stemmer algorithm (Porter 1980) to reduce words to their morphological roots. Then, the cosine similarity is computed between the vectors defined by the presence or absence of different words in the tweets of each user (Salton and Mcgill 1986).

3.2   Balancing Technique

As stated in Sect. 2, one of the main issues of approaching the user recommendation problem in microblogging systems as a classification problem is the highly imbalanced class distribution. This class imbalance negatively influences the classification results since they will be skewed to the predominant class. For example, if we have a class distribution with a relation of 95:5, a classifier that simply output the predominant class will have 95% accuracy.

One of the main approaches to solve this problem is to under-sample the dataset by keeping all the instances of the minority class and randomly selecting an equal number of the predominant class. This approach has the advantage of being simple and efficient. However, as in any under-sampling method, much information is lost in the procedure and the reported results will be dependent on the subset of negative instances that were selected to be included in the balanced dataset.

In this work, we propose an alternative approach that we call *cross balancing*. Following the idea of the cross validation method, we divide the number of negative examples in N disjoint subsets. Each of these subsets is then combined with all the positive instances to obtain N different datasets that are used to test the performance of the classification algorithms. Results are then averaged to measure the performance of the algorithms.

3.3   Classification Algorithms

We compare different classification algorithms in order to obtain the one that better fit the problem of user recommendation in microblogging systems. Classification algorithms build a function or model that describes and distinguishes data classes or concepts, in order to use the model to predict the class of objects whose class label is unknown (Han et al. 2011). In this research, we consider four well-known classification algorithms: Naïve Bayes, Decision Trees, Random Forest, and Support Vector Machines.

Naïve Bayes Classifier (NBC): NBC is a learning algorithm that simplifies learning by assuming that features (observable variables) are independent given the class (inferred variable). Despite this unrealistic assumption, the resulting classifier is remarkably successful in practice, often competing with much more sophisticated techniques (Rish 2001).

Decision Trees (DT): Decision trees classify instances by sorting them down in a tree-like structure from the root to some leaf node, which provides the classification of the instance (Mitchell 1997). Learned trees can also be represented as sets of if-then rules to improve human readability.

Random forests (RF):   are an ensemble learning method that operate by constructing a set of decision trees at training time and outputting the class that is the mode of the classes of the individual trees (Ho 1995). Random forests correct the decision trees' habit of overfitting to their training set.

Support Vector Machines (SVM):   are a family of algorithms that blend linear modelling and instance-based learning. Support vector machines select a small number of critical boundary instances called support vectors from each class and build a linear discriminant function that separates them as widely as possible (Witten et al. 2016).

4   Experiments

4.1   Settings

We have evaluated our approach with a dataset crawled from Twitter. The dataset was composed of 42032 nodes and 54372 directed edges, where each node represented a user of Twitter, and each edge, a following relationship among users. Moreover, each node had an associated set of Tweets posted by the user. Since the number of possible connections among nodes is higher than the number of edges, the graph was highly disconnected. The average degree of the graph was 1.29.

4.2   Procedure

Before running the experiments, we first preprocessed the tweets of each user. To do this, we eliminated dates, number, symbols and links, since this information was not relevant from the point of view of content-based metric. Then, the rest of the words were processed with the "EnglishAnalyzer" of Lucene (McCandless et al. 2010) by removing stopwords and running the Porter stemming algorithm.

After preprocessing the tweets, we computed the topological and content-based metrics for each pair of nodes, and created a new dataset indicating whether a relationship exists between the nodes or not. The resulting dataset contained 99.96% instances belonging to the negative class (no link) and 0.04% instances belonging to the positive class (existent links). Then, we trained the classifiers by using 10-folds cross-validation. This training was carried out by using the WEKA API in two experiments: without balancing the dataset and by applying the balancing technique detailed in Sect. 3.2.

On the other hand, we ran experiments by applying the topological and content-based metrics individually. To do so, we determined the set of 500 users with more relationships. Then, from that set, we randomly selected a set of 100 target users. For each target user, we hid 20% of her links and tried to predict them by using the individual metrics. To carry out these predictions, a ranking of candidate users is built according to the value of the metric in a descendant order. Finally, the top-k ranked nodes were taken as recommendation to the target user (where k is the number of real links that were hidden).

Finally, we computed several measures in order to compare the results of our approach and the individual metrics. These measures were Precision (average, positive and negative), Recall (average, positive and negative), the Relative Absolute Error (RAE) and the Area Under the Receiver Operating Curve (AUC).

5 Results

Table 1 shows the results obtained by several classifiers with and without balance. Moreover, Table 2 shows the results obtained by the topological and content-based metrics individually. We analyze next the results from the point of view of the research questions.

RQ1 Does the balancing of the dataset affect the classification performance?

As we can see in Table 1, when the classifiers were trained without a balanced dataset the results were considerably poor. This fact is due to the high difference

Table 1. Performance of different classification algorithms with and without class balance

Classifier	Balance	AUC	RAE	Precision (+)	Precision (−)	Precision (−)	Recall	Recall (+)	Recall (−)
Naïve Bayes	No	0.974	6424.67	99.9%	0.5%	99.9%	97.8%	57.9%	97.8%
	Yes	0.973	10.54	95%	92.17%	97.64%	94.73%	94.89%	94.59%
DT	No	0.948	82.82	99.9%	85.9%	99.9%	99.9%	15.7%	99.9%
	Yes	0.962	11.37	96.54%	94.76%	98.31%	96.46%	98.37%	94.56%
Random forest	No	0.740	64.76	99.9%	67.6%	99.9%	99.9%	26.2%	99.9%
	Yes	0.979	10.39	96.52%	94.86%	98.19%	96.46%	98.25%	94.67%
SVM	No	0.5	49.95	99.9%	0%	100%	99.9%	0%	100%
	Yes	0.95	9.93	92.26%	92.26%	98.21%	95.03%	98.3%	91.74%

between the positive and negative values of precision and recall and a high RAE. For example, the positive recall values obtained by DT and SVM were very low. This was because the classifier predicted most instances as no-links due to the high imbalance of the dataset. Without balancing, Naive Bayes got the highest positive recall, but its precision was very low. This occurred because many negative instances were classified as positive.

In contrast, when the dataset was balanced with the proposed technique, positive and negative values were similar and the RAE was drastically reduced. Moreover, the AUC measure improved using the balancing approach by most of the classification algorithms.

Although the results obtained by the four classifier are similar, we can distinguish a small improvement using Random Forest. This classifier achieved the highest positive precision, AUC and a low RAE.

RQ2 Does a supervised learning approach outperform individual metrics for link prediction in microblogging systems?

By comparing the results of Tables 1 and 2, we can clearly see that the learning-based approach outperform the individual metrics for link prediction. By using the topological metrics, the highest precision value did not exceed 1%. On the other hand, the content-based metric also reached low precision and recall, though these values were higher than those reached by the topological metrics were. Notice that negative precision and negative recall values are not reported in Table 2 because they was close to 1.

Table 2. Performance of different local topological metrics

Individual metric	AUC	Precision (+)	Precision (−)
AA	0.422	0.1%	0.05%
CN	0.488	0.4%	0.15%
HD	0.377	0.5%	0.18%
HP	0.529	0.3%	0.10%
JC	0.462	0.4%	0.19%
LHN	0.516	0.5%	0.30%
RA	0.413	0.9%	0.25%
SC	0.422	0.0%	0.00%
SI	0.401	0.4%	0.14%
CB	0.589	1.6%	0.50%

6   Discussion and Conclusions

In this work, we used a learning-based approach to user recommendation in microblogging systems with a proposed balancing approach to cope with the highly class imbalance, which is a common issue to deal with when approaching user rec-ommendation using learning-based techniques in low degree networks. We compared different classification algorithms using different local topological metrics and a content-based metric as features for describing when two users are connected in the network.

The experiments showed that our approach outperformed the individual metrics for link prediction, even though these individual metrics were used as the input of the classifiers. In the context of link prediction, it is particularly important to evaluate the ability of the recommender system to predict elements in the "positive" class, which correspond to links in the network. In this sense, random forests and decision tree algorithms performed better than Naïve Bayes and SVM.

Furthermore, it is worth noticing that the poor performance obtained by the individual metrics was caused by the low degree of the network. In this context, the traditional recommendation mechanism using individual metrics does not allow us to deal with this factor, since all the nodes of the network are candidate to be recommended, once a set of real links have been hidden.

A problem with the evaluation methodology used in this study is that we are not able to affirm that a user classified as "no link" will not be interesting to the target user. Since we are using a snapshot of the network, we are only able to consider as positive those instances representing links that are already present in the network. Then, some links recommended by the classification algorithm might be indeed good recommendations that we consider false positives with our approach. However, experiments with real users are need to evaluate the real interest of users in the recommended connections.

Future work will focus on adding new individual metrics as input of the learning based approach. Moreover, we will study the use of assemble classifiers to compare with the results obtained by the individual ones.

Acknowledgements. This work was partially supported by research project PICT-2014-2750.

References

Ahmed, C., ElKorany, A., Bahgat, R.: A supervised learning approach to link prediction in Twitter. Soc. Netw. Anal. Min. **6**(1), 24 (2016)

Al Hasan, M., Zaki, M.J.: A survey of link prediction in social networks. In: Aggarwal, C.C. (ed.) Social Network Data Analytics, pp. 243–275. Springer, US (2011). https://doi.org/10.1007/978-1-4419-8462-3_9

Armentano, M.G., Godoy, D., Amandi, A.: Topology-based recommendation of users in micro-blogging communities. J. Comput. Sci. Technol. **27**(3), 624–634 (2012)

Armentano, M.G., Godoy, D., Amandi, A.A.: Followee recommendation based on text analysis of micro-blogging activity. Inf. Syst. **38**(8), 1116–1127 (2013)

Bhattacharyya, P., Garg, A., Wu, S.F.: Analysis of user keyword similarity in online social networks. Soc. Netw. Anal. Min. **1**(3), 143–158 (2011)

Chawla, N.V., et al.: SMOTE: synthetic minority over-sampling technique. J. Artif. Intell. Res. **16**, 321–357 (2002)

Chen, H., Jin, H., Cui, X.: Hybrid followee recommendation in microblogging systems. Sci. China Inf. Sci. **60**(1), 012–102 (2017)

Ertekin, S., Huang, J., Giles, C.L.: Active learning for class imbalance problem. In: Proceedings of the 30th Annual International ACM SIGIR Conference on Research and Development in Information Retrieval, SIGIR 2007, pp. 823–824. ACM, New York (2007)

Han, J., Pei, J., Kamber, M.: Data Mining: Concepts and Techniques. Elsevier, Amsterdam (2011)

Han, S., Xu, Y.: Link prediction in microblog network using supervised learning with multiple features. JCP 11(1), 72–82 (2016)

Ho, T.K.: Random decision forests. In: Proceedings of 3rd International Conference on Document Analysis and Recognition, vol. 1, pp. 278–282 (1995)

Karakoulas, G., Shawe-Taylor, J.: Optimizing classifiers for imbalanced training sets. In: Proceedings of the 11th International Conference on Neural Information Processing Systems, NIPS 1998, pp. 253–259. MIT Press, Cambridge (1998)

Liben-Nowell, D., Kleinberg, J.: The link-prediction problem for social networks. J. Assoc. Inf. Sci. Technol. 58(7), 1019–1031 (2007)

McCandless, M., Hatcher, E., Gospodnetic, O.: Lucene in Action, Second Edition: Covers Apache Lucene 3.0. Manning Publications Co., Greenwich (2010)

Mitchell, T.M.: Machine Learning, vol. 45, no. 37, pp. 870–877. McGraw Hill, Burr Ridge (1997)

Porter, M.F.: An algorithm for suffix stripping. Rossiiskaya Akademiya Nauk. Programmirovanie 14(3), 130–137 (1980)

Rattigan, M.J., Jensen, D.: The case for anomalous link discovery. SIGKDD Explor. Newsl. 7(2), 41–47 (2005)

Rish, I.: An empirical study of the naive Bayes classifier. In: IJCAI 2001 Workshop on Empirical Methods in Artificial Intelligence, pp. 41–46. IBM, New York (2001)

Salton, G., Mcgill, M.J.: Introduction to Modern Information Retrieval. McGraw-Hill, New York (1986)

Wang, P.: Link prediction in social networks: the state-of-the-art. Sci. China Inf. Sci. 58(1), 1–38 (2015)

Witten, I.H., et al.: Data Mining: Practical Machine Learning Tools and Techniques. Morgan Kaufmann, Burlington (2016)

Volcanic Anomalies Detection Through Recursive Density Estimation

Jose Eduardo Gomez[1,2], David Camilo Corrales[2,3], Emmanuel Lasso[2(✉)], Jose Antonio Iglesias[3], and Juan Carlos Corrales[2]

[1] Observatorio Vulcanológico y Sismológico, Popayán, Colombia
`jegomez@sgc.gov.co`
[2] Telematics Engineering Group, University of Cauca, Popayán, Colombia
`{jeduardo,dcorrales,eglasso,jcorral}@unicauca.edu.co`
[3] Computer Science Department, Carlos III University of Madrid, Leganes, Spain
`davidcamilo.corrales@alumnos.uc3m.es`, `jiglesia@inf.uc3m.es`
`http://www.sgc.gov.co/`, `http://www.unicauca.edu.co`, `http://www.uc3m.es`

Abstract. The volcanic conditions of Latin America and the Caribbean propitiate the occurrence of natural disaster in these areas. The volcanic-related disasters alter the living conditions of the populations compromised by their activity. We propose to use Recursive Density Estimation (RDE) method to detect volcanic anomalies. The different data used for the design and evaluation of this method are obtained from Puraće volcano of two surveillance volcanic areas: Geochemistry and Deformation. The proposed method learns quickly from data streams in real time and the different volcanic anomalies can be detected taking into account all the previous data of the volcano. RDE achieves good performance in the outliers detection; 82% of precision for geochemestry data, while 77% of precision in geodesy data.

Keywords: Recursive Density Estimation (RDE) · Geochemistry Deformation · Outlier

1 Introduction

Volcanoes have always generated emergency situations for those who live in their surroundings. People living in areas close to volcanoes coexist with a complex combination of benefits and risks. The risks affect the health of a population (earthquakes, flows, explosions, emissions of gases and ash, and so on), and can cause morbidity and high mortality due to large eruptions size. Indirectly, vulcanological events can cause socioeconomic deterioration, the damage of vital lines or infrastructures and, in general, alter the living conditions of the populations compromised. For this reason, volcano monitoring is a key task to detect volcanic anomalies and act accordingly. In this paper, we propose a method for detecting any volcanic anomaly based on data density. The different data used for the design and evaluation of this method are obtained from the seismological

© Springer Nature Switzerland AG 2018
I. Batyrshin et al. (Eds.): MICAI 2018, LNAI 11288, pp. 299–314, 2018.
https://doi.org/10.1007/978-3-030-04491-6_23

stations which are close to a volcano. However, an important aspect to consider in this (volcanic) environment is that these data need to be analysed in real-time to produce a rapid and specific response. Thus, the proposed method needs to learn quickly from data streams in real time. This cause that the common calculation of the data density cannot be used in the proposed scenario because it is computationally very complex since it requires storing all the data, which imposes strict limitations on the memory and computational power. For these reasons, we propose in this paper the use a recursive approach which was introduced in 2001 [1,2] called RDE (Recursive Data Estimation) and was also patented in 2016 [3]. The results of the proposed method will be analysed by 2 experts to evaluate its success and validity.

The remainder of this paper is organized as follows: Sect. 2 describes the background of this research; Sect. 3 refers to the dataset description, Sect. 4 relates the outliers detection approach; Sect. 5 contains the results; and Sect. 6 presents the conclusions and future work.

2   State of the Art

In this section are described the volcanology areas used for building of dataset, the algorithm for volcano anomalies and the related works.

2.1   Background

2.1.1   Volcanology

Over the years, volcanoes have generated emergency situations for the human being who lives in their surroundings. There are many populations located in areas close to volcanoes that coexist with a complex combination of benefits and risks. The benefits are several: agricultural, tourist, therapeutic, etc.; while the risks in an active volcano can affect the health of a population directly, because of their earthquakes, flows, explosions, emissions of gases and ash, among others, causing morbidity due to different pathologies and high mortality due to large eruptions size. Indirectly, volcanological events can cause socioeconomic deterioration, the damage of vital lines or infrastructures and, in general, alter the living conditions of the populations compromised. For this reason, volcano monitoring is a key element to be able to study its behavior and establish alert levels.

Volcano monitoring involves the constant measurement of different variables related to volcanic activity through various stations making use of a certain methodology. Dependent on the frequency of obtaining the data, a station is classified as high and low sampling rate. Generally, the only stations that have a high sampling rate are seismological stations, while geochemistry, deformation, magnetometry and climatological stations belong to stations with a low sampling rate. According to experts, the areas that provide most information on possible changes of the volcano, in addition to seismology, are geochemistry and geodesy (deformation) [4–6].

– Geochemistry: geological science that studies the chemistry of the planet; In volcanology, geochemistry specializes in the collection and analysis of volcanic fluids. The chemical composition of the water or gases present in the emissions of a volcano is a reflection of its activity. The main components of volcanic gases are water vapor (H_2O), carbon dioxide (CO_2) and sulfur dioxide (SO_2) [7].
– Geodesy (Deformation): in volcanology, it studies the deformation in the surface of the volcanic cone as expansions or contractions due to the internal activity of the volcano when it is subjected to the pressure generated by the magma during its ascent towards the surface [4].

2.1.2   Outlier Detection

An anomaly is an observation that is significantly different from the rest of observations, so that it is suspected to have been generated by a different mechanism. The detection of anomalies (outliers) is a critical task at the moment for the detection of frauds, the detection of intrusions or the cleaning of noisy data. Outlier detection algorithms can be based on different techniques such as statistical distributions, distance or angle between objects and density of objects [8].

Recursive Density Estimation
Data density is a key measure in the outliers detection and other related problems [9]. However, the (common) estimation of the data density is very complex since it requires storing all which restricts the limitations on the memory and the computational power.

To address these restrictions, a recursive approach was introduced in 2001 [1,2] called RDE (Recursive Data Estimation) and was also patented in 2016 [3]. In [10], RDE is used to obtain the result of the data density expression by storing (and updating) only a very small amount of data (mean of all the samples, and the scalar product averages quantity, $\sum_k$ at each moment k). Thus, this recursive calculation can be carried out very fast, in real time. In addition, it is important to remark that the obtained value is exactly the same (not an approximation).

The recursive expression is as follows [11,12]:

$$D(x_k) = \frac{1}{1 + \|x_k - \mu_k\| + \sum_k -\|\mu_k\|^2}$$

where both the mean, μ_k and the scalar product mean, $\sum_k$ are updated recursively as follows:

$$\mu_k = \frac{k-1}{k}\mu_{k-1} + \frac{1}{k}x_k; \mu_1 = x_1$$

$$\sum_k = \frac{k-1}{k}\sum_{k-1} + \frac{1}{k}\|x_k\|^2; \sum_1 = \|x_k\|^2$$

One of the important aspects to remark by using RDE is that the outliers detection task can work in online mode and in real-time.

Local Outlier Factor
Local Outlier Factor (LOF) [13,14] is an algorithm created for the detection of outliers of a data flow that is based on the density calculation and results in a value of an object p representing the degree in which p is an anomaly. Points with high LOF value have smaller local densities than their neighbors and represent stronger outliers. In other words, the LOF algorithm calculates a value for each data record inserted in the data set and instantly determines whether the inserted data record is atypical. The LOF values for the existing data records are updated if necessary, however, their performance depends on the algorithm's own indexing structures, for this reason LOF has limitations in the data dimensionality and does not it is applicable when it is required to process a large number of dimensions of a data set.

Angle Based Outlier Detection
In [15] an approach called ABOD (Angle Based Outlier Detection) where the variance is evaluated by the difference of the angles between points in a vector space but mainly the directions of distance vectors, is proposed. Comparing the angles between pairs of distance vectors to other points helps to distinguish between points similar to other points and outliers. The performance in high dimension datasets are improved compared to purely distance-based approaches. A main advantage of this approach is that this method does not depend on any selection of parameters that influences the quality of the classification obtained.

2.2   Related Works

Geochemistry. In [16] the effectiveness of several supervised classification techniques for detecting transient geophysical phenomena is experimentally evaluated using volcanic plume detection methods in planetary satellites. Four sets of mission data were used with accuracy ranging from 73 to 95 percent (73%–95%). Additionally, the same techniques are applicable to other geological characteristics. The authors of [17] make use of a hybrid computational model for the characterization of oilfields in the oil industries. This requires highly accurate predictions for the exploration, analysis and efficient management of oil and gas resources. Some techniques as fuzzy logic, support vector machines and artificial neural networks are used. Artificial Neural networks were used for feature selection, while support vector machines were used for final predictions in conjunction with fuzzy logic. The research presented in [18] aims to monitor soil radon anomalies of 3 drilling wells in the Orlica fault in the Krško basin (Slovenia). In order to distinguish the anomalies caused by the environmental parameters (air and soil temperature, barometric pressure, rainfall) related to seismic activity, the following approaches were considered: *(i)* deviation of the radon gas concentration from the seasonal average; *(ii)* correlation between the time gradients of the radon concentration and the barometric pressure, and *(iii)* an automatic learning process through regression trees. The first approach is less successful in predicting anomalies caused by seismic events, while *(iii)* obtains a much smaller number of false seismic anomalies than *(ii)*. When radon gas is influenced only by environmental parameters, this correlation is significantly reduced in the seismically active periods. In [19], radon gas is also studied, this time as a precursor of earthquakes, using anomaly detection methods. The objective is to identify radon anomalies using supervised learning techniques (such as artificial neural networks, regression and decision trees) to get a prediction model of radon concentration based on some environmental parameters such as barometric pressure, rainfall and soil temperature.

Deformation. In [4] volcanic deformation is considered as a precursor event to an eruption, which occurs a few days or even months later. The detection of deformation events provides a better understanding of magmatic processes and long-term eruption predictions. A study of geodetic deformations in bridges is presented in [20] from artificial neural networks. Variables such as temperature, pressure, humidity, water level variations and traffic volume are taken as input variables (forces) and displacements of points components (deformations) as output variables. The results show that it is possible to predict a deformation up to 180 days before it occurs.

The research presented in [21] aims to find signs of deformation events in the region near the Chiles-Cerro Negro volcanoes. Additionally, analytical models (least squares) and Markov models were used to determine the source of the deformation. Also, two model inversion methods were tested, the first of a pressurized spherical reservoir and the second corresponding to an oblique displacement. The results of the inversions are modeled with the Finite Element Method (FEM) taking into account the region topography. Finally, in [22], a set of algorithms to detect abrupt changes on the data obtained by GPS instruments are presented. The data represent measurements of deformations in real time continuously. In addition, the algorithms make use of a statistical cumulative sum technique.

In Geochemistry and Deformation, few works are closely linked with the domain of application of volcanology. In addition, these focus on the detection and classification of alerts on a data set that may become obsolete due to the dynamic nature of volcanic phenomena [23]. On the other hand, the outlier detected in some of the related works, process the whole dataset in each iteration, which becomes a task of high resource consumption and inefficient.

2.3 Outlier Detection with RDE

RDE was first introduced as part of an autonomous real-time anomaly detector for Russian flights [24]. The authors propose and implement a fault detection application that automatically determines flight phases (eg. takeoff, landing) using an RDE-based clustering algorithm. The result allows to detect the problem or anomaly in the flight data in real time. In [25], the authors make use of RDE for real-time fault detection of industrial plants based on the analysis of control and error signals. The approach is based on the concept of density of the data space, which are obtained from the monitoring of a production plant. The detected outliers allow continuous monitoring of the process, especially in parameters such as pressure, temperature, flow and level in the tanks of the plant. The authors of [26] propose an algorithm called Autoclass, based on a classifier of an auto-adaptive system, which in turn makes use of RDE to estimate the density of the data space. Taking the characteristics of RDE, this algorithm allows the classifier to start learning from scratch, without having to specify fuzzy rules or the number of Autoclass labels. The algorithm is validated in a level control process where control signals and errors are used as attributes for the fault detection system. Finally, in [27] an algorithm that integrates three techniques for automatic detection and identification of objects in video streams is presented. One of these techniques is based on the inclusion of the RDE method. The main difference compared to other approaches is that it works on data set per frame without requiring a set of frames stored and processed in RAM.

3   Dataset Description

The data used in this work were collected by Geological Survey (Servicio Geológico Colombiano - SGC) in the Puracé volcano, which is located in the department of Cauca, at the geographical coordinates: 2° 19' 01 N and 76° 23' 53 W, at a distance of 26 km, southeast of the city of Popayán [28]. We built 7 data sets among periods: June 2015–July 2016 with 480.865 instances, which belong two surveillance volcanic areas:

- **Geochemistry:** contains data of carbon dioxide concentration (CO_2) and temperature (°C) through geochemistry stations in two points: Cocuy and Crater. The dataset built is composed by 4 attributes.
- **Deformation:** contains data of landslides in north and vertical axis, and temperature for each deformation station in five points: Agua Blanca, Cocuy, Curiquinga, Lavas Rojas and Guañarita. The dataset built is composed by 15 attributes.

Figure 1 shows the location of the stations: Crater (CRADC), Cocuy (CO3DC), Guañarita (GN2IN), Lavas Rojas (LARIN), Curiquinga (CURIN), Cocuy (CO_2IN) and Agua Blanca (ABLIN) around of Puracé volcano:

4   Outliers Detection Approach

The RDE application in order to find outliers were organized according to the volcanic monitoring area (geochemistry and volcano deformation).

4.1   Geochemistry

The result of applying the RDE method to the carbon dioxide stations belonging to the OVSPOP (*Observatorio Vulcanológico y Sismológico de Popayán* - Seismological and Vulcanological Observatory of Popayán) is shown in Fig. 2.

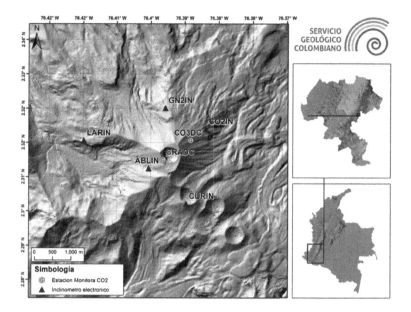

Fig. 1. Location map of CO_2 monitoring stations and electronic inclinometers. Source: OVSPOP (Seismological and Vulcanological Observatory of Popayán)

These stations have a sampling rate of one hundred samples per second (100 m/s). The horizontal axis of the figure corresponds to the number of instances (480865 records that make up the data set of a year of volcanic monitoring); the vertical axis represents the density given by the algorithm (values range from zero to one). Additionally, two curves can be identified, where the upper one (orange) represents the average of the data and the second one (yellow) represents the threshold that delimits the outliers, that is, the values that are below the second curve are the anomalies detected.

23 outliers from RDE applied to the aforementioned dataset were detected, for which the corresponding date, time and instance are extracted. For example, the first outliers detected are in the range from instance 5082 to 8030, corresponding to the dates of 2015-07-05 21:10 a 2015-07-08 02:54. The analysis of the variation of CO_2 concentration and temperature of each one of the geochemistry stations allows to find the relation of these variations with the outliers detected.

Figures 3 and 4, the first one hundred thousand instances are shown (for a better visualization, only the previously instances were presented.) for each of the stations (Crater and Cocouy). The concentration is measured in CO_2 Berkelium (color) and the temperature measured in degrees centigrade (red). Instances of the entire dataset. Finally, the vertical bars of gray color represent the outliers detected and are located in their corresponding instances of the horizontal axis.

In the Crater station the increases in the external temperature of the sensor are clearly identified (Fig. 3), such as the outlier marked approximately between

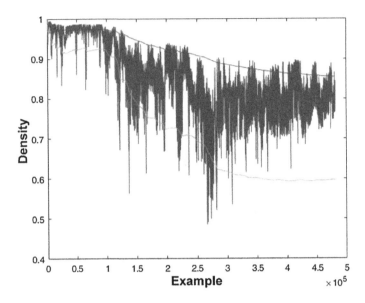

Fig. 2. Density in geochemistry stations (Color figure online)

instances number 23000 and 26000. This station is very close to the crater of the Puracé volcano. Only in the last period (outlier around instance 95000) a start of a continuous degassing of CO_2 gas that lasts approximately between 7 and 8 days is identified.

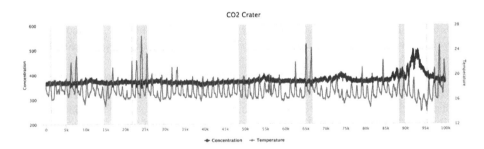

Fig. 3. Data of geochemistry stations: Crater. (Color figure online)

For Cocuy station (Fig. 4), changes in temperature rise and associated with the movement of volcanic gas carbon dioxide can also be identified, as are the precursor pulses of degassing phenomenon in small amounts of gas concentration (around instances 15000, 49000, 66000, 89000, 99000).

Additionally, in the year-round data of these two stations, other types of changes were identified such as: low, medium, and high temperature degassing,

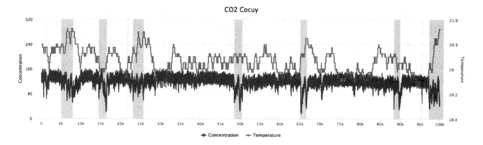

Fig. 4. Data of geochemistry stations: Cocuy.

phenomena precursor pulse, continuous degassing, small intermittent precursor degassing.

4.2 Deformation

Similar to Sect. 4.1, in the Fig. 5 the data density thrown by the RDE method for the dataset of the volcanic deformation area (5 inclinometers) is shown, where outliers were identified with date, time and associated instance information.

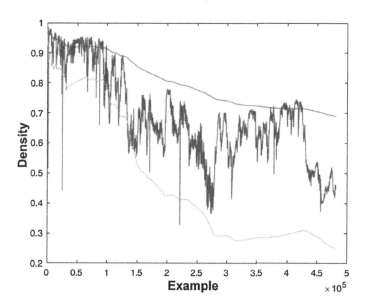

Fig. 5. Density in deformation stations

In Figs. 6, 7, 8, 9 and 10, the components: north, east (whose unit of measure is micro radians) and temperature (measured in degrees centigrade) are shown, from the deformation stations: Guañarita, Lavas Rojas, Curiquinga, Cocuy and

Agua Blanca, respectively. As in the Geochemistry stations, the outliers detected by RDE correspond to one or several instances of each of the graphs (gray bars).

In the Guañarita station (Fig. 6), around the instances number 25000 and 66000 a phenomenon of thermal contraction is observed, which indicates in a deflation on one of the flanks of the volcano associated with volumetric changes in the magma.

Fig. 6. Data of deformation stations: Guañarita.

For the Lavas Rojas station (Fig. 7), some problems of digitalization of the sensor (noise) are identified (instances between 24072 and 25633), which do not belong to any change of the volcano, then these data can not be taken into account in the decision making. On the other hand, changes in temperature and increase in the northern component of the digitizer are identified (around the instances 26000 and 95000), indicator of heat effect expansion which is associated with inflation on one of the flanks of the volcano. In this station there are some false negatives (instances between 42000 and 47000), corresponding to changes in the volcano that were not detected by the algorithm used.

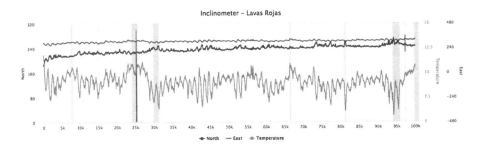

Fig. 7. Data of deformation stations: Lavas Rojas.

In Fig. 8, for the Curiquinga station, one of the outlier (around instance 81000) is related to a distant earthquake that occurred at a later time (instance

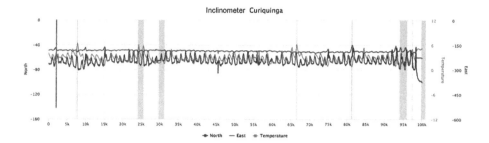

Fig. 8. Data of deformation stations: Curiquinga.

around 98000). For the sampling rate of these stations (1 sample per minute), the earthquakes detected are generally those that last more than one minute.

In the Cocuy station (Fig. 9), no significant behavior was observed indicating a change in the volcano, so the outliers obtained by the algorithm are false positives. However, the technique used indicates that at some of the inclinometer measuring stations something happened and a side effect was recorded at the Cocuy station.

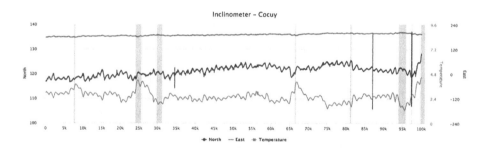

Fig. 9. Data of deformation stations: Cocuy.

Figure 10 presents the outliers detected in the Agua blanca station (around instance 8000 and 25000), changes in the inclinometer associated with deformation are observed, which are indicators of inflation phenomena and deflation of the volcanic cone.

Some stations register the change better than others; this is due to the geographical location of the inclinometer where the volcanic deformation is measured. However, the volcano monitoring from all its flanks represents a better response to the changes generated in its structure.

As in Geochemistry, for the dataset of one year monitoring, in the five monitoring stations of volcanic deformation other types of changes were identified such as: problems in digitalization, contraction, dilation, distant earthquakes, probability of solar storm and cycles belonging to day-night phenomena.

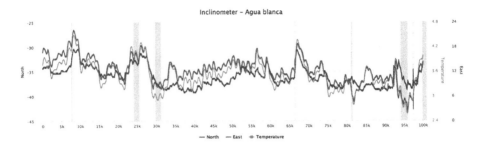

Fig. 10. Data of deformation stations: Agua Blanca.

5   Results

The Outliers detected in each dataset (13 for deformation and 23 for geochemistry) by RDE, were analyzed by OVSPOP experts, both in the area of Geochemistry and Volcanic Deformation. The information of experts who helped us with the interpretation of the outlier to be able to label the results obtained is presented below:

- **Geochemistry:** Luisa Fernanda Mesa, e-mail: lmesa@sgc.gov.co, chemical engineer.
- **Deformation:** Jorge Armando Alpala, e-mail: jalpala@sgc.gov.co, civil engineer.

Additionally, the anomalies detections from the outliers identified by the RDE algorithm are analyzed from the relationship between true positives (TP), true negatives (TN), false positives (FP) and false negatives (FN).

5.1   Geochemistry Expert Analysis

The confusion matrix for the results of the RDE application, mentioned in the previous section, is presented in Table 1.

Table 1. Confusion matrix for Geochemistry

Number of instances = 480888		Predicted value	
		Anomaly	No anomaly
Real value	Anomaly	19 (TP)	4 (FP)
	No anomaly	6 (FN)	480859 (TN)

The expert's analysis of the anomalies found by the algorithm is described below:

- For 4 of the instances where the algorithm has detected an anomaly, the expert has classified them as false positives, since the values of the sensors are in normal ranges.
- 3 anomalies correspond to low degassing.
- 1 anomaly corresponds to high temperatures.
- 9 anomalies correspond to continuous degassing.
- 4 anomalies correspond to medium degassing.
- 2 anomalies correspond to small low degassing imminent precursor.

5.2   Geodesy (deformation) Expert Analysis

The confusion matrix for the results of the RDE application in deformation dataset is presented in Table 2.

Table 2. Confusion matrix for Geodesy (deformation)

Number of instances = 480865		Predicted value	
		Anomaly	No anomaly
Real value	Anomaly	10 (TP)	3 (FP)
	No anomaly	2 (FN)	480850 (TN)

The expert's analysis of the anomalies found by the algorithm is described below:

- For 3 of the instances where the algorithm has detected an anomaly, the expert has classified them as false positives, since the values of the sensors are in normal ranges.
- 2 anomalies correspond to shrinkage by low temperature thermal phenomenon.
- 2 anomalies correspond to expansion of the thermal effect heat.
- 5 anomalies correspond to digitization. 2 of them have to do with solar storms and 3 are part of a combination between digitization and contraction of deformation phenomenon.
- One anomaly corresponds to a detection of a distal earthquake that was recorded by one of the stations close to the epicenter of this event.

Additionally, from confusion matrices presented above, a comparison of algorithm performance measures such as: Precision, Recall and F-measure; were extracted for each monitoring type and are presented in the Fig. 11.

For Geochemistry, the performance measures are 82% Precision, 86% Recall and F-measure value around 0.845. In the case of Geodesy, the results are 77% Precision, 83% Recall and F-measure value around 0.8. The Geochemistry area presented better results in the three calculated measures, where Recall was the best value obtained.

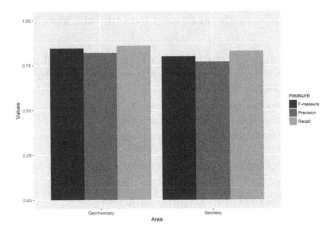

Fig. 11. RDE algorithm performance comparison for Geochemistry and Geodesy

6   Conclusions and Future Work

Volcano monitoring is a key task to detect volcanic anomalies and act accordingly. The fast and accurate detection of these anomalies is essential in those areas close to volcanoes which are very affected for its behaviour.

In this paper, we have proposed and efficient method for detecting any volcanic anomaly based on data density. One of the most important aspects of this paper is that the proposed method learns quickly from data streams in real time. In order to obtain this, RDE was taking into account and successfully used.

The experimentation results obtained have been analyzed considering the knowledge of several experts. This analysis shows that, in general, the different volcanic anomalies can be detected taking into account all the previous data of the volcano.

We propose as future works:

- Build time series based on data from Geochemistry and Deformation stations and Recursive Density Estimation to predict outliers.
- Use active learning for semi-automatic labeling of the outliers detected by Recursive Density Estimation.
- Consider data from Seismic stations as types of earthquakes: Vulcano tectonics, Long period, Tremors, Hybrids, Screw to validate the outliers detected by Recursive Density Estimation.
- Compare the results obtained (precision, overall, F-measure) with RDE applied to the geochemistry and deformation datasets against other anomaly detection techniques such as the ABOD and LOF algorithms. This will allow comparing the efficiency of an online anomaly detection algorithm (RDE) in volcanic environments against a density-based offline algorithm (LOF) and an algorithm based on angle distance (ABOD).

Acknowledgments. We are grateful to the Colombian Geological Survey (SGC) - especially to the volcanological and seismological observatory located in Popayán (OVSPOP) - for giving us the necessary data and helping us with this paper. In addition, we are grateful to Colciencias (Colombia) for PhD scholarship granted to MsC. David Camilo Corrales. This work has been also supported by:

– Project: "Alternativas Innovadoras de Agricultura Inteligente para sistemas productivos agrícolas del departamento del Cauca soportado en entornos de IoT - ID 4633" financed by Convocatoria 04C-2018 "Banco de Proyectos Conjuntos UEES-Sostenibilidad" of Project "Red de formación de talento humano para la innovación social y productiva en el Departamento del Cauca InnovAcción Cauca".
– The Spanish Ministry of Economy, Industry and Competitiveness (Projects TRA2015-63708-R and TRA2016-78886-C3-1-R).

References

1. Angelov, P., Buswell, R.: Evolving rule-based models: a tool for intelligent adaptation. In: Joint 9th IFSA World Congress and 20th NAFIPS International Conference, vol. 2, pp. 1062–1067. IEEE (2001)
2. Angelov, P., Buswell, R.: Identification of evolving fuzzy rule-based models. IEEE Trans. Fuzzy Syst. **10**(5), 667–677 (2002)
3. Angelov, P.: Anomalous system state identification, 12 July 2016. US Patent 9,390,265
4. Dzurisin, D.: A comprehensive approach to monitoring volcano deformation as a window on the eruption cycle. Rev. Geophys. **41**(1) (2003)
5. Dzurisin, D.: Volcano Deformation: New Geodetic Monitoring Techniques. Springer, Heidelberg (2006). https://doi.org/10.1007/978-3-540-49302-0
6. Caselli, A., Vélez, M., Agusto, M., Bengoa, C., Euillades, P., Ibáñez, J.: Copahue volcano (Argentina): a relationship between ground deformation, seismic activity and geochemical changes. In: The Volume Project., Volcanoes: Understanding subsurface mass movement, Jaycee Printing, Dublin, Ireland, pp. 309–318 (2009)
7. McSween, H.Y., Richardson, S.M., Uhle, M.E.: Geochemistry: Pathways and Processes. Columbia University Press, New York (2003)
8. Aggarwal, C.C.: Outlier analysis. In: Aggarwal, C.C. (ed.) Data mining, pp. 237–263. Springer, Heidelberg (2015)
9. Angelov, P., Ramezani, R., Zhou, X.: Autonomous novelty detection and object tracking in video streams using evolving clustering and Takagi-Sugeno type neurofuzzy system. In: 2008 IEEE International Joint Conference on Neural Networks (IEEE World Congress on Computational Intelligence), pp. 1456–1463, June 2008
10. Angelov, P.: Autonomous Learning Systems: From Data Streams to Knowledge in Real-Time. Wiley, Hoboken (2012)
11. Angelov, P.P.: Evolving rule-based models: a tool for design of flexible adaptive systems. Physica **vol**, 92 (2013)
12. Angelov, P.: Typicality distribution function-a new density-based data analytics tool. In: 2015 International Joint Conference on Neural Networks (IJCNN), pp. 1–8. IEEE (2015)
13. Breunig, M.M., Kriegel, H.-P., Ng, R.T., Sander, J.: LOF: identifying density-based local outliers. In: ACM Sigmod Record, vol. 29, pp. 93–104. ACM (2000)
14. Kuna, H.D., et al.: Avances en procedimientos de la explotación de información con algoritmos basados en la densidad para la identificación de outliers en bases de datos. In: XIII Workshop de Investigadores en Ciencias de la. Computación (2011)

15. Kriegel, H.-P., Zimek, A., et al.: Angle-based outlier detection in high-dimensional data. In: Proceedings of the 14th ACM SIGKDD International Conference on Knowledge Discovery and Data Mining, pp. 444–452. ACM (2008)
16. Lin, Y., Bunte, M., Saripalli, S., Greeley, R.: Autonomous detection of volcanic plumes on outer planetary bodies. In: 2012 IEEE International Conference on Robotics and Automation (ICRA), pp. 3431–3436. IEEE (2012)
17. Helmy, T., Fatai, A., Faisal, K.: Hybrid computational models for the characterization of oil and gas reservoirs. Expert Syst. Appl. **37**(7), 5353–5363 (2010)
18. Zmazek, B., Živčić, M., Todorovski, L., Džeroski, S., Vaupotič, J., Kobal, I.: Radon in soil gas: how to identify anomalies caused by earthquakes. Appl. Geochem. **20**(6), 1106–1119 (2005)
19. Gregorič, A., Zmazek, B., Džeroski, S., Torkar, D., Vaupotič, J.: Radon as an earthquake precursor-methods for detecting anomalies. In: Earthquake Research and Analysis-Statistical Studies, Observations and Planning. InTech (2012)
20. Miima, J.B., Niemeier, W.: Adapting neural networks for modelling structural behavior in geodetic deformation monitoring. Zfv Heft **3**, 1–8 (2004)
21. Angarita Vargas, M.F., et al.: Procesos de deformación en la región de los volcanes Chiles-Cerro Negro por medio de imágenes InSAR. Ph.D. thesis, Universidad Nacional de Colombia-Sede Bogotá (2016)
22. Mertikas, S., Rizos, C.: On-line detection of abrupt changes in the carrier-phase measurements of GPS. J. Geodesy **71**(8), 469–482 (1997)
23. Canon-Tapia, E., Szakács, A.: What is a Volcano? vol. 470. Geological Society of America, Boulder (2010)
24. Kolev, D., Angelov, P., Markarian, G., Suvorov, M., Lysanov, S.: ARFA: automated real-time flight data analysis using evolving clustering, classifiers and recursive density estimation. In: 2013 IEEE Conference on Evolving and Adaptive Intelligent Systems (EAIS), pp. 91–97. IEEE (2013)
25. Costa, B.S.J., Angelov, P.P., Guedes, L.A.: Real-time fault detection using recursive density estimation. J. Control Autom. Electr. Syst. **25**(4), 428–437 (2014)
26. Costa, B.S.J., Angelov, P.P., Guedes, L.A.: Fully unsupervised fault detection and identification based on recursive density estimation and self-evolving cloud-based classifier. Neurocomputing **150**, 289–303 (2015)
27. Angelov, P., Sadeghi-Tehran, P., Ramezani, R.: An approach to automatic real-time novelty detection, object identification, and tracking in video streams based on recursive density estimation and evolving takagi-sugeno fuzzy systems. Int. J. Intell. Syst. **26**(3), 189–205 (2011)
28. Gómez, J.E., Corrales, D.C., Corrales, J.C., Sanchis, A., Ledezma, A., Iglesias, J.A.: Monitoring of vulcano puracé through seismic signals: description of a real dataset. In: Evolving and Adaptive Intelligent Systems (EAIS), pp. 1–6. IEEE (2017)

A Rainfall Prediction Tool for Sustainable Agriculture Using Random Forest

Cristian Valencia-Payan$^{(\boxtimes)}$ (iD) and Juan Carlos Corrales (iD)

Universidad del Cauca, Popayán, Colombia
{chpayan, jcorral}@unicauca.edu.co

Abstract. In recent years world's governments have focused its efforts on the development of the Sustainable Agriculture were all resources, especially water resources, are used in a more environmentally friendly manner. In this paper, we present an approach for estimating daily accumulated rainfall using multi-spatial scale multi-source data based on Machine Learning algorithms for three HABs in the Andean Region of Colombia where the agricultural activities are one of the main production activities. The proposed approach uses data from different rain-related variables such as vegetation index, elevation data, rain rate and temperature with the aim of the development of a rain forecast, able to respond to local or large-scale rain events. The results show that the trained model can detect local rain events event when no meteorological station data was used.

Keywords: Rainfall · Machine learning · Cubist · CART · Random forest
Multiscale data · High Andean Basin · Sustainable agriculture

1 Introduction

Several authors have defined a High Andean Basin (HAB) as a basin located in the upper regions of the Andes Mountain, mainly located between the 2800 and the 4200 m above mean sea level. The precipitations in the Andean Region in Colombian are above the 4000 mm per year and it affects the levels on two of the most important rivers in Colombia which together support more than 80% of the population. Therefore, the main challenges in the Andean Region Basins of Colombia are related to the efficient use of water resources.

In recent years worlds governments have made efforts to move its agriculture into a more environmentally friendly system, the sustainable agriculture. The sustainable agriculture goal is to meet society's foods and textile needs in the present without compromising the ability to meet the needs of future generations [1]. One of the main topics in sustainable agriculture is the water use efficiency, it refers to the correct use of water in the agriculture processes.

The HABs located in the department of Cauca, Colombia are mainly used for agriculture, livestock production and pisciculture activities which demand a heavy use of water. On these HABs the agriculture processes located area mainly for use in potatoes and to a lesser extent other vegetable crops. According to [2] water supply is

© Springer Nature Switzerland AG 2018
I. Batyrshin et al. (Eds.): MICAI 2018, LNAI 11288, pp. 315–326, 2018.
https://doi.org/10.1007/978-3-030-04491-6_24

important to avoid deformations and diseases in the potato crops in dry periods or in rainy periods, but in HAB most of the crops depend mainly on the amount of rain due to the lack of industrialization of agriculture in the region. It's clear that the forecast of precipitation is very important for these HABs.

This paper is organized as follows: the first part outlines studies related to the main topics addressed around rain forecasting using ML; subsequently, the dataset construction showing the selected data sources and the data processing; ML architectures are validated by experimental evaluation results; and finally, the conclusions of this research are presented.

2   Related Work

In [3], an approach for extreme rainfall events using Artificial Neural Networks (ANN) is proposed, using data from multiple site observations of the study zone. This approach was used to estimate the rainfall events with up to 12 months lead time with 25% to 80% skill score. In [4] a one-year daily rainfall forecasting is presented using Singular Spectrum Analysis (SSA) to decompose a meteorological time series from 1961 to 2013.

In [5], data mining and machine learning techniques such as Support Vector Regression (SVR) and ANN for weather forecasting (rain and temperature) using a dataset constructed from a single meteorological station. In [6], predictive capacity of meteorological data is evaluated using multiple machine learning methods to predict the range of the amount of rain expected the next day using information not only on rain but on temperature and wind.

Meanwhile [7] proposed an approximation for the estimation of the climate properties based on a large images database related to multiple climatological conditions using Decision Trees (DT). In [8] a method using a modified version of K-means to identify storm patrons is proposed, this method performs a mining process on all the storm-related data available in the studied zone.

In [9] a methodology to estimate precipitation using multiple infrared and radar satellite images is proposed. The authors use a machine learning algorithm and morphological operations to detect clouds and estimate the precipitation. In [10] a Deep Learning approach to predict heavy rainfall within 6 to 48 h before it's occurrence is proposed, but this proposal leave aside normal rainfall events.

According to the prior literature review, it is important to consider multi-spatial scale data, to obtain rain estimates with a precision comparable to those using rain gauge data, but with a wider range. Moreover, no studies using multiple climatic data at multiple spatial scales were found. The closest studies to this proposal use only rain information and have not been applied in the High Colombian Andes.

3   Dataset Construction

For this study we selected three HABs located in the northeastern Cauca, these HABs were selected based on the proximity and the importance of the resources extracted and generated in the most populated city in the department. Using the approach proposed by [11] we got the land cover classification for the three selected HABs. This approach uses Landsat and a Digital Elevation Model (DEM) to obtain the expected land cover, processing all Landsat data to obtain multiple vegetation indexes that allow them to differentiate between plants, water, soil and more [12], and the DEM data. All the data is processed using a Machine Learning Model. An example of the obtained classification used in this study is included in the following (Fig. 1).

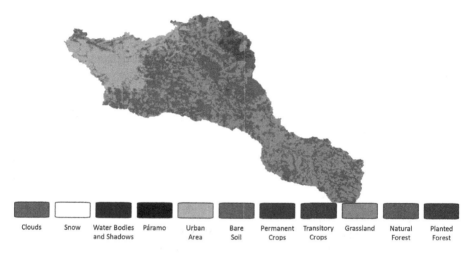

Clouds   Snow   Water Bodies   Páramo   Urban   Bare   Permanent   Transitory   Grassland   Natural   Planted
                        and Shadows   Area   Soil   Crops   Crops   Forest   Forest

Fig. 1. Landsat classification for the Molino River HAB using the [11] approach.

This classifications are useful to see that agriculture is important in the selected HAB considering the amount of land used in every HAB, also other natural coverage of great importance, which has a great dependence on the amount of available water.

This proposal is important due the implications the water resources available for agricultural activities has on the HABs located in the department of Cauca which demand 42% of its resources [13], and the world governments are putting their efforts in the development of sustainable agriculture promoting a better use of natural resources.

3.1   Data Sources

In this study, different multiscale climatological databases were pre-processed to estimate the amount of daily rain. To achieve a predictor strong enough to estimate with the highest accuracy using multiscale data, five free access and one non-free access databases were used as training data. Data sources are as follows.

TRMM Data

The Tropical Rainfall Measuring Mission (TRMM) is a joint NASA and JAXA mission launched in 1997 to study rainfall [14]. The TRMM product used in this study was the 3B42 3-Hourly Rainfall Data with a spatial resolution of 25 km by 25 km.

GOES Data

The Geostationary Operational Environmental Satellite Program (GOES) is a joint mission of NASA and the National Oceanic and Atmospheric Administration (NOAA). In this study, GOES-13 was the satellite used to gather the climatic data. To obtain the rainfall data from the GOES data, the model proposed in [15] called the auto-estimator was used.

Meteoblue Data

Initially developed at the University of Basel, based on NOAA and the National Centers for Environmental Prediction (NCEP) models [16]. It delivers local weather information for any point in the world, with a spatial resolution of 5 km by 5 km, providing temperature, wind direction, and speed, relative humidity, solar radiation, cloud cover, and precipitation.

Modis NDVI Data

The Moderate Resolution Imaging Spectroradiometer (MODIS) vegetation indices, produced at 16-day intervals at multiple spatial resolutions, provide consistent spatial and temporal comparisons of vegetation canopy greenness, a composite property of leaf area, chlorophyll and canopy structure [17] using daily data.

Digital Elevation Map (DEM)

The Shuttle Radar Topography Mission [18] of NASA produced a DEM with a spatial resolution of 30 m by 30 m with global coverage and free access. It has reported a vertical error to be less than 16 m.

Oceanic el Nino Index (ONI)

Is a measure of the El Nino - Southern Oscillation or ENSO [19], produced by the Climate Prediction Center (CPC) of NOAA. This index is an indicator of the presence of the El Niño (warm periods) or La Niña (cold periods) in the ocean temperature. The data are present monthly since 1950, although they are produced and updated on a quarterly basis.

3.2   Data Selection and Cleaning

In [20] the authors concluded that the amount of precipitation increases with elevation, following the characteristics of the topography. In [21] a relationship between NDVI and the amount of precipitation was found. In [22] a high correlation was found between humidity and rainfall. Finally, in [23] found that Temperature and

Table 1. Selected variables and their corresponding units

Attribute name	Unit	Source
Year	Numeric (dimensionless)	[16]
Month	Numeric (dimensionless)	[16]
Day	Numeric (dimensionless)	[16]
MaxTemp	Numeric (°C)	[16]
MinTemp	Numeric (°C)	[16]
MeanTemp	Numeric (°C)	[16]
Relative_Humidity	Numeric (%)	[16]
TRMM	Numeric (mm h^{-1})	[14]
NDVI	Numeric (dimensionless)	[17]
ONI	Numeric (dimensionless)	[24]
AMSL	Numeric (m)	[18]
GOES	Numeric (mm h^{-1})	[15]
Actual_Total_Precipitation	Numeric (mm)	[16]

precipitation have a positive or negative correlation depending on spatial location and other climatic variables. Based on the previous studies the final dataset has the following attributes.

Table 1 shows all the selected variables and the objective variable, with its corresponding name on the dataset, the unit, and the source. In this case, Actual Total Precipitation is the objective variable.

After this selection, we processed the TRMM and GOES databases to give it a more adequate georeferenced information, in our case the World Geodetic System 1984 (WGS 84) using its reference geolocation points and changed the HDF and NetCDF format to GeoTIFF format for both cases. To achieve these changes an automatic program was developed using OpenCV and Gdal Libraries.

Once this process was completed, we proceeded to extract all the information with the same date from the databases, for this task an automatic program was developed using OpenCV and Gdal libraries in C++ language, to load all the information and generate a file with all the available data. Due to the temporal coverage in the Meteoblue data we decide to work with a temporal window, from January 1, 2012 to December 31, 2016. Once the data was collected we notice the existence of a data gap between 2012 and 2014 in the GOES data. The next phases were applied initially for the Molino HAB, after finishing all phases the same process was applied to Las Piedras and Cajibío HABs.

Following the procedure that is shown at [25] where a Data Cleaning guide for regression models is proposed. We proceeded to perform a data cleaning process to achieve the best results in the forecasting. This cleaning process is resumed in Fig. 2.

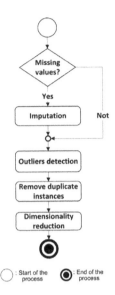

Fig. 2. Process for data cleaning in regression models [25].

To impute the missing values three imputation methods were used to fill in the missing values (MV): Predictive Mean Matching (PMM) [26]; HotDeck [27], Expectation Maximization (EM) [28] and missForest [29] in the R software tool. To achieve this, TRMM data was used, due to the close relationship between this data. The GOES data is used in the process to obtain TRMM data. The summary of this process can be seen in Table 2.

Table 2. Results from the imputation methods

Data	Min.	Median	Mean	Max.	NA
Original Data	0.0006	0.0315	0.4209	10.5148	873
PMM Imp.	0.000584	0.035935	0.458953	10.5148	-
HotDeck Imp.	0.000584	0.036811	0.410911	10.5148	-
EMImputation	−3.57588	0.09893	0.43213	62.08591	-
missForest Imp.	0.000584	0.256726	0.433130	10.5148	

As can be seen in Table 2, the PMM and HotDeck methods give similar results compared with the Original data, the missForest data differs a little from the behavior of the original data but its stats are close. The EMImputation method produces different results, which must be discarded due to the negative values that have been used to impute the missing values on the dataset. Negative values make no sense for rainfall rate $(mm*h^{-1})$. Finally, the PMM, HotDeck and missForest imputed GOES variables were used in this study after performing a correlation test between them, using parametric and non-parametric correlation test.

Table 3. Results from the correlation test

Correlation model	PMM/HotDeck	PMM/missForest	HotDeck/missForest
Pearson	0.42	0.67	0.57
Kendall's Tau	0.38	0.5	0.4
Spearman's Rho	0.44	0.6	0.48

As can be seen in Table 3, PMM and missForest have a strong correlation, while with HotDeck the correlation is slightly lower but considerable, in all cases, PMM and missForest has a strong correlation. Taking this into account a t-test was performed to determine which dataset will be used in the modeling process. The t-test results showed that the null hypothesis must be rejected according to the p-values greater than 0.05 for all cases, so none could be rejected at this point. According to these results, the three constructed datasets were used in the modeling process.

To perform the Outliers and Duplicates detection we use the R software tool function called outlierTest included in the Caret package, and the duplicated function included in the base R packages. Due to the size of the dataset, there was no need to perform dimensionality reduction

4 Modeling the Rainfall Accumulation Forecast

Machine learning was developed to learn and perform specific tasks for example to predict the amount of daily rain based on multiscale data. The predictor should be able to estimate the total amount of rain based on parameters such as temperature, rainfall rate, and humidity, among others. This task is formally called regression. The regressors showed in this study were the one with the best results on the experimentation process. All the models were trained using R package Caret.

Random Forest (RF)
This ensemble learning method is the combination of tree predictors. Each one depends on the values of a random vector of samples obtained independently from the training samples and with the same distribution for all the trees [30].

CART
It is a method for the construction of prediction models from a set of input data. These are obtained by partitioning the space of the input data and adjusting a prediction model to each partition [31].

Cubist
It is an extended version of the M5 model, it is a model of regression trees based on rules [32]. This model generates a list of decisions for regression problems using the design paradigm of divide and conquers algorithm.

RF, CART and Cubist, predictors in Table 4 were used to evaluate the consolidated dataset imputed with PMM, HotDeck, and missForest. To apply this test, R tool and the Caret package were used, and all the predictors were configured using the default parameters, all datasets were randomly divided to use 80% of the data as training samples and the remaining 20% as test samples.

Table 4. Results of the first test using the imputed datasets

Predictor	Correlation	MAE	RMSE
$CART_{PMM}$	0.55	2.92	4.91
$Cubist_{PMM}$	0.5	2.75	5.93
RF_{PMM}	0.6	2.78	4.58
$CART_{HotDeck}$	0.55	2.91	4.86
$Cubist_{HotDeck}$	0.46	2.84	6.08
$RF_{HotDeck}$	0.6	2.80	4.59
$CART_{missForest}$	0.56	2.91	4.85
$Cubist_{missForest}$	0.46	2.85	6.11
$RF_{missForest}$	0.59	2.81	4.63

As can be seen in Table 4 the correlation coefficient in CART and RF predictors suggests a strong correlation between the selected variables and Total Precipitation, while the Cubist predictor has a moderate correlation. The MAE and RMSE suggest that the predicted values are closest to the real values in the predictors with a strong correlation compared with the one that has the moderate correlation. The high RMSE value could be due to some points where the predicted value has a large difference from the real value. The worst results were obtained using the Cubist model.

Figure 3 shows the predicted value vs the real precipitation value, for the RF machine learning model using the PMM imputed dataset that has the best results in all tests.

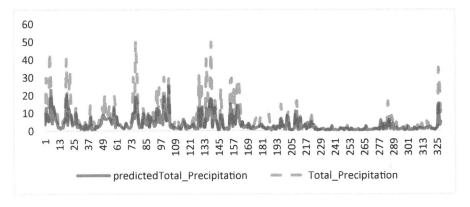

Fig. 3. RF predicted total precipitation vs Actual_Total_Precipitation (using PMM imputed dataset) in the Molino HAB.

As can be seen, the Fig. 3, confirms the results in Table 4 for the correlation value in RF model using the PMM imputation method, the predicted and real precipitation have a very similar behavior. The RF model predicts the amount of rain correctly in most cases, except for those where the *Total_Precipitation* has high values or where the values are closest to the 10 mm of rain accumulation. But the overall behavior is good. More information on different spatial scales is probably needed to adjust the predictions when the amount of rain has high values or to apply a correction after the prediction model to adjust the predicted value in these conditions.

Finally, a comparison was carried out of the *predictedTotal_Precipitation* from the RF model and the accumulated rainfall measured by a meteorological station within the study area during the first 100+days of 2016.

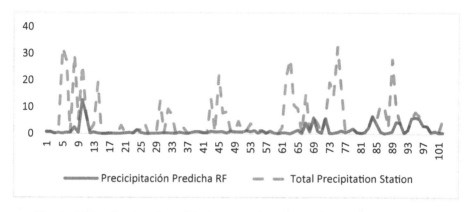

Fig. 4. RF predicted total precipitation vs station precipitation in the Molino HAB.

Figure 4 shows the comparison between the RF predicted precipitation and the data collected by the meteorological station data. As can be seen, the RF model has a better behavior at a local level, giving outputs that in most cases underestimate the local precipitation. The use of multiple spatial scale data has given the RF model the ability to predict rainfall at large spatial scales while been able to react to local precipitations events.

After completing this test, we applied the same process for the remaining selected HAB, using only the PMM imputation method and the RF algorithm. The results can be seen in Table 5.

Table 5. Results for the remaining HAB using the RF model

Model	Correlation	MAE	RMSE
RF$_{Cajibio}$	0.41	4.48	8.11
RF$_{Las\ Piedras}$	0.78	1.62	3.03

As can be seen in Table 5 the RF model response for Cajibío and Las Piedras HABs. The values for the Cajibío HAB shows that it has a moderate correlation and higher MAE and RMSE values when compared to the Molino HAB, these results can be explained because for this HAB we found more missing data values present on other attributes on the cleaning process, so the instances number was reduced for this case. In Las Piedras HAB we got best results with the lowest MAE and RMSE and the highest correlation, in this case, due to the mayor size of the HAB we get more data from the sources without instance duplication.

The three HAB the underestimation problem is far greater than the over-estimation, but in general terms, the models have a good performance.

5   Conclusions

This paper explains the importance of water resources in the HABs located in the department of Cauca. Most of the crops present on this HABs depends mostly on precipitation for the irrigation process so an adequate use of the water resources is highly important. A reliable prediction tool would allow the decision-making processes related to the planting time, irrigation and harvest to be more accurate, and for watershed management authorities aiming to develop activities aiming to the development of agricultural practices in concordance with the Sustainable Agriculture, will be a helpful tool.

As can be seen in the Landsat Classifications crops are one the mayor land cover on these HABs making this daily rainfall accumulation prediction model a suitable tool to the development of a more adequate crop planning considering the amount of rain expected so the crop planting happens at the right moment with the correct planting considerations.

In the tests, the RF model was best, according to the results shown in the cross validated test and the realistic approach test. The results have shown that RF model was able to reproduce local rain events even when no local data such as the meteorological station data was used to build the model, although it underestimates the real amount of precipitation registered at the station.

The results suggest that more data on highest spatial scales or more data related with the processes involved in the rain generation such as other vegetation index, more geomorphological data or more climatological data requires to be taken into account to be able to achieve the forecast of the amount of rain in local and large rainfall events, although these local events were able to be successfully detected. Future work would focus on the improvement of the prediction of the amount of rain in these two scenarios.

Acknowledgments. The authors are grateful to the Telematics Engineering Group (GIT) and the Optics and Laser Group (GOL) of the University of Cauca, The University of Cauca, Meteoblue, RICCLISA Program, and the AgroCloud project for supporting this research, as well as the AQUARISC program for the PhD support granted to Cristian Valencia-Payan.

References

1. Feenstra, G.: What is sustainable agriculture? — UC SAREP, UC Sustainable Agriculture Research and Education Program (2017). http://asi.ucdavis.edu/programs/sarep/about/what-is-sustainable-agriculture. Accessed 23 May 2018
2. Ministerio De Agricultura y Desarrollo Rural and Departamento Administrativo Nacional de Estadística: El cultivo de la papa, Solanum tuberosum Alimento de gran valor nutritivo, clave en la seguridad alimentaria mundial. Insumos y Factores asociados a la producción agropecuaria, no. 15, p. 92 (2013)
3. Abbot, J., Marohasy, J.: Forecasting extreme monthly rainfall events in regions of Queensland, Australia using artificial neural networks. Int. J. Sustain. Dev. Plan. **12**(7), 1117–1131 (2017)
4. Unnikrishnan, P., Jothiprakash, V.: Daily rainfall forecasting for one year in a single run using singular spectrum analysis. J. Hydrol. **561**, 609–621 (2018)
5. Rasel, R.I., Sultana, N., Meesad, P.: An application of data mining and machine learning for weather forecasting. In: Meesad, P., Sodsee, S., Unger, H. (eds.) IC2IT 2017. AISC, vol. 566, pp. 169–178. Springer, Cham (2018). https://doi.org/10.1007/978-3-319-60663-7_16
6. Ahmed, B.: Predictive capacity of meteorological data: will it rain tomorrow? In: Proceedings of the 2015 Science and Information Conference, SAI 2015, pp. 199–205 (2015)
7. Chu, W.T., Zheng, X.Y., Ding, D.S.: Image2weather: a large-scale image dataset for weather property estimation. In: Proceedings - 2016 IEEE 2nd International Conference on Multimedia Big Data, BigMM 2016, pp. 137–144 (2016)
8. Gupta, U., Jitkajornwanich, K., Elmasri, R., Fegaras, L.: Adapting k-means clustering to identify spatial patterns in storms. In: Proceedings - 2016 IEEE International Conference on Big Data, Big Data 2016, pp. 2646–2654 (2016)
9. Hong, Y., Chiang, Y.M., Liu, Y., Hsu, K.L., Sorooshian, S.: Satellite-based precipitation estimation using watershed segmentation and growing hierarchical self-organizing map. Int. J. Remote Sens. **27**(23), 5165–5184 (2006)
10. Gope, S., Sarkar, S., Mitra, P., Ghosh, S.: Early prediction of extreme rainfall events: a deep learning approach. In: Perner, P. (ed.) ICDM 2016. LNCS (LNAI), vol. 9728, pp. 154–167. Springer, Cham (2016). https://doi.org/10.1007/978-3-319-41561-1_12
11. Pencue-Fierro, E.L., Solano-Correa, Y.T., Corrales-Muñoz, J.C., Figueroa-Casas, A.: A semi-supervised hybrid approach for multitemporal multi-region multisensor landsat data classification. IEEE J. Sel. Top. Appl. Earth Obs. Remote Sens. **9**(12), 5424–5435 (2016)
12. Ho, P.-G.P.: Geoscience and Remote Sensing. INTECH (2009)
13. Ministerio de Ambiente Vivienda Y Desarrollo Teritorial, Ministerio de Hacienda y Crédito Público: CONPES 3624 - Programa para el saneamiento, manejo y recuperación ambiental de la cuenca alta del río Cauca, p. 60 (2009)
14. National Aeronautics and Space Administration: TRMM Home Page | Precipitation Measurement Missions. https://pmm.nasa.gov/trmm. Accessed 29 Mar 2018
15. Vicente, G.A., Scofield, R.A., Menzel, W.P.: The operational GOES infrared rainfall estimation technique. Bull. Am. Meteorol. Soc. **79**(9), 1883–1893 (1998)
16. Meteoblue. https://content.meteoblue.com/en/about-us. Accessed 28 Jul 2017
17. MODIS Vegetation Index Products. https://modis.gsfc.nasa.gov/data/dataprod/mod13.php. Accessed 28 Jul 2017
18. NASA: Shuttle Radar Topography Mission 2017. https://www2.jpl.nasa.gov/srtm/. Accessed 28 Jul 2017
19. N. C. P. Center. NOAA's Climate Prediction Center

20. Sasaki, H., Kurihara, K.: Relationship between precipitation and elevation in the present climate reproduced by the non-hydrostatic regional climate model. SOLA **4**, 109–112 (2008)

21. Purevdorj, T., Hoshino, B., Ganzorig, S., Tserendulam, T.: Spatial and temporal patterns of NDVI response to precipitation in Mongolian Steppe. J. Rakuno Gakuen Univ. **35**(2), 55–62 (2011)

22. Umoh, A.A.: Rainfall and relative humidity occurrence patterns in Uyo metropolis, Akwa Ibom State, South-South Nigeria. IOSR J. Eng. **03**(08), 27–31 (2013)

23. Trenberth, K.E., Shea, D.J.: Relationships between precipitation and surface temperature. Geophys. Res. Lett. **32**(14) (2005)

24. NOAA Center for Weather and Climate Prediction: Climate Prediction Center (CPC), Madden Jullian Oscillation (MJO) 2013. http://www.cpc.ncep.noaa.gov/products/analysis_monitoring/ensostuff/ensoyears.shtml. Accessed 28 Jul 2017

25. Corrales, D.C., Corrales, J.C., Ledezma, A.: How to address the data quality issues in regression models: a guided process for data cleaning. Symmetry (Basel) **10**(4), 1–20 (2018)

26. Vink, G., Frank, L.E., Pannekoek, J., van Buuren, S.: Predictive mean matching imputation of semicontinuous variables. Stat. Neerl. **68**(1), 61–90 (2014)

27. Andridge, R.R., Little, R.J.: A review of hot deck imputation for survey non-response. Int. Stat. Rev. **78**(1), 40–64 (2010). NIH Public Access

28. Dempster, A.P., Laird, N.M., Rubin, D.B.: Maximum likelihood from incomplete data via the EM algorithm. J. R. Stat. Soc. Ser. B Methodol. **39**(1), 1–38 (1977)

29. Stekhoven, D.J., Bühlmann, P.: Missforest-Non-parametric missing value imputation for mixed-type data. Bioinformatics **28**(1), 112–118 (2012)

30. Breiman, L.: Random forests. Mach. Learn. **45**(1), 5–32 (2001)

31. Loh, W.-Y.: Classification and regression trees. Wiley Interdiscip. Rev. Data Min. Knowl. Discov. **1**(1), 14–23 (2011)

32. Quinlan, J.R.: An overview of Cubist. Retrieved June 2017. http://rulequest.com/cubist-win.html. Accessed 30 Mar 2018

Kolb's Learning Styles, Learning Activities and Academic Performance in a Massive Private Online Course

Mario Solarte[1]([✉]), Raúl Ramírez-Velarde[2], Carlos Alario-Hoyos[3], Gustavo Ramírez-González[1], and Hugo Ordóñez-Eraso[4]

[1] Universidad del Cauca, Popayán, Colombia
{msolarte,gramirez}@unicauca.edu.co
[2] Instituto Tecnológico de Estudios Superiores de Monterrey,
Monterrey, Mexico
rramirez@itesm.mx
[3] Universidad Carlos III de Madrid, Madrid, Spain
calario@it.uc3m.es
[4] Universidad San Buenaventura, Bogotá, Colombia
hugoeraso@gmail.com

Abstract. Massive Open Online Courses (MOOC) have been considered an "educational revolution". Although these courses were designed to reach a massive number of participants, Higher Education institutions have started to use MOOCs technologies and methodologies as a support for educative traditional practices in what has been called Small Private Online Courses (SPOCs) and Massive Private Online Courses (MPOCs) according to the proportion of students enrolled and the teachers who support them. A slightly explored area of scientific literature is the possible correlations between performance and learning styles in academic value courses designed to be offered in massively environments. This article presents the results obtained in the MPOC "Daily Astronomy" at University of Cauca in terms of the possible associations between learning styles according to Kolb and the results in the evaluations and the activity demonstrated in the services of the platform that hosted the course.

Keywords: MOOC · MPOC · Kolb´s learning styles · Assessment
PCA

1 Introduction

MOOCs appeared in 2008 as an evolution of Open Educational Resources (OERs), as a proposal to universalize education and offer free, quality education to people living in remote or disadvantaged areas, inspired by Connectivism [1], an innovative Educational Theory proposed by George Siemens in 2005. They are characterized by the offer of free courses accessible through the Internet [2], from which a certificate of approval can be issued after the respective payment [3], usually of short duration [4], focusing on content - which should be open - basically video type [5], with relatively simple evaluative activities [6] with no limit on the number of registrations [7].

© Springer Nature Switzerland AG 2018
I. Batyrshin et al. (Eds.): MICAI 2018, LNAI 11288, pp. 327–341, 2018.
https://doi.org/10.1007/978-3-030-04491-6_25

Over the years, MOOCs have begun to play an important educational role in higher education [8], hand in hand with the so-called bMOOCs or combined MOOCs that merge traditional education strategies with the advantages of MOOCs in improving teaching and learning [9]. This new approach uses MOOC content and activities as part of courses also supported by face-to-face sessions leading to various combinations of hybrid methodologies such as the inverted classroom where students must first watch MOOC videos at home, then take advantage of classroom sessions to resolve concerns with the teacher or develop the practical component of the content [10].

It is also possible to use MOOC's own strategy as an instrument for the development of formative processes with academic recognition in higher education without the need to mix it with face-to-face activities. Terms such as SPOCs (Small Private Online Courses) and MPOCs (Massive Private Online Courses) [11, 12] reflect the academic community's concern to use the essence of the MOOC movement in formal higher education processes.

This use of MPOCs in higher education (formal education), basically online courses with a relatively high number of students and a small number of teachers and assistants who do not make it possible to follow one by one the learning activities [13], should become an interesting research scenario in the classroom due to the large amount of data that can be captured through the web platforms that offer them and the data analysis techniques used in the studies of the MOOCs.

This article aims to present the results obtained in the MPOC "Daily Astronomy" of the University of Cauca, offered to 414 students during the second academic period of 2016 and hosted in an Open edX instance [14], in order to determine the influence on the learning outcomes of variables such as the learning style of students and demographic variables.

From a methodological point of view, the large amount of data collected during the MPOC implies an initial difficulty that must be resolved before the analysis itself, from which another research question also arises. What variables contain more information regarding student performance in an MPOC?

In [15, 16] Principal Component Analysis (PCA) is applied for the reduction of dimensionality applied to verify the accuracy of a course's evaluation system and also seeking the correlation between performance in the learning activities of a course and the results of the final evaluation. But such research has been carried out with face-to-face surprisies and has not taken into account the students' learning style according to Kolb [17]. Nor did they analyse courses with a high number of students enrolled.

This article describes the structure of the data set collected from the MPOC's "Daily Astronomy" data set that collects the pedagogical strategy with which it was designed, illustrates the results of reducing the dimensionality of the dataset through the PCA, student groupings through observation points (evaluation and activities on the MPOC platform), to end with the presentation of conclusions and future work.

2 Study Case and Data

The purpose of this study is to determine if on-line learning activities influence student learning at MPOC in University of Cauca.

The research questions are multiple: (a) Determine if there are difference between different learning styles according to Kolb. (b) Determine if there are correlations between activities carried out on Open edX, such as number of accesses, videos played, web resources accessed, participation in forums, on-line practice labs, etc. (c) Classify students by their platform activities and determine such classification is related to Kolb's learning styles. (d) Determine if demographic factors influence student performance.

The main tool used to analyse the courses is Principal Component Analysis (PCA) descrito [18]. PCA, by finding the eigenvalues and the eigenvector of the correlation matrix, determines orthogonal axis of maximum variance, called principal component or PC. The first principal component or factor (F1) captures the largest amount of variability, the second (F2) the next largest variability and so on. Variables are projected into those axis calling these "factor loadings". Whereas observations, when projected into the axis are called "factor scores". Factor loadings show how variables group together. The dimension of a dataset can be reduced when factor loadings of different variables are rather close. Factor scores show how observations group together. Clusters of observations may eventually be observed suggesting similitudes and differences between groups. Furthermore, if factors resulting from PCA are adequate and interpretable, factor scores can also be used to analyse correlations and build predictive models of desired outcomes. The following principal component analysis were carried out:

1. PCA1: Analysis of grade related variables
2. PCA2: Analysis of platform based activities with some profile grade data
3. PCA3: Analysis of platform based activities without grade profile data

The date is a consolidation of platform delivered activities, surveys and off-line activities and considerations too. For example, the computation for the 70% course grade, the best 6 grades of the best 8 exams were considered, whereas the grade for the final 30% is integrated using the last five tests. Also, the classification in Kolb's Learning Styles was done through a Google Docs survey which as automatically processed to determine 9 possible combinations that were reduced to 4 major learning styles.

The combined data contains 54 variables described for 284 samples out of an original cohort of 414 students enrolled in the MPOC. The variables included in the study are described in Table 1:

3   Dimension Reduction

In this section we aim to determine which variables are not giving useful information and can be eliminated. We do this by reviewing in PCA1, PCA2 and PCA3 those groups of variables (student activities on MPOC) that give the same information. This is shown in the biplots for F1 and F2. These biplots show the correlation between variables and two PCA axis, mainly F1 and F2. They are equivalent to plotting the factor loadings. We show this in Figs. 1, 2 and 3. Biplot sticks that are rather horizontal

Table 1. Variables and their descriptions

Variable	Description
ES01… ES13	Exam week 1… Exam week 12 (between 0 and 1)
R1… R2	Remedial exam 1 and 2 (voluntary)
Parcial.70	Grade for first 70% course (between 0 and 1)
Parcial.30	Grade for last 30% course (between 0 and 1)
Definitiva	Final grade (between 0 and 1)
Pro6p	Average of the first 6 exams
Estilo.segun.Kolb	Style according to Kolb
Kolb	(C = Accomodating S = Assimilative, O = Convergent I = Divergent)
Kolb.Num	Kolb style number (1 = Accomodating, 2 = Assimilative, 3 = Convergent, 4 = Divergent)
Genero	Gender 1 (Femenine), 2 (Masculine)
Programa	Academic program
Semestre	Academic semester
Promedio.2016.2	Academic average
Diferencia. Aprendizaje	Learning difference = Definitiva - Promedio.2016.2
Puntaje.Ingreso	Entrance examination average
Prioridad	Academic program priority
Estrato	Social-economic strata
Nacimiento	Year of birth
Edad	Age
Municipio	Current residence municipality
Departamento	Current residence department
Tipo.Bachillerato	Public or private high school
Ingresos	Number of accesses to online course
Examenes. Presentados	Number of exams solved
Accesos. Diapositivas	Number of hits on course slides
Accesos.Paginas. Video	Number of hits on course videos
Play.Video	Number of pauses while viewing videos
Stop.Video	Number of videos completed without pauses
Accesos.Recurso. Web	Number of hits on pages with course resources
Acesos. Simuladores	Number of hits on simulators
Accesos.Practicas	Number of hits on labs
Accesos.Talleres	Number of hits on workshops
Foros.Leidos	Number of read fora
Foros.Creados	Number of created fora

are correlated to F1 and vertical ones are correlated to F2. Diagonal sticks are correlated to both. Short sticks are more prominent in other dimensions (F3, F4, etc.).

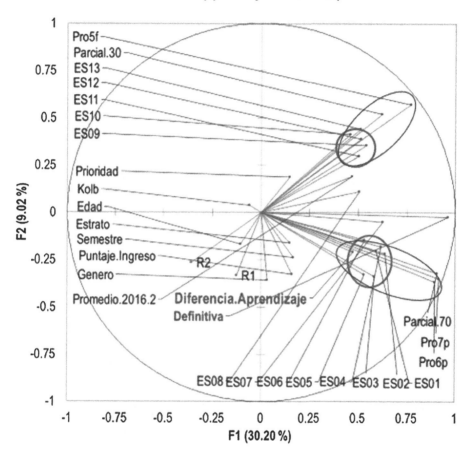

Fig. 1. Biplot of PCA1 showing clusters of variables indicating different evaluations.

Variables with sticks grouped together are highly correlated regardless of orientation. Therefore, in any such close group of variables, all can be dropped but one. Since each variable represents either an action or a grade, we can reduce either the number of evaluations or discard activities that are not relevant (and possible make room for other activities).

To determine is the number of course evaluations can be reduced, PCA 2, containing only information about the evaluations was carried out. The biplot is shown in Fig. 1. In this figure we see two closed clusters of evaluation as shown in Table 3. The first close cluster shown in green is formed by Evaluations 1 to 8. The second cluster in blue consists of evaluations 9 to 13. The purple ovals also show that the green cluster is

Variables (ejes F1 y F2: 30.90 %)

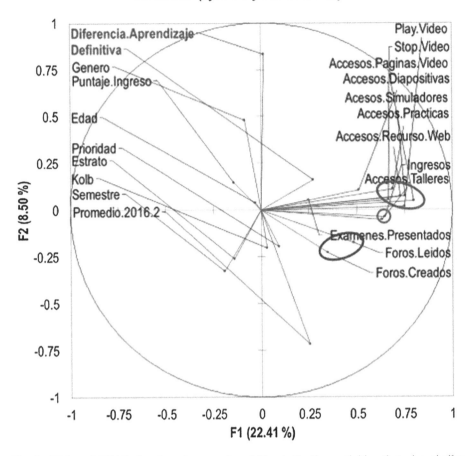

Fig. 2. Biplot of PCA2 showing clusters of variables indicating activities that give similar results. (Color figure online)

related to grades Pro7p, Pro6p and Parcial.70 whereas the blue cluster is related to grades Prof5f and Parcial.30. It means that the course is divided en two sections. It also means that the course is over evaluated. From evaluation 1 to 8, 7 can be dropped as they give the same information. From evaluations 9 to 13 4 can be dropped as they give the same information also. Course evaluation should consist of two partial evaluations and one final global evaluation. Since final grade is a linear combination of all the grades, obviously the final grade is correlated to the grades variables. It also seems, that these variables are also correlated with learning difference variable.

Figure 1 also shows that socio-economic fata is unrelated to grades, as the short sticks indicate prominance in other dimensions and in general this sticks are oriented orthogonally to grades. Nevertheless, a few correlations remain that we will explore later.

Variables (ejes F1 y F2: 54.19 %)

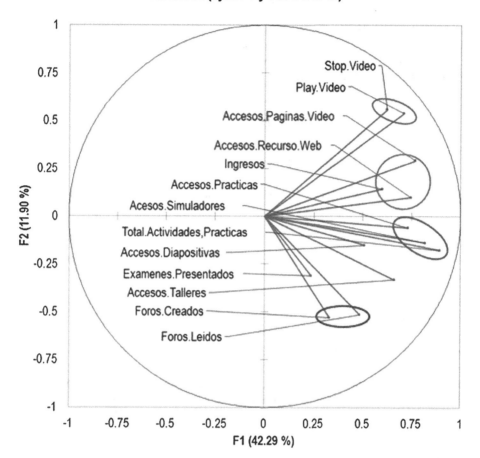

Fig. 3. Biplot of PCA3 showing clusters of variables indicating activities that give similar results. (Color figure online)

Figure 2 shows the biplot for PCA2 which is related to student actions. We see three mini clusters show in Table 2. Blue cluster is about accessing resources. Green cluster is about general platform access. Purple cluster is about accessing online forums. In green cluster several variables can be dropped. In blue cluster up to 5 variables could be dropped.

Nevertheless, once grade information and demographic data are removed, we see a somewhat different picture shown in Fig. 3, where the biplot for PCA3 is shown (only practical activities included without any grades data).

In Fig. 3 three clusters of activities are shown (see Table 3). Cluster 1 in blue relates to video access. Cluster 2 in green relates to accessing online resources. Cluster 3 in red relates to accessing practise labs. And cluster 4 in purple relates to accessing discussion fora.

Table 2. PCA3 biplot clusters

Cluster	Members
Green	Ingresos, Accesos.Talleres
Blue	Play.Video, Stop.Video, Accesos.Paginas.Video, Accesos. Simuladores, Accesos. Practicasm Accesos.Recurso.Web
Purple	Foros.Leidos, Foros.Creados

Table 3. PCA 3 biplot clusters

Cluster	Members
Blue	Play video, stop video
Green	Ingresos, Acceso.Paginas.Web, Acceso.Paginas.Video
Red	Acceso.Practicas, Acceso.Simuladores, Total.Actividades
Purple	Foros.Creado, Foros.Leidos

Therefore, one variable may be dropped from cluster 1, two from cluster 2, two from cluster 3 and one from cluster 4.

4   Clustering and Discussion

Whereas dimension reduction is achieved by grouping factor loadings, or variable projections towards component axis, groupings of students can be discovered by plotting the projections of the samples, called observation scores. If we appropriately label those projections, clusters based on different characteristics appear.

4.1   Clustering by Grades

David Kolb's Model of Learning Styles [17] known as Experiential Learning Theory affirms that learning is the result of how people perceive and then process what they have perceived. He also identified that some people prefer to perceive through concrete experience (CE) while others prefer abstract conceptualization and generalizations (AC). Also that some people prefer to process through active experimentation (AE) while others prefer to process through reflective observation (RO).

The juxtaposition of the ways of perceiving and processing creates a model of four quadrant, which in turn characterizes each of the learning styles proposed by Kolb: Accommodator (CE-AE), Assimilator (AC-RO), Convergent (CE-RO), Divergent (AC-AE). By the algorithm defined for the definition of styles, it is possible that a person may present a combination of styles or present characteristics of all of them simultaneously without any particular preference, so that a student may belong to one of nine different groups.

PCA1 uses student grade data. In Fig. 4 we show a scatterplot in which the horizontal axis corresponds to F1 scores and the vertical axis corresponds to F2 sample scores from PCA2, which uses grade information. The colour is based on Kolb's Learning Style classification (students take surveys to determine this classification). In Fig. 5 we show the same scatterplot as Fig. 4 with colours corresponding to learning difference.

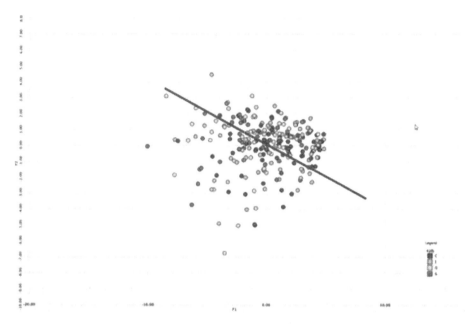

Fig. 4. PCA 2 (Grades) scatterplot of F1 and F2 scores. Colour is based on Kolb classification. (Color figure online)

Considering that the University of Cauca is an eminently face-to-face higher education institution and that 47% of the students enrolled in the MPOC had no experience in online courses, it is important to determine the "learning difference" the difference between the assessments of the course in line with the average grades of the face-to-face courses obtained during the same academic period. Therefore, the learning gap provides information on the performance of students in virtual courses with respect to their performance in face-to-face courses.

In Fig. 4 we see that there seems to be a line (in red) separating the quadrant in which both F1 and F2 are above zero in which C (Accommodator) and S (Assimilator) learning style profiles seem dominant. As both F1 and F2, as seen by the squared cosines statistic are related to students general grades performance, students for whom the sample is placed in this quadrant would be higher performance students.

In Fig. 5 there seems to be a line (red line) dependent only on F1 separating mostly positive learning difference from mostly negative learning difference.

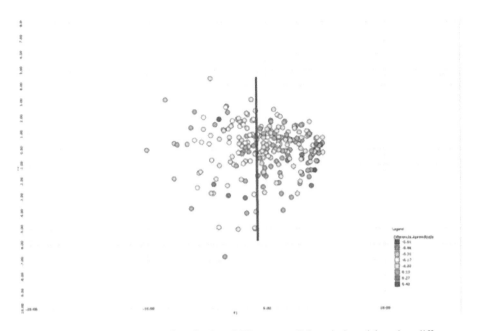

Fig. 5. PCA 2 (Grades) scatterplot of F1 and F2 scores. Colour is based learning difference. (Color figure online)

To investigate what Figs. 3 and 4 seem to show we need to know the data in table which the distribution of the learning styles (C = 29%, I = 29%, O = 23%, S = 19%).

We show the count of students in said quadrant by learning style. It shows that even though by count profiles C and I (Divergent) are dominant (23 both), in percentage by style, they are not. We also see that 23% of all C style samples are on the quadrant whereas the figure for styles I, O (Convergent) and S are 32%, 24% and 31%. Thus we can only conclude that style O is weak in overall grade performance as measured by F1 and all the other styles are the same.

In Table 4 we show the averages on F1 by learning style and by gender. We see that males outperform females and that C and I profiles have averages in the quadrant (positive F1 values) whereas O and S profiles do not.

Nevertheless, statistical tests show with α = 0.05 and p = 0.001 that the means for learning styles are in fact statistically different, whereas with p = 0.661, the means between genders are not different with any statistical significance. Therefore, males outperform females in grades, but only slightly. Whereas Profiles C and I clearly outperform profiles O and S.

Table 5 shows the average in learning difference considering both profile and gender. In this case, the differences between learning difference in females and males is statistically significant with p = 0.003 and between learning styles with p = 0.013. Therefore, males have wider learning difference as females and profiles C and I have wider learning difference than profiles O and S.

Table 4. F1 averages by learning style and gender.

Avg of F1	Column labels		
Row labels	1	2	Grand total
C	0.7031	0.8605	0.8217
I	0.2447	0.7250	0.4849
O	−0.9254	−1.0681	−1.0041
S	−0.3492	−1.0731	−0.7481
Grand total	−0.1007	0.0685	0

Table 5. Learning difference by learning style and gender

Avg of dif.aprend	Column labels		
Row labels	1	2	Grand total
C	0.0800	0.1127	0.1046
I	0.0764	0.1132	0.0948
O	0.0212	0.0776	0.0523
S	0.0274	0.0661	0.0487
Grand total	0.0524	0.0969	0.07889

4.2 Clustering by Activities

Activities are analysed in PCA3. PCA3 includes only platform activities. According to squared cosines statistics, F1 can be associated to 9 out of 13 activities and F2 to only two, which are access to fora. Therefore, F1 serves as a global measure of student platform activities and F2 as supplementary relating to fora activity.

In Fig. 6 is a scatter plot of F1 vs F2 by PCA3, that is, activities as recorded by platform. Colours are by Kolb's learning style profile. There seems to be a line for F1 > 0 and F2 > 0 that seems to be dominated mainly by profile C, with presence of I and S. This indicates that different profiles will carry out different activities in platform. Also, there is a cluster (blue circle) in F1 < 0 ≪ 1 in which profile I is dominant. That is, I profile students are more likely to carry no learning activities at all as recorded by platform.

In Fig. 7 we show same scatter plot as Fig. 6 but colours are based on learning difference. There seems to be no obvious clustering according to learning differences as shown by learning activities.

Figure 6 seems to indicate that different learning style profiles will behave differently when carrying out activities as recorded by platform. But, Fig. 7 seems to indicate that those learning activities have no influence on learning outcomes.

To investigate these matters further we present Table 6, indicating which proportions of Kolb's learning styles have F1 > 0 and F2 > 0 indicating higher platform activity level.

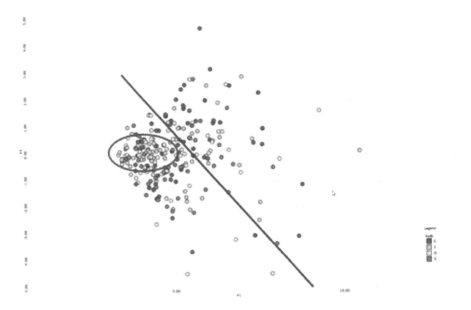

Fig. 6. PCA3 (Activities) scatterplot of F1 and F2 scores. Colour is based on Kolb classification. (Color figure online)

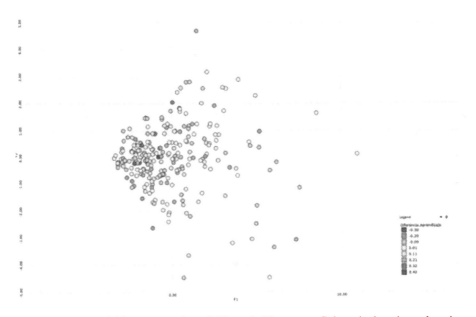

Fig. 7. PCA3 (Activities) scatterplot of F1 and F2 scores. Colour is based on learning difference.

Table 6. Count of samples in higher activity level quadrant and percentages

Kolb style	Count	% of total	% of Kolb style
C	25	10	35
I	15	06	21
O	13	05	24
S	7	03	15

We see that C profile has 35% of all participants in the high activity quadrant. Profile O is second place with 24% followed by profile I in third place with 21%. The least active profile is profile S with only 15%.

In Table 7 we show F1 averages (general activity measure) by learning style and gender. It seems to show that females tend to have more platform activity than males and that profile C by far the most active profile of all. ANOVA analysis shows that with $p = 0.483$ with $\alpha = 0.05$ the perceived differences in activity by learning style are not statistical significant, neither are differences by gender with $p = 0.5$.

Table 7. F1 (general activity measure) averages by learning style and gender.

Avg of F1	Column labels		
Row labels	1	2	Grand total
C	−0.2164	0.4912	0.3118
I	−0.0711	−0.5015	−0.2894
O	0.4303	−0.2376	0.0660
S	0.3731	−0.4836	−0.1088
Grand total	0.1233	−0.0836	0

Is this relevant? Unfortunately, Fig. 7 seems to indicate that learning activities as recorded by platform are not relevant to the learning difference or learning outcomes.

5 Conclusions

After the reduction of dimensionality and clustering using PCA to the dataset built through the offer of the MPOC "Daily Astronomy" at the University of Cauca, with 284 records of a total of 414 students enrolled, the following can be concluded:

- Students Accommodator and Assimilator present better results on assessments as well as a greater learning difference.
- Divergent students are more likely to not engage in learning activities at the MPOC.
- Males slightly outperform in grades and have a larger learning difference than females.
- There is no statistical difference in overall grade performance or learning difference by age, socioeconomic strata or academic semester.

- There is no statistical difference in platform use between different learning styles, neither between males and females.
- The number of learning activities recorded in Open edX has no effect on the learning difference or assessment results.

PCA found that the dataset variables can be grouped into two sets, one group relating to school performance and the other to learning activities recorded by the platform. It was found that the interest variables are mainly linked to the overall performance of the rating. It was also found that these variables of interest are apparently not correlated with the learning activities registered by the platform.

But one interesting question remains: Is getting a certain final grade related to a certain learning difference?

As Fig. 8 shows, the answer is yes, but with high variability. The R^2 statistic of the model is not very high, only 0.2802, but the linear tendency is clear.

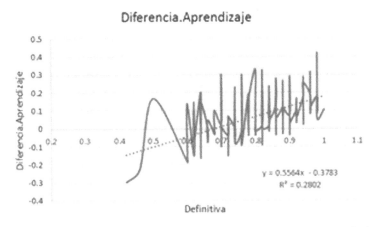

Fig. 8. Final grade (Definitiva) vs learning difference (Diferencia.Aprendizaje)

As a work in the future it is proposed to find predictive models of both the final qualification and the learning difference according to the different variables that record the students' activity with the Open edX.

References

1. Siemens, G.: Connectivism: a learning theory for the digital age. http://er.dut.ac.za/handle/123456789/69. Accessed 15 Dec 2017
2. Wiley, D.: The MOOC misnomer. http://opencontent.org/blog/archives/2436. Accessed 15 June 2018
3. McAuley, A., Stewart, B., Siemens, G., Cormier, D.: The MOOC model for digital practice. http://elearnspace.org/Articles/MOOC_Final.pdf. Accessed 15 June 2018
4. Liyanagunawardena, T., Adams, A., Williams, S.: MOOCs: a systematic study of the published literature 2008–2012. Int. Rev. Res. Open. Distrib. Learn. **14**(3), 202–227 (2013)

5. Guo, P.J., Kim, J., Rubin, R.: How video production affects student engagement: an empirical study of MOOC videos. In: Memorias Proceedings of the First ACM Conference on Learning@ Scale Conference, pp. 41–50. ACM (2014)
6. Roig, R., Mengual, S., Suarez, C.: Evaluación de la calidad pedagógica de los MOOC. Profesorado **18**(1), 27–41 (2014)
7. Jansen, D., Schuwer, R.: Institutional MOOC strategies in Europe. Status Report Based on a Mapping Survey Conducted in October–December 2014 (2014). https://www.surfspace.nl/media/bijlagen/artikel-176322974efd1d43f52aa98e0ba04f14c9f3.pdf. Accessed 15 June 2018
8. Adone, D., Michaescu, V., Ternauciuc, A., Vasiu, R.: Integrating MOOCs in traditional higher education. In: Memorias Third European MOOCs Stakeholders Summit, pp. 71–75 (2015)
9. Mohamed, A., Yousef, F., Chatti, M., Schroeder, U., Wosnitza, M.: A usability evaluation of a blended MOOC environment: an experimental case study. Int. Rev. Res. Open Distance Learn. **16**(2), 69–93 (2015)
10. Bishop, J., Verleger, M.: The flipped classroom: a survey of the research. In: IASEE National Conference, pp. 1–18 (2013)
11. Fox, A.: From MOOCs to SPOCs. Commun. ACM **56**(12), 38–40 (2013). https://doi.org/10.1145/2535918. Accessed 15 June 2018
12. Guo, W.: From SPOC to MPOC–the effective practice of Peking University online teacher training. In: International Conference of Memorias Educational Innovation through Technology, pp. 258–264. IEEE (2014)
13. Solarte, M., Ramírez, G., Jaramillo, D.: Access habits and assessment results in massive online courses with academic value. Ingeniería e Innovación **5**(1), 1–10 (2017)
14. (n.d.). https://open.edx.org/. Accessed 15 June 2018
15. Ramirez-Velarde, R., Alexandrov, N., Sanhueza-Olave, M., Perez-Cazares, R.: The impact of learning activities on the final grade in engineering education. Procedia Comput. Sci. **80**, 1812–1821 (2015)
16. Ramirez-Velarde, R., Olave, M.S., Alexandrov, N., de Marcos Ortega, L.: Do learning activities matter? In: International Conference on Interactive Collaborative and Blended Learning, pp. 76–82. IEEE (2016)
17. Kolb, D.: Experiential Learning: Experience as the Source of Learning and Development. FT press, Upper Saddle River (2014)
18. Jolliffe, I.T.: Principal Component Analysis. Springer, Heidelberg (2002). https://doi.org/10.1007/978-1-4757-1904-8

Tremor Signal Analysis for Parkinson's Disease Detection Using Leap Motion Device

Guillermina Vivar-Estudillo, Mario-Alberto Ibarra-Manzano[✉],
and Dora-Luz Almanza-Ojeda

Department of Electronics Engineering, Universidad de Guanajuato,
Carr. Salamanca -Valle de Santiago Km. 3.5+1.8 Comunidad Palo Blanco,
36885 Salamanca, Guanajuato, Mexico
{g.vivarestudillo,ibarram,dora.almanza}@ugto.mx

Abstract. Tremor is an involuntary rhythmic movement observed in people with Parkinson's disease (PD), specifically, hand tremor is a measurement for diagnosing this disease. In this paper, we use hand positions acquired by Leap Motion device for statistical analysis of hand tremor based on the sum and difference of histograms (SDH). Tremor is measured using only one coordinate of the center palm during predefined exercises performed by volunteers at Hospital. In addition, the statistical features obtained with SDH are used to classify tremor signal as with PD or not. Experimental results show that the classification is independent of the hand used during tests, achieving 98% of accuracy for our proposed approach using different supervised machine learning classifiers. Additionally, we compare our result with others classifiers proposed in the literature.

Keywords: Tremor analysis · Parkinson detection · SDH method
Classification

1 Introduction

Parkinson's disease (PD) is a progressive degenerative disorder of the central nervous system that generally affects 1% of the population above 60 years [17]. PD is commonly associated by motor impairment that treming, bradykinesia (slowness of movement), postural instability, gait difficulty and rigidity of the limbs. Furthermore, non-motor symptoms, such as depression, speech difficulties are also characteristic of this disease [16].

Tremor in motor control is defined as involuntary, approximately rhythmic and roughly sinusoidal movements [18], being the hand tremors the major symptom observed in people diagnosed with PD. Measurement of tremors is a significant parameter for diagnosing and continuously monitoring patients with PD [1,15].

© Springer Nature Switzerland AG 2018
I. Batyrshin et al. (Eds.): MICAI 2018, LNAI 11288, pp. 342–353, 2018.
https://doi.org/10.1007/978-3-030-04491-6_26

The diagnosed of the movements are evaluate by a neurologist based on his/her clinical experience using the Unified Parkinson's Disease Rating Scale (UPDRS) [5].

Many methods have been developed to monitor the tremor associated with PD [4,7,13,14], however, these methods have been implemented in a very invasive form for patients.

In [10] the principal idea was consider a no-invasive, portable and economical method for acquiring patients measurements.

The statistical analysis gives a representative number for interpreting and estimate the data tendency. One of the most common features used for data description is the texture, which represents an approximation that measures the uniform distribution of patrons in the dataset. For proper analysis of any signal using texture features, it is essential to extract a set of discriminative features to provide better classification; first and second order statistics are commonly used for this purpose. Sum and difference histograms (SDHs) provide an efficient approximation of the joint probability function by counting the frequencies of data pairs of sums of the radial distance and the angular displacement, respectively [12]. In [6,19], the authors use SDHs to extract characteristics instead of Gray Level Co-occurence Matrices (GLCM). SDHs is a powerful technique for feature texture analysis that let obtain essential texture information in an optimized way.

The purpose of this paper is to detect tremor in patients and to group measured data of voluntaries as Parkinson patient or not, using a leap motion sensor. Hand positions are modeled by texture features using the sum and difference of histograms (SDH). The classification is performed using supervised methods obtaining two groups: with PD or without PD.

This document is organized as follows. In Sect. 2, we describe the data acquisition procedure and the feature extraction method. Subsequently, in Sect. 3 different features are evaluate using supervised classifiers. In Sect. 4, we provide the score of the classifiers and a comparison with similar classifiers proposed in the literature. Finally, conclusions are provided in Sect. 5.

2   Methodology

The proposed strategy measures hand motion for predicting Parkinson disease, from real data acquired during routine exercises performed by patients in a Hospital. Overall block diagram is shown in Fig. 1 consisting in the following general stages: (1) Data Acquisition, (2) Coordinates Selection, (3) Features computation using sum and difference of histograms (SDH), (4) Classification (with PD or not) and (5) Comparison result.

2.1   Data Acquisition

Leap motion controller (LMC), shown in Fig. 1, is an interactive device mainly used for hand gestures and finger position detection. As described in [2], the

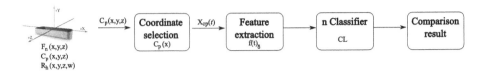

Fig. 1. Block diagram of the analysis of tremor for Parkinson detection

LMC consists of three infrared light (IR) emitters and two IR cameras used as depth sensors. Moreover, introduces a new gesture and position tracking system with sub-millimeter accuracy. This sensor tracks hands and fingers in its field of view (about 150°) and it provides updates of data. The information regarding the user's hand, fingers and gestures is captured between 25 to 600 mm above the sensor. LMC was used in [10] to implement a natural interaction system through Virtual reality interface as a medical tool with the aim of acquiring data from 20 patients. The study is conducted at the Edmonton Kaye Alberta Clinic, and all patients provided a prior consent to participate in the tests. Thus, the data is collected for both, right and left hand, 9 males and 11 females, aged in 30–87 years. During the measurement of kinetic tremor, the patient extends the right or left hand at initial position (indicated on the screen) and tries to move or push a sphere located in the center of the scene to the final goal located in the right.

The duration of the test was 15 s, because the idea is to induce the patient in a stressed situation and to monitor how he focus in when performing a specific activity of the daily life. The acquired data is illustrated in Table 1 and is divided into two groups (with PD and without PD), by gender and by the hand used for the test. We want to point out that patients could have tremor in only one hand (left or right). Therefore we found that even if we have 5 patients with PD they manifest tremor only in left or right hand as shown in the first row of the Table 1. For the test, LMC worked at the rate of 40 fps.

Table 1. Information of patients divided as with and without PD.

Patients	Gender (F/M)	Hand		Subtotal
		Right	Left	
with PD	F	2	3	5
	M	6	7	13
without PD	F	7	6	13
	M	5	4	9
Total		20	20	40

2.2 Coordinate Selection

The data base described above is used in this work and consists in a collection of 600 samples per hand of the volunteers that includes: position and velocity of the fingertips $F_n(x, y, z)$, the palm center position $C_p(x, y, z)$ and the rotation hand $R_h(x, y, z, w)$. Each acquired sample corresponds to 40 variables in the 3D space (see Table 2).

Table 2. Hand measurements captured with the LMC

Data	Acquired data	Coordinates	Total
Position, Velocity	Thumb, index, middle, ring, pinky, center palm	(x, y, z)	36
Rotation	Hand	(x, y, z, w)	4
		Total	40

The hand movements in patients with PD are oscillatory, rhythmic and involuntary. Figure 2 shows the displacement in $C_p(x, y, z)$ of two different subjects recorded by LMC. We analyze the movement (tremor) as a change in the position of a hand (center of the palm) over time with respect to the LMC reference system. In order to determinate such displacement, we analyze the $C_p(x)$ coordinate of the right and left center palm over time $(X_{cp}(t))$.

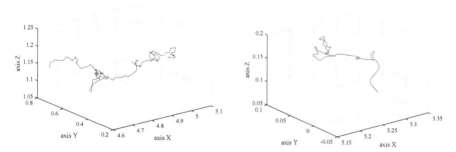

(a) Hand Left of Subject A (without PD)   (b) Hand left of Subject B (with PD)

Fig. 2. Original displacement in $C_p(x, y, z)$ coordinates of the center palm.

2.3 Features Extraction Using Sum and Difference of Histograms

Feature extraction transform raw signal in attributes. Statistical analysis is used for texture feature extraction, because provides high relevant and distinguishable features. The most common features used for data description is texture. Texture is as a property in a neighborhood and is generally computed using

methods extract statistical information like spatial gray level dependences matrices (SGLDM). Besides, sum and differences of histograms are used in images for analysis of texture and it is an alternative accuracy for classification. This method provide an efficient approximation for the joint probability by counting the frequencies of sums respectively of data pairs with the radial distance and the angular. The method sum and difference of histograms (SDH) employ in [6,19], where the let obtain essential texture information in a simplified and optimized way. In SDHs a rectangular grid that contains the analyze discrete texture image is defined on $K \times L$ and is denoted by $I(k, l)$ where ($k \in [0, k-1]$; $l \in [0, L-1]$). And assumes that gray level at each pixel is quantified to N_g levels, so let $G \in [0, ...N_g - 1]$ be the set of these N_g levels. Next, for a given pixel (k, l), let $(\delta_k, \delta_l) = \{(\delta_{k_1}, \delta_{l_1}), (\delta_{k_2}, \delta_{l_2}), ...(\delta_{k_M}, \delta_{l_M})\}$ be the set of M relative displacements. The sum and difference images, I_S and I_D respectively, associated with each relative displacement (δ_k, δ_l), are defined as:

$$I_S(k, l) = I(k, l) + I(k + \delta_k, l + \delta_l) \tag{1}$$
$$I_D(k, l) = I(k, l) - I(k + \delta_k, l + \delta_l)$$

Thus, the range of the I_S image is $[0, 2(N_g - 1)]$, and for the I_D image is $[-N_g + 1, N_g - 1]$. From this, let define i and j as two any gray levels in the I_S and I_D image range respectively. Then, let D be a subset of indexes which specifies a region to be analyzed, so, the SDHs with parameters (δ_u, δ_v) over the domain $(k, l) \in D$ are. respectively define as

$$h_S(i; \delta_u, \delta_v) = h_S(i) = \#\{(k, l) \in D, I_S(k, l) = i\} \tag{2}$$
$$h_D(i; \delta_u, \delta_v) = h_D(i) = \#\{(k, l) \in D, I_D(k, l) = j\}$$

where, the total number of count is

$$N = \#\{D\} = K \times L = \sum_i h_S(i) = \sum_j h_D(j) \tag{3}$$

The normalized SDHs is given by

$$\hat{P}_S(i) = \frac{h_S(i)}{N} , \hat{P}_D(j) = \frac{h_D(j)}{N} \tag{4}$$

The sum and difference of data is the first task executed during the texture analysis. Addition and subtraction operations are carried out between the current pixel of data and the displaced pixels. And now, using the approach in [6], addition and subtraction operations are carried out between coordinate positions x instead pixels.

To achieve the purported goal we must have consider a vector L to extract features, and will preserve the principle of SDH.

$$\mu = \frac{1}{d} \sum_i i \cdot \hat{P}_S(i) \tag{5}$$

from this, $\hat{P}_S$ could be replaced by the sum of histogram given in Eq. 4 where h_S represent the sum histogram.

The sum and difference of data whit respect to relative displacement are calculated from input data positions x-coordinate of center palm $(X_{cp}(t))$ over time. The distance used in [6] was $d = 1, 2, 3, 4$ pixels with $0°, 45°, 90°$ and $135°$ as directions. However in our case, we use displacement $\delta_k = 1$ and $0°$ direction.

The features most frequently used are: mean, variance, energy, correlation, entropy, contrast, homogeneity, cluster shade and cluster prominence showed in Table 3.

Table 3. Statistical textures calculated $f(t)_\delta$

Texture	Sum and difference histograms
Mean	$\frac{1}{2}\sum_i (i) \cdot \hat{P}_S(i) = \mu$
Variance	$\frac{1}{2}(\sum_i (i - 2\mu)^2 \cdot \hat{P}_S(i) + \sum_j j^2 \cdot \hat{P}_D(i))$
Energy	$\sum_i \hat{P}_S(i)^2 \cdot \sum_j \hat{P}_D(j)$
Correlation	$\frac{1}{2}(\sum_i (i - 2\mu)^2 \cdot \hat{P}_S(i) - \sum_j j^2 \cdot \hat{P}_D(i))$
Entropy	$-\sum_i \hat{P}_S(i) \cdot log\{\hat{P}_S(i)\} - \sum_j \hat{P}_D(j) \cdot log\{\hat{P}_D(j)\}$
Contrast	$\sum_j j^2 \cdot \hat{P}_D(j)$
Homogeneity	$\sum_j \frac{1}{1+j} \cdot \hat{P}_D(j)$
Cluster shade	$\sum_i (i - 2\mu)^3 \cdot \hat{P}_S(i)$
Cluster prominence	$\sum_i (i - 2\mu)^4 \cdot \hat{P}_S(i)$

Figure 3, show the result applying SDH in coordinate palm $(X_{cp}(t))$ (raw data). The result was obtained by total 9600 samples **without PD** and **PD** group, **left** and **right** hand. In the figure, (a) show the features: mean, variance and correlation, (b) show contrast, homogeneity and cluster shade, and finally (c) illustrate entropy, energy and cluster prominence. This features are grouped by without PD (o) and PD group(+).

3   Tremor Classification: Parkinson Disease Detection

Nowadays, there are different approaches used in machine learning to evaluate tremors and its performance. However, first it is important to detect PD in patients, that is a binary classification (with PD or not).

In the first experiment, the training and validation process were performed using the x-coordinate $(X_{cp}(t))$ of the center right palm, for all classifiers (CL) of the Table 4. The parameters used to compute statistical features are:

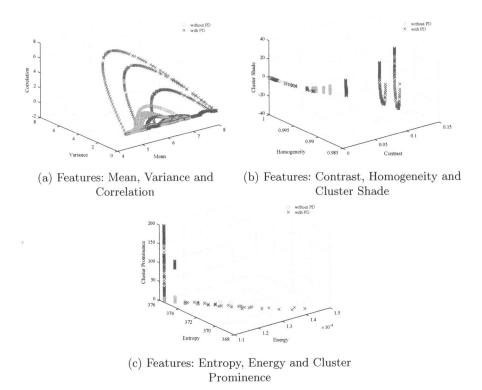

(a) Features: Mean, Variance and Correlation

(b) Features: Contrast, Homogeneity and Cluster Shade

(c) Features: Entropy, Energy and Cluster Prominence

Fig. 3. Result features extract $f(t)_\delta$ applying SDH in x-coordinate of center palm $(X_{cp}(t))$ over time.

(i) $\delta_l = 1$ (displacement), (ii) coordinates x to right hand (PD and without PD group), (iii) the window sizes tested were in the range of $\delta_u = 3$ to $\delta_v = 501$. For each window size the classification performed was repeat 30 times for each classifier $C(t)$ and the obtained results allow us to choose the most representative window. Figure 4 illustrates the performance for every size window for all the classifiers of the Table 4, (a) show classifiers based in KNN that Co-KNN, Cs-KNN, Cu-KNN, F-KNNm M-KNN, W-KNN. Similarly in (b) show performance of SVM classifiers that CGSVM, CSVM, FGSVM, FGSVM, LSVM, MGSVM, QSVM. At the same time in (c) show BgT, Sb-KNN, Bt, RT, SD, LR. Simultaneously in (d) show the performance of ST, MT, CT, LD, QD. Finally in (e) show the better performance classifier (F-KNN, FGSVM, BgT, CT) and show that the grow most significant was in the first 50 window size in the majority of the classifier and the tendencies are the same as those observer from window size 95, classifiers with better perform have a monotonous growth.

In the second experiment, the $(X_{cp}(t))$ coordinate of the left hand center palm were used, the window size was fixed to 95 for all the classifiers CL in Table 4. The average accuracy obtained for the **BgT** classifier was 0.9838. Indeed, the

Table 4. List of Classifiers CL used during the experimental tests

Classifier	
Bagged Tree (BgT)	**Logistic Regression (LR)**
Boosted Tree (BT)	**Medium Gaussian SVM (MGSVM)**
Coarse Gaussian SVM(CGSVM)	Medium KNN (M-KNN)
Coarse KNN (Co-KNN)	Medium Tree (MT)
Complex Tree (CT)	Quadratic Discriminant (QD)
Cosine KNN (Cs-KNN)	**Quadratic SVM (QSVM)**
Cubic KNN (Cu-KNN)	RUSBoosted Tree (RT)
Cubic SVM (CSVM)	Simple Tree (ST)
Fine Gaussian SVM (FGSVM)	Subspace Discriminant (SD)
Fine KNN (F-KNN)	Subspace KNN (Sb-KNN)
Linear Discriminant (LD)	Weighted KNN (W-KNN)
Linear SVM (LSVM)	

performance of classifiers is expected to be the same (left and right hand), and after both experiments, the difference between both obtained accuracies in BgT is 0.0079, therefore we consider that the measures obtained for left or right hand could be used equally. After that, the last classification performed for every classifier using left and right hand position data and the parameters described above. However in this case, the coordinate x was right and left hand from the two group and this were combined and balanced. The results of the performance classification are presented in the next section.

4   Result and Discussion

The Table 5 illustrates the rank rate for the eleven classifiers using window size of 95 in order: (Acc) Accuracy (maximum, minimum), (Pr) precision, (Sp) specificity and (Sens) sensitivity. The values were obtained by performing classification test repeated 30 times and the each column in this table specifies the mean. The rank rate was obtained to convolution of the measurements m (Acc, Pr, Sp, Sens) normalized with $\prod_{n=1}^{4} \frac{m_i}{max(m_i)}$. Thus, Bagged Tree got the best classification percentages. Afterward, F-KNN classify closer to first, specifically, the mean in Acc of F-KNN (0.9824) is higher than minimum the BgT (0.9813). In like manner, BgT has the better specificity and sensitivity whereas CSVM got better precision than BgT by 0.005. In addition, these tree classifiers (BgT, F-KNN, CT) have the best classification percentages, in contrast, CSVM and QSVM are the less stable due to the higher variation in accuracy with 12% and 16% respectively. On the other hand, the best SVM was Fine Gaussian SVM, and finally, the Logistic Regression classifier shows the worst performance.

The results obtained in this work were compared with similar classifiers proposed in the literature, as is shown in Table 6. In [8] also the sensor LMC is used

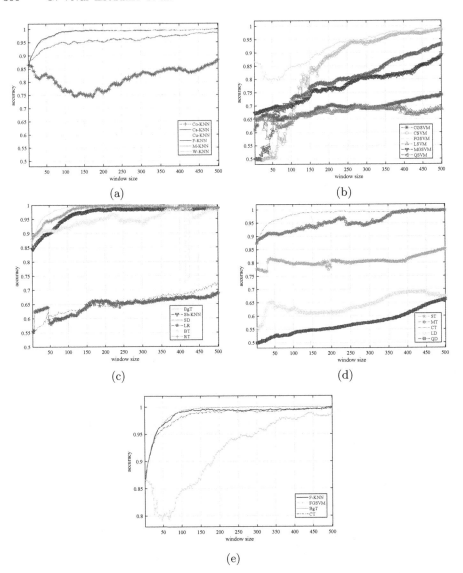

Fig. 4. Performance of classifiers CL tested (Table 4) for each window size. (a) KNN classifiers, (b) SVM classifiers, (c) Ensemble classifiers, (d) Decision tree and linear discriminant (e) Better classifiers grouped: BgT, F-KNN, CT, FGSVM.

to acquire data therefore it can be compared with the proposed approach. The overall accuracy of classification using K-means clustering was 77.5%, sensitivity with 60% and specificity with 95%. And using SVM classifier was 85% accuracy, sensitivity with 75% and specificity with 95%. In this case, Bgt behaves better than SVM in Acc 13% and Sens in 23%.

Table 5. Descriptive results for all evaluation classifier using window size of 95. Meaning: (CL) Classifier Learner, (Acc) Accuracy (max, min), (Pr) Precision, (Sp) Specificity, (Sens.) Sensitivity.

CL	Acc (max, min)	Pr.	Sp.	Sens.	False	Rank
BgT	**0.9862** $(0.9896, 0.9813)$	0.9880	0.9880	0.9843	98	**1**
F-KNN	0.9824 $(0.9857, 0.9786)$	0.9819	0.9819	0.9833	123	2
CT	0.9548 $(0.9689, 0.9409)$	0.9577	0.9576	0.9570	305	3
FGSVM	0.7945 $(0.8065, 0.7672)$	0.8781	0.8531	0.7530	1473	4
CSVM	0.5045 $(0.5568, 0.4301)$	0.9885	0.5958	0.5025	3503	5
MGSVM	0.6431 $(0.6800, 0.6215)$	0.5137	0.6196	0.6894	2557	6
CGSVM	0.5771 $(0.5905, 0.5606)$	0.6194	0.5831	0.5735	2490	7
QSVM	0.5458 $(0.6135, 0.4477)$	0.6911	0.5871	0.5291	3481	8
LSVM	0.5687 $(0.5902, 0.5486)$	0.5796	0.5718	0.5667	3071	9
LD	0.5573 $(0.5766, 0.5464)$	0.5941	0.5617	0.5529	3146	10
LR	0.5457 $(0.5708, 0.5349)$	0.5202	0.5437	0.5482	3671	11

Table 6. Comparative performance evaluation of the classifier with similar classifiers.

Author	Technology device	Acc	Sp.	Sens.	Classifier
Our approach	LMC	**0.9858** Avg	0.9884	0.9835	Bagged Tree
Johnson [8]	LMC	0.775	0.95	0.60	K-means
		0.85	0.95	0.75	SVM
Butt et al. [3]	LMC	0.8571	0.835	0.875	SVM
Kostikis et al. [9]	Smartphone	–	0.90	0.82	Random forest of decision trees
Manzanera et al. [11]	Uniaxial accelerometer	0.95	0.98	0.69	Welch(2)

In the same way, in [3], LMC was used to recorder data. They report using three supervised learning methods to classify PD and healthy subjects. The accuracy obtained using SVM was 85.71% with sensitivity 83.5%, and specificity in 87.5% from SVM. In comparison with the proposed approach it can be seen that BgT is better than [3] in accuracy for 12% and specificity in 15%.

On the other hand, Kostikis et al. in [9] use signals from an iPhone (acquired by its accelerometer and gyroscope) worn for each volunteer as a glove. Authors use six different machine learning approaches to classify the tremor detected as Parkinson. The classification by BgT is better in 8% specificity and 12% in sensitivity than Random forest of decision tree.

Whereas, Manzanera et al. in [11] using an uniaxial accelerometer on the dorsal side of the hand to test no-parametric and parametric methods to tremor detection in segment using classical and parametric power spectral density estimation techniques. BgT behaves slightly better for 3% in accuracy, while in specificity is de same. Although Bgt delivers slightly higher in sensitivity in terms of classification performance.

5   Conclusions and Future Work

This paper proposes an algorithm based in SDH to extract features for classify tremor signal as PD patients or not. Furthermore, using LMC that is non-invasive, portable and economical system to acquisition signal data.

The proposed approach has two advantages: first, extract features with SDH in a signal and second one coordinate was fairly successful to classify. Classification between PD group and without PD group was performed based on supervised machine learning classifiers. The result show that the approach based in SDH is wide considerable successful result. Furthermore, the result is significant because the classification is independent of hand patient. The best classification result were obtained with BgT and F-KNN. This represent, that is possible to use fine KNN to classify PD and healthy group for our case. Finally, this work can be used for detecting early onset PD and monitoring progression. In the future, it is interesting to look for the behavior of the classifiers with less features u others and overcome the performance of classification. Analogous to this, investigate the behavior with sub-classification tremor.

Acknowledgments. Authors gratefully acknowledge all the volunteers at Edmonton Kaye Alberta Clinic, Canada, the Research and Postgraduate studies Support Program (DAIP) by the Universidad de Guanajuato and the Universidad Autónoma "Benito Juárez "de Oaxaca.

References

1. Alam, M.N., Johnson, B., Gendreau, J., Tavakolian, K., Combs, C., Fazel-Rezai, R.: Tremor quantification of Parkinson's disease - a pilot study. In: 2016 IEEE International Conference on Electro Information Technology (EIT), pp. 0755–0759, May 2016
2. Bachmann, D., Weichert, F., Rinkenauer, G.: Evaluation of the leap motion controller as a new contact-free pointing device. Sensors **15**(1), 214–233 (2015). http://www.mdpi.com/1424-8220/15/1/214
3. Butt, A.H., Rovini, E., Dolciotti, C., Bongioanni, P., Petris, G.D., Cavallo, F.: Leap motion evaluation for assessment of upper limb motor skills in Parkinson's disease. In: 2017 International Conference on Rehabilitation Robotics (ICORR), pp. 116–121, July 2017
4. Chen, K.H., Lin, P.C., Chen, Y.J., Yang, B.S., Lin, C.H.: Development of method for quantifying essential tremor using a small optical device. J. Neurosci. Methods **266**, 78–83 (2016). http://www.sciencedirect.com/science/article/pii/S0165027016300206
5. Goetz, C.G., et al.: Movement disorder society-sponsored revision of the unified Parkinson's disease rating scale (MDS-UPDRS): scale presentation and clinimetric testing results. Mov. Disord. **23**(15), 2129–2170 (2008). https://doi.org/10.1002/mds.22340
6. Ibarra-Manzano, M.A., Devy, M., Boizard, J.L.: Real-time classification based on color and texture attributes on an FPGA-based architecture. In: 2010 Conference on Design and Architectures for Signal and Image Processing (DASIP), pp. 250–257, Oct 2010

7. Jilbab, A., Benba, A., Hammouch, A.: Quantification system of Parkinson's disease. Int. J. Speech Technol. **40**, 1–8 (2017)
8. Johnson, M.J.: Detection of Parkinson disease rest tremor. Master thesis, Washington University, Department of Electrical and Systems Engineering. School of Engineering and Applied Science (2014)
9. Kostikis, N., Hristu-Varsakelis, D., Arnaoutoglou, M., Kotsavasiloglou, C.: A smartphone-based tool for assessing Parkinsonian hand tremor. IEEE J. Biomed. Health Inform. **19**(6), 1835–1842 (2015)
10. Lugo, G., Ibarra-Manzano, M., Ba, F., Cheng, I.: Virtual reality and hand tracking system as a medical tool to evaluate patients with Parkinson's. In: Proceedings of the 11th EAI International Conference on Pervasive Computing Technologies for Healthcare, PervasiveHealth 2017, pp. 405–408. ACM, New York, NY, USA (2017). https://doi.org/10.1145/3154862.3154924
11. Martinez Manzanera, O., Elting, J.W., van der Hoeven, J.H., Maurits, N.M.: Tremor detection using parametric and non-parametric spectral estimation methods: a comparison with clinical assessment. PLOS ONE **11**(6), 1–15 (2016). https://doi.org/10.1371/journal.pone.0156822
12. Münzenmayer, C., Wilharm, S., Hornegger, J., Wittenberg, T.: Illumination invariant color texture analysis based on sum- and difference-histograms. In: Kropatsch, W.G., Sablatnig, R., Hanbury, A. (eds.) DAGM 2005. LNCS, vol. 3663, pp. 17–24. Springer, Heidelberg (2005). https://doi.org/10.1007/11550518_3
13. Oung, Q.W., et al.: Technologies for assessment of motor disorders in Parkinson's disease: a review. Sensors **15**(9), 21710–21745 (2015). http://www.mdpi.com/1424-8220/15/9/21710
14. Perumal, S.V., Sankar, R.: Gait and tremor assessment for patients with parkinson's disease using wearable sensors. ICT Express **2**(4), 168–174 (2016). http://www.sciencedirect.com/science/article/pii/S2405959516301382. special Issue on Emerging Technologies for Medical Diagnostics
15. van der Stouwe, A., et al.: How typical are "typical" tremor characteristics? Sensitivity and specificity of five tremor phenomena. Parkinsonism Relat. Disord. **30**, 23–28 (2016). http://www.sciencedirect.com/science/article/pii/S1353802016302280
16. Thomas, B., Beal, M.F.: Parkinson's disease. Hum. Mol. Genet. **16**(R2), R183–R194 (2007). https://doi.org/10.1093/hmg/ddm159
17. Tysnes, O.B., Storstein, A.: Epidemiology of Parkinson's disease. J. Neural Trans. **124**(8), 901–905 (2017). https://doi.org/10.1007/s00702-017-1686-y
18. Vaillancourt, D.E., Newell, K.M.: The dynamics of resting and postural tremor in Parkinson's disease. Clin. Neurophysiol. **111**(11), 2046–2056 (2000). http://www.sciencedirect.com/science/article/pii/S1388245700004673
19. Villalon-Hernandez, M.-T., Almanza-Ojeda, D.-L., Ibarra-Manzano, M.-A.: Color-texture image analysis for automatic failure detection in tiles. In: Carrasco-Ochoa, J.A., Martínez-Trinidad, J.F., Olvera-López, J.A. (eds.) MCPR 2017. LNCS, vol. 10267, pp. 159–168. Springer, Cham (2017). https://doi.org/10.1007/978-3-319-59226-8_16

Fuzzy Logic and Uncertainty Management

Modeling Decisions for Project Scheduling Optimization Problem Based on Type-2 Fuzzy Numbers

Margarita Knyazeva[1], Alexander Bozhenyuk[1(✉)],
and Janusz Kacprzyk[2] (ID)

[1] Southern Federal University,
Nekrasovskiy Street, 44, 347928 Taganrog, Russia
margarita.knyazeva@gmail.com, avb002@yandex.ru
[2] Systems Research Institute Polish Academy of Sciences,
Newelska 6, 01-447 Warsaw, Poland
janusz.kacprzyk@ibspan.waw.pl

Abstract. This paper examines type-2 fuzzy numbers implementation to resource-constrained scheduling problem (RCSP) for agriculture production system based on expert parameter estimations. Some critical parameters in production system are usually treated as uncertain variables due to environmental changes that influence agriculture process. Implementation of type-2 fuzzy sets (T2FSs) can handle uncertain data when estimating variables for solving decision-making problems. The paper focuses on estimation procedure of uncertain variables in scheduling that reflect level of preference or attitude of decision-maker towards imprecise concepts, relations between variables. Special profiles for activity performance allow to consider uncertainty in time variables, expert estimations, flexibilities in scheduling, resource levelling problem and combinatorial nature of solution methodology. An example of activities for agriculture production system is introduced. Heuristic decision algorithm based on enumeration tree and partial schedules is developed. It can handle both resource-constrained optimization problem under uncertain variables and activity profile switching. As initial activity profile we consider expert decision about best activity execution profile on each level of enumeration tree.

Keywords: Type-2 fuzzy number · Fuzzy graph · Combinatorial optimization
Scheduling

1 Introduction

In practice, decision-makers are usually required to choose the best alternative among several alternatives. When modelling agriculture processes experts evaluate each alternative under several production criteria, and then the best execution profile for activity is chosen as a basic one at first initialization step. Scheduling problem is considered as an optimization problem of operations research study that investigates optimal or near-optimal solutions for complex decision-making problems. The main idea of the problem is to construct special plan for activity performance with respect to

© Springer Nature Switzerland AG 2018
I. Batyrshin et al. (Eds.): MICAI 2018, LNAI 11288, pp. 357–368, 2018.
https://doi.org/10.1007/978-3-030-04491-6_27

precedence relations, activity duration, resources leveling profile and objective function [1]. Allocation of resources, variables estimation, sequencing activities and other constraints and objective function in diverse search space represents decision making process in scheduling problems. In practice, the agriculture scheduling analysis considers the entire agriculture cycle process within a year. It includes the annual cycle of activities related to the growth and harvest of crops. Activities include loosening and tillage, transportation and insertion of farming inputs in soil (e.g. fertilizers, minerals) and all agricultural operations in the field (e.g. seeding, watering and harvesting are among them). Forecasting analysis for scheduling agriculture process, indicators of environmental effects, which include resource depletion, land use, climate changes, toxicity and acidification levels of soil, and other factors are usually analyzed by experts in order to estimate variables. Consumption of resources connected to the different agriculture processes is leveled and after normalization and weighting of the indicator values it becomes possible to calculate summarizing indicators for resource usage and environmental impacts (EcoX) [2]. Thus decision-makers are usually able to evaluate each alternative under several criteria (time, resource usage) and choose the best or eligible alternative (profile) for activity performance to be scheduled. This type of problem can also be related to multi-criteria decision making problem to some extend (MCDM).

As a result, with the increasing complexity of decision-making environment in scheduling, limitation of knowledge at some specific planning stages and necessity to consider preference attitude of decision-maker towards some variables the estimation process can't be performed under exact crisp numbers. When applying fuzzy numbers it is relevant to investigate and analyze complex fuzzy interactions between variables affecting scheduling process and performance measure under current operating conditions. Different types of flexibilities appear as a result of multiple criteria nature of decision process [3, 4]. Flexibilities in scheduling can reflect uncertain expert estimations, preferences or attitude of decision makers towards uncertain parameters and precedence relations between activities of project. In this case decision making process can be represented as a necessity to choose best profile for activity performance including uncertain starting times of activities, resource requirements as a function of environmental influence, and finally preference of decision maker towards the fuzzy time estimation (ready-time and deadline time) and resource usage. Flexibility in activity execution profile considers trade-offs between time and resource usage, costs, overlaps between starting times of activities, product performance variety and other degree of freedom available during operational step of planning [5, 6].

2   Type-2 Fuzzy Numbers for Scheduling Flexibilities

In scheduling procedure fuzzy numbers are usually associated with imprecise or approximate time variables, which can stay for a measure of preference or attitude of decision-maker and thus reflect flexibilities. A number of approaches towards fuzzy multi-criteria decision making and fuzzy TOPSIS method based on alpha level sets have been developed [7, 8] to cope with problems based on type-1 fuzzy sets (T1FSs). Different extensions of T1FSs allow to present fuzzy and imprecise information as close to practice as possible. These approaches differ from each other in the way they describe the membership degree of an element and they can handle more types of flexibilities in fuzzy decision making problems. Among them are type-2 fuzzy sets (T2FSs) [9], type-n fuzzy sets [9], interval-valued fuzzy sets [9], Atanassov's intuitionistic fuzzy sets [10–12], interval-valued intuitionistic fuzzy sets [12, 13], hesitant fuzzy sets [14, 15], neutrosophic sets [16] in decision making and other.

Trapezoidal fuzzy numbers of type-1 (T1TFNs) and type-2 trapezoidal fuzzy numbers (T2TFNs) [17] are usually used when it is necessary to handle uncertainty. In order to consider activity profiling in scheduling problem with type-2 trapezoidal fuzzy numbers (T2TFNs), a new scheduling model including fuzzy operations and a ranking method is proposed in this paper. Fuzzy sets here can handle first and second order uncertainties in scheduling by introducing a general type-2 fuzzy set Ã, as a three-dimensional profile (Fig. 1). The third dimension here is the value of the membership function, i.e. satisfaction degree of expert towards each activity execution profile, for example time variable at each point on its two-dimensional domain that stands for a footprint of uncertainty (FOU).

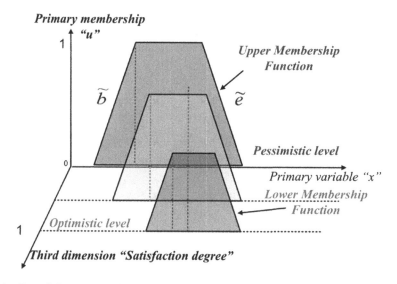

Fig. 1. Type-2 fuzzy trapezoidal start time estimation of activity j ready-time and due-date.

Figure 1 illustrates type-2 fuzzy trapezoidal start time of activity (as well as ready time $\tilde{b}_j$ and due date $\tilde{e}_j$ according to expert estimation). A footprint of uncertainty (FOU) here is the value of the membership function at each point on its two-dimensional domain, which reflects a decision-maker satisfaction/attitude towards optimistic and pessimistic profile.

Here we refer to classical terminology and definition of type-2 fuzzy sets [17].

Definition 1. A type-2 fuzzy set (T2FS) is defined on the universe of disclosure X and can be characterized by its membership function $\mu_{\tilde{A}}(x, u)$ and represented by Eq. (1):

$$\tilde{A} = \{((x,u), \mu_{\tilde{A}}(x,u))/\forall x \in X, \forall u \in J_x \subseteq [0,1]\}, \tag{1}$$

where $0 \leq \mu_{\tilde{A}}(x,u) \leq 1$, the subinterval J_x called *primary membership function* of x, and $\mu_{\tilde{A}}(x,u)$ is called *secondary membership function* that defines the possibilities of primary membership function. Uncertainty in J_x of type-2 fuzzy number $\tilde{A}$ is usually called footprint of uncertainty (FOU) and defined as the union of all primary memberships [17, 18].

Definition 2. An interval type-2 fuzzy set (IT2FS) is defined on the universe of disclosure X and can be characterized by upper membership function (UMF) and lower membership function (LMF) and denoted as follows:

$$\tilde{A} = \{\langle \mu_{A^U}(x), \mu_{A^L}(x) \rangle / \forall x \in X\}, \tag{2}$$

where A^U is an upper T1FS and its membership function is $\mu_{A^U}(x) = max\{J_x\}$ for $\forall x \in X$, and A^L is a lower T1FS whose membership function is equal to $\mu_{A^L}(x) = min\{J_x\}$ for $\forall x \in X$.

Let $\tilde{A} = \langle A^U, A^L \rangle = \langle a_1^U, a_2^U, a_3^U, a_4^U, h(A^U); a_1^L, a_2^L, a_3^L, a_4^L, h(A^L) \rangle$ be an IT2FS on the set of real numbers R. Then its formal presentation of upper and lower membership function [18] can be defined as pessimistic and optimistic attitude of decision-maker towards imprecise variable $\tilde{A}$ as follows:

$$\mu_{A^U}(x) = \begin{cases} h(A^U) * \frac{x-a_1^U}{a_2^U-a_1^U}, a_1^U \leq x < a_2^U \\ h(A^U), a_2^U \leq x \leq a_3^U \\ h(A^U) * \frac{a_4^U-x}{a_4^U-a_3^U}, a_3^U < x \leq a_4^U \\ 0, otherwise. \end{cases} \tag{3}$$

$$\mu_{A^L}(x) = \begin{cases} h(A^L) * \frac{x-a_1^L}{a_2^L-a_1^L}, a_1^L \leq x < a_2^L \\ h(A^L), a_2^L \leq x \leq a_3^L \\ h(A^L) * \frac{a_4^L-x}{a_4^L-a_3^L}, a_3^L < x \leq a_4^L \\ 0, otherwise. \end{cases} \tag{4}$$

Figure 2 illustrates interval type-2 fuzzy trapezoidal parameter estimation of activity duration according to pessimistic and optimistic attitude towards execution profile of activity.

Such kind of first-order uncertainty presupposes modelling decision-maker attitude towards fuzzy estimation that is represented as a footprint of uncertainty (FOU). Each upper and lower point a^L and a^U shows optimistic and pessimistic evaluation of the same variable within a fuzzy number representation. Such kind of aggregation allows to formalize an activity profile description with fuzzy representation for MCDM problem.

Traditionally when a decision-maker needs to solve RCPSP he uses classical forward propagation to calculate fuzzy earliest time and fuzzy total end-time of a project,

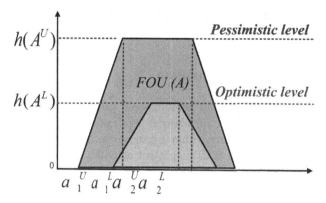

Fig. 2. IT2FS for scheduling problem with respect to decision-maker attitude towards fuzzy variable.

however backward propagation is inadequate due to the nature of uncertainty. Otherwise such uncertainty would be considered twice. Here we need to rank fuzzy numbers (activity start times) all the time when precedence relations constraints should be checked [5]. Finally fuzzy subtraction increases the imprecision in the sense that the result of "*duedate*" $\tilde{e}_j$ – "*duration*" $\tilde{p}_j$ is more imprecise that $\tilde{e}_j$ or $\tilde{p}_j$ separately. Therefore, variables $\tilde{e}_j$ or $\tilde{p}_j$ are non-interactive. D.Dubois and H.Prade introduced the concept of weakly non-interactive fuzzy numbers and operations there are based on the extension principle corresponding to each t-norm in place of the minimum operator [19].

A computational profile of fuzzy numbers including T1TFNs and IT2TFNs types usually involves their arithmetic operations and a ranking method. When applying ranking method in scheduling we operate with fuzzy durations, estimated by experts and precedence relations between activities. Such relations presuppose that in some technological cases an activity can't be performed unless the predecessor is finished. Thus it is necessary to make some assumptions and check out practical efficiency of ranking method. Arithmetic operations are also important when justifying the reasonability of the ranking method [21, 22].

However, considering specific scheduling constraints, such as precedence relations, we need to take into account overlaps between fuzzy variables and combine variable estimation within activity profile with the ranking process. Dynamical evaluation of alternatives in algorithm allows to obtain feasible/optimal solution. The paper also considers an inadmissible overlaps or absence of overlapping between fuzzy "finish-start times" or "start-start times" of predecessors while ranking them to check the Eq. (6) performance. Types of overlaps between activities were introduced in paper [6] and can have configurations depending on intersection degree between left- and right-spread of trapezoidal fuzzy numbers.

In order to provide technique based on simultaneous dealing with fuzzy resource-constrained scheduling problem with non-preemptive activity without splitting, activity execution profiles and precedence relations let's consider mathematical formulation of classical multi-mode resource-constrained project scheduling problem MRCPSP with respect to type-2 interval fuzzy numbers (problem statement m, $1T|cpm$, $mu|C_{max}$ in notation of Herroelen et al. [20]).

3 MRCPSP Problem Formulation

Many MCDM problems, especially combinatorial problems can be modeled using graph theory. It has a wide range of practical applications and intuitive visual representation. Any activity sequencing within a project can be characterized by Activity-on-Node representation, where nodes represent activities of a project; directed arcs represent precedence relations between activities. Each activity execution profile $p \in P_j$ is a combination of certain expert estimated fuzzy duration and certain resource requirement level as a specific trade-off.

Definition 3. *A resource-constrained schedule* $(P, \tilde{S})$ *consists of profile* P, *which assigns start time to each activity* $\tilde{S}$ *within resource usage compromise, where* $P_j = (p_j), j \in V$ *assigns to each activity only one execution profile* p_j. *A start time vector* $\tilde{S} = (\tilde{s}_j), j \in V$ *assigns to each activity exactly one point in time* $\tilde{t} \geq 0$ *as start time* $\tilde{s}_j$ *with* $\tilde{s}_0 := 0$; $\tilde{s}_{n+1}$ *stands for fuzzy project duration so that:*

$$\tilde{t} \in \left[\tilde{s}_j, \tilde{s}_j + \tilde{d}_{jp}\right). \tag{5}$$

Definition 4. *A Schedule is a combination of four sets* $\{A, R, Pr, P_j\}$, *where* A – *the set of activities* j *of a project;* R – *the set of renewable resources available at each moment of a project performance horizon* $\tilde{T}$; Pr – *precedence constraints in set* A; P_j – *the set of alternative execution profiles for each activity* $j \in A$. *There are* n - *activities with discrete resource requirements at every moment* $\tilde{t}$. *For each activity* $j \in A$ *experts estimate the following variables:* $\tilde{d}_{jp}$ – *duration time of activity* j *executed in profile* $p \in P_j$; $\tilde{b}_j$ - *ready time and* $\tilde{e}_j$-*deadline or due date for activity performance;* y_{jp} - *is a binary variable that shows if activity* $j \in A$ *is performed under profile* $p \in P_j$ *or not;* x_{jpt} – *is a binary variable that shows whether activity* $j \in A$ *is performed in profile* $p \in P_j$ *and starts at fuzzy moment* $\tilde{t}$ *or not.*

$$\sum_p \sum_{\tilde{t}=\tilde{s}_j}^{\tilde{c}_j} x_{jpt} = 1, \ \forall j \in A, \tag{6}$$

where $\tilde{S}_j$ earliest starting time of activity j and latest completion time $\tilde{C}_j$ of activity j;

$$x_{jpt} = \begin{cases} 1, & \textit{if activity } j \textit{ starts at moment } \tilde{t} \\ & \textit{with } p - \textit{execution profile} \\ & 0, \textit{ otherwise.} \end{cases} \tag{7}$$

Let's R_{kt} be a renewable resource availability in every fuzzy moment $\tilde{t}$, r_{jkp} - number of units of renewable resources when activity j is performed with p-profile.

$$\sum_j \sum_p \sum_{q=\tilde{t}}^{\tilde{t}+\tilde{p}_{jpq}-1} r_{jkp} * x_{jp\tilde{t}} \leq R_{kt}, \forall k, \tilde{t}. \tag{8}$$

Finally the start time of dummy last activity $\tilde{S}_{|n|+1}$ shows project makespan. Data inputs are introduced in resource profiles $\left[0, U_{jkp}\right]$ for renewable resources [23]. Mathematical formulation for multi-profile resource-constrained project scheduling problem MRCPSP (m, 1T|cpm, mu|C_{max} problem) can be stated as follows:

$$Min\ \tilde{S}_{|n|+1}, \tag{9}$$

subject to

$$\sum_p \sum_{\tilde{t}=\tilde{s}_j}^{\tilde{c}_j} x_{jp\tilde{t}} = 1, \quad \forall j \in A, \tag{10}$$

$$\tilde{c}_i \leq \tilde{s}_j - 1, \ \forall (i,j) \in V; \tag{11}$$

$$\tilde{s}_j \leq x_{jp\tilde{t}} * \tilde{t} + P\left(1 - x_{jp\tilde{t}}\right), \ \forall j \in A, \ p \in P_j, \ \tilde{t} = 1, \ldots, \tilde{T}; \tag{12}$$

$$\tilde{c}_j \geq x_{jp\tilde{t}} * \tilde{t}, \ \forall j \in A; \ p \in P_j; \ \tilde{t} = 1, \ldots, \tilde{T}; \tag{13}$$

$$\tilde{s}_j \geq b_j, \quad \forall j \in A; \tag{14}$$

$$\tilde{c}_j \leq e_j, \quad \forall j \in A, \tag{15}$$

$$\sum_j \sum_p \sum_{q=\tilde{t}}^{\tilde{t}+\tilde{p}_{jpq}-1} r_{jkp} * x_{jp\tilde{t}} \leq R_{kt}, \forall k, \tilde{t}; \tag{16}$$

$$x_{jp\tilde{t}} \in \{0, 1\}, \quad \forall j \in A, \ p \in P_j, \ \tilde{t} = 1, \ldots, \tilde{T}; \tag{17}$$

$$\tilde{s}_j \geq 0, \quad \forall j \in A, \tag{18}$$

$$\tilde{c}_j \geq 0, \quad \forall j \in A; \tag{19}$$

$$\tilde{c}_j - \tilde{s}_j = \sum_{p \in P_j} \sum_{\tilde{t}=1}^{\tilde{T}} x_{jp\tilde{t}} - 1, \quad \forall j \in A. \tag{20}$$

The idea is to construct a schedule for activity performance, i.e. activity performance within best selected execution profile, so that precedence and resource constraints are satisfied and the project make span $\tilde{S}_{|n|+1}$ is minimized. Constraint (10) assigns exactly one execution profile to perform each activity. Constraints (11)–(13) correspond to precedence relations, where $\tilde{s}_j - 1$ eq. (11) removes strict inequality given integer time units and there is no activity splitting Eq. (20). Activity ready time and due-dates constraints are presented in (14), (15). Equation (16) specifies resource availability for renewable resources for each fuzzy moment $\tilde{t}$. Finally to check out if an activity j performed in profile p, $p \in P_j$ is feasible to be included in current project we introduced Eq. (21) as follows:

$$U_{jkp\tilde{\imath}} = 1_{[0,R_{kt}]}\left(r_{jkp\tilde{\imath}}\right) := \begin{cases} 1, & if\ r_{jkp\tilde{\imath}} \in [0, R_{kt}] \\ 0, otherwise, \end{cases} \qquad (21)$$

where $U_{jkp\tilde{\imath}}$ denotes resource usage to perform activity j within profile p at each fuzzy moment $\tilde{\imath}$.

Definition 5. *An activity execution profile* $p \in P_j$ *is composed of fuzzy estimated activity duration* $\tilde{d}_{jp}$ *and corresponding resource requirements of a certain type* $r_{jkp\tilde{\imath}}$ needed for activity execution. Usually activity duration depends on resources available for each scheduling moment.

4 Solution Methodology Based on IT2TFN for Agriculture Scheduling

This paper is based on idea of Lee and Chen [24] that presented the TOPSIS method for handling fuzzy multiple attributes group decision-making problems. On the first step decision-maker is required to choose the best execution profile for each activity. This expert choice will be used as initial expert estimated activity profile from set P_j at l-level of decision tree. During the optimization procedure it may happen that it will be switched off algorithmically to another profile.

If we have $\tilde{A} = \langle A^U, A^L \rangle = \langle a_1^U, a_2^U, a_3^U, a_4^U, h(A^U); a_1^L, a_2^L, a_3^L, a_4^L, h(A^L) \rangle$, then index of pessimism is set as $h(A^U)$ and illustrates the worst activity duration execution variant; and index of optimism is set as $h(A^L)$ and illustrates the attitude of decision-maker towards a preferable activity duration. Data have been aggregated from 5 experts and given after detailed environmental and economic cost/benefit analysis.

Detailed aspects of the wheat production system investigated in this example (e.g. P, K, Mg fertilization, plant protection, use of agricultural machinery etc.) have been defined according to good agricultural practice in Western Europe [2]. Depending on environmental and forecast factors and specific regional aspects timing period for each activity may vary significantly from year to year.

Table 1 illustrates the example of basic agriculture activities.

Different approaches based on resource-constrained scheduling apply heuristic search, tree-searching algorithms and their modifications. Some of them propose a precedence tree-based solution that consider enumeration scheme and estimates all possible partial schedules. In this research the precedence tree-based approach is based on enumeration scheme and estimates all possible partial combinations of activity profiles and consider a possibility to change activity profile, i.e. choose another alternative in case of fuzzy numbers overlapping.

Overlapping here means that precedence relations are not followed. Let's consider some notations for resource-constrained problem formulation (Table 2):

Modified pseudo-code, that was first introduced by Cheng et al. in [25], was improved in this paper by fuzzy performance and profile-switching and illustrates precedence tree branch-and-bound approach for solving fuzzy MRCPSP.

Table 1. Basic agricultural operations for winter wheat production system

Activity	Resource requirement/timing	Expert timing (duration) estimation (days). aggregated data (IT2TFN)
1. Fertilizer application: P as triple super phosphate (TSP, 46% P2O5), K as potassium sulfate (50% K2O), Mg as kieserite (26% MgO)	October, fertilizer spreader, tractors	<5,6,7,10,**1**; 5.5,6.5,7,8,**0.4**>
2. Soil preparation: ploughing	October, plow (5 shares), tractors	<7,8,9,10,**0.8**; 5.5,6.5,7,8,**0.3**>
3. Sowing: seedbed preparation, drilling.	October, seedbed combination, drill, tractors	<11,13,17,21,**0.9**; 12,13,14,15,**0.3**>
4. Fertilizer application: N as ammonium nitrate (AN, 33.5% N)	March/April, fertilizer spreader, tractors	<5,6,7,10,**1**; 5.5,6.5,7,8,**0.4** >
5. Plant protection: 1st herbicide application	May, sprayer, tractors	<3,5,8,10,**0.8**; 4,5,7,9,**0.6**>
6. Plant protection: 1st fungicide application	May/June, sprayer. tractor	<4,5,8,10,**0.7**; 4,5,7,9,**0.6**>
7. Plant protection: 2nd herbicide application	June/July, sprayer, tractor	<2.5,5,8,10,**0.8**; 4,5,7,9,**0.4**>
8. Plant protection: 2nd fungicide application	June/July, together with 2nd herbicide application	<1.5,3,8,10,**1**; 4,5,7,9,**0.6**>
9. Harvest: combine harvesting, bale pressing	August, combine, baling press, tractor	<30,35,40,50,**1**; 15,25,35,40,**0.6**>

Table 2. Resource-constrained notations for solution methodology

Symbol	Description	Comments
l, j_l	Level l of enumeration tree, Activity j at l-level	
p_{j-l}	Selected profile* for activity j at l level	*Initial expert estimated activity profile from set P_j at l-level
P_{j-l}	Set of available profiles for activity j performance at l level	
$\tilde{s}_{j-l}, \tilde{c}_{j-l}$	Start and finish time of activity j_l	Finish time of activities are set according to $\tilde{c}_j = \tilde{s}_j + \tilde{d}_{jp}$
$\tilde{d}_{jp-l}$	Activity duration within profile j_l	
$SchJ_l$	Set of already scheduled activities at l-level	

(continued)

Table 2. (*continued*)

Symbol	Description	Comments
$SuitJ_l$	Set of suitable (active) to scheduling activities at l-level	Activities whose immediate predecessors were completely finished
EST_j	Earliest precedence feasible start time of activity j	Earliest Start Time of activity is set according to CPM, forward propagation
$\gamma_{i_p j_p}^{1,2,3}$	Types of fuzzy overlaps between activity (i,j)	Types of fuzzy overlaps between activities were introduced in [6]
$Pred(j)$	Set of predecessor activities of j-activity	For dummy start $Pred(0) = \varnothing$
$Succ(j)$	Set of successor activities of j-activity	For dummy finish $Succ\ (|n|+1) = \varnothing$

Step 1:Initialization.
Set $g=1$; Dummy activity $j_1=0; \tilde{t}:=(0,0,0,0)$; $p_{j1}=1; \tilde{s}_{j1}=0; SchJ\text{-}l=\varnothing$;
Step 2: Updating sets of activities.
Increase tree level $g=g+1$. Update set of scheduled activities:
$SchJ_l = SchJ_{l-1} \cup \{j_{l-1}\}$.
Compose the set of suitable activities: $SuitJ_l = \{j \in A \setminus SchJ_l | pred(j) \subseteq SchJ_l\}$.
If the last dummy activity is active: $n+1 \in SuitJ_l$, then store the current solution and go to Step 5.
Else go to step 3.
Step 3:Selecting the next activity from $SuitJ_l$ to be scheduled.
If there is no untested activity left in $SuitJ_l$, then go to step 5, Else select next activity $j_l \in SuitJ_l$.
Step 4:Selecting a profile for selected activity and fuzzy scheduling.
If there is no untested profiles left in $\{1,...,P_{j-l}\}$ for activity j, then go to step 3,
Else select an untested profile$p_{j-l} \in \{1,...,P_{j-l}\}$.
Compute the earliest precedence feasible start time:
$EST_{j-l} = max\{\tilde{c}_{j-l}|i \in pred(j_l)\} + 1$.
Compute the start time$\tilde{s}_{j-l}$ and completion time $\tilde{c}_{j-l}$ satisfying equation (20).
For each pair (i,j):
If time overlapping is$\gamma_{i_m j_m}^1 := 0$, then there is no switch between profiles;
If $\gamma_{i_m j_m}^2 := 1$, then set $d_A := a_B$;If $\gamma_{i_m j_m}^3 := -1$, then select untested profile for j.
Go to Step 2.
Step 5: Backtracking.
Decrease the precedence tree level by 1: $g=g\text{-}1$. If the precedence tree level $l=1$, then STOP. Else go to Step 4.

The precedence tree-based approach algorithm consider enumeration scheme and evaluate all possible partial schedules unless all the activities will be scheduled.

5   Conclusion

This paper examines type-2 fuzzy numbers and its implementation to resource-constrained fuzzy scheduling. The main idea is to estimate and formalize uncertain variables in scheduling process and chose best execution profile under preference attitude of decision-maker and flexible operating conditions. The developed precedence tree-based approach relies on enumeration scheme and evaluates all possible partial schedules. As a future research an efficient ranking method need to be developed. It can improve execution performance of enumeration strategy and solution methodology using type-2 fuzzy numbers performance.

Acknowledgments. This work has been supported by the Ministry of Education and Science of the Russian Federation under Project "Methods and means of decision making on base of dynamic geographic information models" (Project part, State task 2.918.2017).

References

1. Blazewicz, J.: Scheduling subject to resource constraints: classification and complexity. Discrete Appl. Math. **5**(1), 11–24 (1983)
2. Brentrup, F., Küsters, J., Lammela, J., Barraclough, P., Kuhlmann, H.: Environmental impact assessment of agricultural production systems using the life cycle assessment (LCA) methodology II. The application to N fertilizer use in winter wheat production systems. Eur. J. Agron. **20**, 265–279 (2004)
3. Billaut, J.-C., Moukrim, A., Sanlaville, E.: Flexibility and Robustness in Scheduling. Control Systems, Robotics and Manufacturing Series. Willey-ISTE, Hoboken (2013)
4. Dubois, D., Fargier, H., Fortemps, F.: Fuzzy scheduling: profilelling flexible constraints vs. coping with incomplete knowledge. Eur. J. Oper. Res. **147**, 231–252 (2003)
5. Brucker, P., Knust, S.: Complex Scheduling, pp. 29–115. Springer, Heidelberg (2012). https://doi.org/10.1007/978-3-642-23929-8
6. Knyazeva, M., Bozhenyuk, A., Rozenberg, I.: Scheduling alternatives with respect to fuzzy and preference modeling on time parameters. In: Kacprzyk, J., Szmidt, E., Zadrożny, S., Atanassov, K.T., Krawczak, M. (eds.) IWIFSGN/EUSFLAT - 2017. AISC, vol. 642, pp. 358–369. Springer, Cham (2018). https://doi.org/10.1007/978-3-319-66824-6_32
7. Liang, G., Wang, M.J.: Personnel selection using fuzzy MCDM algorithm. Eur. J. Oper. Res. **78**(1), 22–33 (1994)
8. Wang, Y., Elhag, T.M.S.: Fuzzy topsis method based on alpha level sets with an application to bridge risk assessment. Expert Syst. Appl. **31**(2), 309–319 (2006)
9. Zadeh, L.A.: The concept of a linguistic variable and its application to approximate reasoning. Inf. Sci. **8**(3), 199–249 (1975)
10. Atanassov, K.T.: Intuitionistic fuzzy sets. Fuzzy Sets Syst. **20**(1), 87–96 (1986)
11. Atanassov, K.T.: Two theorems for intuitionistic fuzzy sets. Fuzzy Sets Syst. **110**, 267–269 (2000)
12. Atanassov, K.T., Gargov, G.: Interval valued intuitionistic fuzzy sets. Fuzzy Sets Syst. **31**(3), 343–349 (1989)
13. Wang, P.J., Wang, X., Cai, C.: Some new operation rules and a new ranking method for interval-valued intuitionistic linguistic numbers. J. Intell. Fuzzy Syst. **32**(1), 1069–1078 (2017)

14. Torra, V.: Hesitant fuzzy sets. Int. J. Intell. Syst. **25**(6), 529–539 (2010)
15. Rodríguez, R.M., Martínez, L., Torra, V., Xu, Z.S., Herrera, F.: Hesitant fuzzy sets: state of the art and future directions. Int. J. Intell. Syst. **29**(6), 495–524 (2014)
16. Majumdar, P.: Neutrosophic sets and its applications to decision making. In: Acharjya, D., Dehuri, S., Sanyal, S. (eds.) Computational Intelligence for Big Data Analysis. Adaptation, Learning, and Optimization, vol. 19, pp. 97–115. Springer, Cham (2015). https://doi.org/10.1007/978-3-319-16598-1_4
17. Mendel, J.M., John, R.I.: Type-2 fuzzy sets made simple. IEEE Trans. Fuzzy Syst. **10**(2), 117–127 (2002)
18. Wang, J., Chen, Q., Zhang, H., Chen, X., Wang, J.: Multi-criteria decision method based on type-2 fuzzy sets. Filomat **31**(2), 431–450 (2017)
19. Dubois, D., Prade, H.: Additions of interactive fuzzy numbers. IEEE Trans. Autom. Control **26**(4), 99–135 (1981)
20. Herroelen, W., De Reyck, B., Demeulemeester, E.: Resource-constrained project scheduling: notation, classification, modes and methods by Brucker et al. Eur. J. Oper. Res. **128**(3), 221–230 (2000)
21. Cheng, C.-H.: A new approach for ranking fuzzy numbers by distance method. Fuzzy Sets Syst. **95**, 307–317 (1998)
22. Liou, T.-S., Wang, M.-J.: Ranking fuzzy numbers with integral value. Fuzzy Sets Syst. **50**, 247–255 (1992)
23. Hartmann, S., Drexl, A.: Project scheduling with multiple modes: a comparison of exact algorithms. Networks **32**, 283–297 (1998)
24. Chen, S.M., Lee, L.W.: Fuzzy multiple attributes group decision making based on the interval type-2 TOPSIS method. Expert Syst. Appl. **37**, 2790–2798 (2010)
25. Cheng, J., Fowler, J., Kempf, K., Mason, S.: Multi-mode resource-constrained project scheduling problems with non-preemptive activity splitting. Comput. Oper. Res. **53**, 275–287 (2015)

Differential Evolution Algorithm Using a Dynamic Crossover Parameter with High-Speed Interval Type 2 Fuzzy System

Patricia Ochoa, Oscar Castillo$^{(\boxtimes)}$, José Soria,
and Prometeo Cortes-Antonio

Division of Graduate Studies, Tijuana Institute of Technology, Tijuana, Mexico
ochoa.martha@hotmail.com, ocastillo@tectijuana.mx,
jsoria57@gmail.com, prometeo.cortes@gmail.com

Abstract. The main contribution of this paper is the use of a new concept of type reduction in type-2 fuzzy systems for improving performance in differential evolution algorithm. The proposed method is an analytical approach using an approximation to the Continuous Karnik-Mendel (CEKM) method, and in this way the computational evaluation cost of the Interval Type 2 Fuzzy System is reduced. The performance of the proposed approach was evaluated with seven reference functions using the Differential Evolution algorithm with a crossover parameter that is dynamically adapted with the proposed methodology.

Keywords: Differential evolution algorithm · Crossover
Dynamic parameter adaptation and interval type 2 fuzzy logic

1 Introduction

Currently the use of metaheuristics is very common for the solution of complex optimization problems [1–6]. In this paper we propose the use of the Differential Evolution algorithm for the set of benchmark mathematical functions of the CEC 2017 competition. We use a subset of functions of the total set that is used in the CEC 2017 competition. In particular we only use the set of simple multimodal functions.

Generally, the algorithms used to optimize functions have static parameters, and in this paper we propose to dynamically change the crossover (CR) parameter, which represents the crossover of individuals in the differential evolution algorithm. In addition, we are using in the proposed approach an analytical approximation to the Continuous Karnik-Mendel (CEKM) method. Some works related to the optimization of parameters in metaheuristics using fuzzy logic are briefly mentioned below [7–11].

The approximation to the CEKM method is inspired on the traditional Fuzzy Sets originally proposed by Zadeh, and this method is based on CEKM as a root-finding problem, which was originally proposed by Liu and Mendel in [12]. This method for type reduction is based on two proposed equations that allow approximation of the output points through numerical methods, like Newton Raphson. The objective is to provide a solution to the type-reduction/defuzzification problem that eliminates the iterations and sampling requirements. In addition, we consider this proposed

I. Batyrshin et al. (Eds.): MICAI 2018, LNAI 11288, pp. 369–378, 2018.
https://doi.org/10.1007/978-3-030-04491-6_28

approximation to achieve the dynamic form of the CR parameter of the Differential Evolution algorithm.

There are works in which the concept of fuzzy logic combined with the Differential Evolution algorithm is used and to mention some related works [13–15, p. 2, 16]. In the same way, there are multiple algorithms that participate in the competition of CEC2017 and for this paper the competition winners are mentioned in [17–20].

The paper is organized in the following form: Sect. 2 describes the Differential Evolution algorithm. Section 3 describes the methodology using the fuzzy logic approach. Section 4 presents the experimentation with the set of CEC 2017 Benchmark functions. Finally, Sect. 5 offers some Conclusions.

2 Differential Evolution Algorithm

The use of the Differential Evolution algorithm is very widespread at present time. The DE algorithm was originally proposed by Storn and Price in 1994, and for this paper we consider the original version of algorithm, which has the following mathematical structure represented in Eqs. 1–10 [21]:

Population Structure

$$P_{x,g} = (x_{i,g}), i = 0, 1, \ldots, Np, g = 0, 1, \ldots, g_{max} \tag{1}$$

$$x_{i,g} = (x_{j,i,g}), j = 0, 1, \ldots, D - 1 \tag{2}$$

$$P_{v,g} = (v_{i,g}), i = 0, 1, \ldots, Np - 1, g = 0, 1, \ldots, g_{max} \tag{3}$$

$$v_{i,g} = (v_{j,I,g}), j = 0, 1, \ldots, D - 1 \tag{4}$$

$$P_{v,g} = (u_{i,g}), i = 0, 1, \ldots, Np - 1, g = 0, 1, \ldots, g_{max} \tag{5}$$

$$u_{i,g} = (u_{j,I,g}), j = 0, 1, \ldots, D - 1 \tag{6}$$

Initialization

$$x_{j,i,0} = rand_j(0, 1) \cdot (b_{j,U} - b_{j,L}) + b_{j,L} \tag{7}$$

Mutation

$$v_{i,g} = x_{r0,g} + F \cdot (x_{r1,g} - x_{r2,g}) \tag{8}$$

Crossover

$$u_{i,g} = (u_{j,i,g}) = \begin{cases} v_{j,i,g} & \text{if } rand_j(0, 1) \leq Cr \text{ or } j = j_{rand} \\ x_{j,i,g} & \text{otherwise} \end{cases} \tag{9}$$

Selection

$$x_{i,g+1} = \begin{cases} u_{i,g} & if \quad f(u_{i,g}) \leq f(x_{i,g}), \\ x_{i,g} & otherwise. \end{cases} \tag{10}$$

3 Methodology

This work is mainly based on the use of the Differential Evolution algorithm by applying a new concept of type reduction, which approximates the Interval Type-2 Fuzzy System, and using this concept we propose to dynamically adapt the CR parameter, which represents the crossover during the execution of the algorithm.

Figure 1 represents the structure of the Differential Evolution algorithm to which we have added the fuzzy system, which gives as the output the crossover parameter.

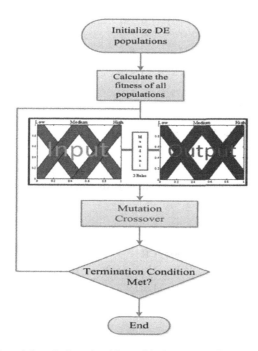

Fig. 1. Differential evolution algorithm with the proposed type-2 fuzzy system

The fuzzy system is constructed with one input which is the generations and is calculated in Eq. 11 and the one output which is the CR, is calculated with Eq. 12.

$$Generations = \frac{Current\ Generations}{Maximun\ of\ Generations} \tag{11}$$

$$CR = \frac{\sum_{i=1}^{r_{CR}} \mu_i^{CR}(CR_{1i})}{\sum_{i=1}^{r_{CR}} \mu_i^{CR}} \qquad (12)$$

Where CR, is the crossover; r_{CR}, is the number of rules of the interval type-2 fuzzy system corresponding to CR; CR_{1i}, is the output result for rule i corresponding to CR; μ_i^{CR}, is the membership function of rule i corresponding to CR. Figure 2 represents the structure for the interval type-2 fuzzy system, is of Mamdani type, and the input and output membership functions are granulated into *Low*, *Medium* and *High*.

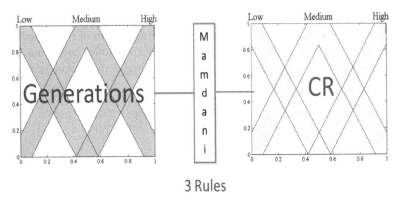

Fig. 2. Structure of the type-2 fuzzy system

Table 1 represents the rules of the interval type-2 fuzzy system for the DFE algorithm, which are on an increase fashion as the number of generations advance.

Table 1. Rules of the FDE algorithm

Generation	CR		
	Low	Medium	High
Low	Low	-	-
Medium	-	Medium	-
High	-	-	High

4   Simulation Results

In this paper we used 7 functions of the set of CEC 2017 benchmark functions, which are described in Table 2. We performed experiments with the original algorithm and later using the proposed method, where a fuzzy system will help to dynamically change the CR parameter.

The parameters used for these experiments are shown in Table 3 where NP is the size of the population, D is the dimension of each individual, F is the mutation, CR is

Table 2. Summary of the CEC'17 learning-based benchmark suite

	No.	Functions	$Fi^* = Fi$ (x^*)
Simple multimodal functions	f1	Shifted and rotated Rosenbrock's function	400
	f2	Shifted and rotated Rastrigin's function	500
	f3	Shifted and rotated expanded Scaffer's F6 function	600
	f4	Shifted and rotated Lunacek Bi_Rastringin function	700
	f5	Shifted and rotated non-continuous Rastrigin's function	800
	f6	Shifted and rotated Levy function	900
	f7	Shifted and rotated Schwefel's function	1000

the crossover and GEN are the generations. For the case of the experiments using the original algorithm CR is assigned a fixed value of 0.4 and subsequently for the following experiments CR is dynamic by using the fuzzy system. Experiments were only carried out with a number of dimensions of 10 and 30 with the same guidelines of the CEC2017 competition.

The results obtained with the original algorithm are presented in Table 4 for a number of dimensions of 10, the table contains the best and the worst results, the means and the standard deviations.

Table 3. Parameters of the experiments

Parameters
NP = 250
D = 10 and 30
F = 0.7
CR = 0.4 and dynamic
GEN = 100000 and 300000

Table 5 shows the results for number of dimensions of 30 and contains the best and the worst results, the means and the standard deviations.

Tables 6 and 7 represent the results obtained with a dynamic CR parameter for 10 and 30 dimensions respectively and contains the best and the worst results, the means and the standard deviations.

The results obtained in our experimentation so far show that the improvement is significant with respect to the original algorithm, the comparison of the results with respect to the mean and the best error show the improvement of the algorithm using the proposed methodology.

Table 4. Results for dimensions D = 10

Differential evolution algorithm for D = 10

	Best	Worst	Mean	Std
f1	3.43E+02	4.74E+03	1.89E+03	8.68E+02
f2	8.51E+01	1.82E+02	1.37E+02	2.24E+01
f3	5.04E+01	1.20E+02	8.61E+01	1.57E+01
f4	2.81E+02	5.57E+02	4.29E+02	7.13E+01
f5	8.06E+01	1.53E+02	1.29E+02	1.46E+01
f6	1.76E+03	6.63E+03	3.73E+03	1.11E+03
f7	1.60E+03	3.25E+03	2.56E+03	2.99E+02

Table 5. Results for dimensions D = 30

Differential evolution algorithm for D = 30

	Best	Worst	Mean	Std
f1	1.45E+04	6.20E+04	4.17E+04	9.36E+03
f2	5.13E+02	7.51E+02	6.54E+02	5.11E+01
f3	1.07E+02	1.58E+02	1.33E+02	1.06E+01
f4	1.88E+03	3.20E+03	2.45E+03	2.77E+02
f5	4.66E+02	6.70E+02	5.85E+02	5.08E+01
f6	2.37E+04	4.60E+04	3.44E+04	4.66E+03
f7	7.69E+03	1.03E+04	9.44E+03	5.12E+02

Table 6. Results with CR dynamic for D = 10

D. E. with CR dynamic for D = 10

	Best	Worst	Mean	Std
f1	2.05E+00	3.48E+03	1.17E+03	9.58E+02
f2	2.86E+01	1.82E+02	1.33E+02	3.66E+01
f3	3.56E+00	1.93E+02	8.50E+01	2.43E+01
f4	4.75E+00	5.88E+02	3.32E+02	1.63E+02
f5	1.21E+00	2.45E+02	1.24E+02	3.84E+01
f6	1.86E+02	7.74E+03	3.73E+03	1.23E+03
f7	2.40E+01	2.97E+03	2.37E+03	6.25E+02

Table 8 represents a comparison between the original method and the proposed method where CR is dynamic, and a comparison of the best results obtained for the number or dimensions of 10 and 30 is made.

Figures 3 and 4 show graphically the comparison of the best results between the original algorithm and the proposed method, and Fig. 3 shows the results for 10 dimensions and Fig. 4 shows the results for 30 dimensions, respectively.

Table 7. Results with CR dynamic for D = 10

D. E. with CR dynamic for D = 30			
	Best	Worst	Mean	Std
f1	5.65E+01	5.61E+04	2.74E+04	1.74E+04
f2	1.52E+00	7.61E+02	5.82E+02	1.98E+02
f3	9.69E+00	1.52E+02	1.24E+02	2.75E+01
f4	1.42E+02	3.11E+03	2.22E+03	9.16E+02
f5	1.11E+02	6.29E+02	5.26E+02	1.41E+02
f6	1.71E+01	4.20E+04	2.76E+04	1.16E+04
f7	3.15E+01	1.04E+04	9.08E+03	2.28E+03

Table 8. Comparison between DE and DE with CR dynamic

Comparison between DE and DE with CR dynamic			
	D = 10		D = 30	
	Original	Proposed	Original	Proposed
f1	3.43E+02	2.05E+00	1.45E+04	5.65E+01
f2	8.51E+01	2.86E+01	5.13E+02	1.52E+00
f3	5.04E+01	3.56E+00	1.07E+02	9.69E+00
f4	2.81E+02	4.75E+00	1.88E+03	1.42E+02
f5	8.06E+01	1.21E+00	4.66E+02	1.11E+02
f6	1.76E+03	1.86E+02	2.37E+04	1.71E+01
f7	1.60E+03	2.40E+01	7.69E+03	3.15E+01

We can notice graphically that the proposed method has better results than the original algorithm for both comparisons, the separation between the two methods is clearly shown (Blue is the proposed method in both Figures).

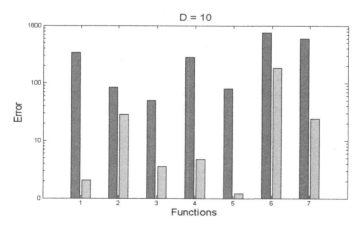

Fig. 3. Comparison of results for D = 10 (Color figure online)

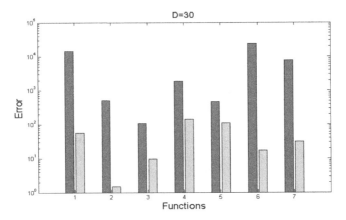

Fig. 4. Comparison of results for D = 30 (Color figure online)

5   Conclusions

In this study we can conclude that the use of the CEKM approximation method combined with the Differential Evolution algorithm produced good results when compared to the original algorithm, we use a set of functions which have a high complexity for the original algorithm but when using the proposed method, the results improved significantly. The fuzzy system used is relatively simple, one input and one output, but the fact of approximating the Interval Type-2 Fuzzy System, demonstrates that the concept of fuzzy logic that is using concepts of Footprint of Uncertainty (FOU) helps to obtain better results, as can be noted in the experimentation for the set of functions of CEC2017 used in this paper.

With respect to the experimentation carried out, it is shown that by increasing the complexity of the problem, for this case from 10 dimensions to 30 dimensions, the fuzzy system improves the results. We need to verify the above with a statistical test, perform more experimentation with more mathematical functions and integrate a more complete fuzzy system, since it is important to note that this study is the beginning of applying the concept of approximation to the CEKM method.

As future works we can mention many possible lines of research, for example: the proposed optimization method can be used in diverse applications, like in the cellular model of [22], in the neural model of [23] or in fuzzy control of [24]. Another application can be in models of time series prediction, like in [25]. Optimization of different types of control structures can also be considered, like in [26, 27]. Pattern recognition systems [28] can also be optimized with the proposed method. Finally, optimizations of intelligent systems in industrial processes, like in [29], or in hybrid models, like in [30], are also possible.

References

1. Barraza, J., Rodríguez, L., Castillo, O., Melin, P., Valdez, F.: A new hybridization approach between the fireworks algorithm and grey wolf optimizer algorithm. J. Optim. **2018**, 1–18 (2018)
2. Caraveo, C., Valdez, F., Castillo, O.: A new optimization meta-heuristic algorithm based on self-defense mechanism of the plants with three reproduction operators. Soft. Comput. **22** (15), 4907–4920 (2018)
3. Peraza, C., Valdez, F., Castillo, O.: Improved method based on type-2 fuzzy logic for the adaptive harmony search algorithm. In: Castillo, O., Melin, P., Kacprzyk, J. (eds.) Fuzzy Logic Augmentation of Neural and Optimization Algorithms: Theoretical Aspects and Real Applications. SCI, vol. 749, pp. 29–37. Springer, Cham (2018). https://doi.org/10.1007/978-3-319-71008-2_3
4. Agrawal, A.P., Kaur, A.: A comprehensive comparison of ant colony and hybrid particle swarm optimization algorithms through test case selection. In: Satapathy, S.C., Bhateja, V., Raju, K.S., Janakiramaiah, B. (eds.) Data Engineering and Intelligent Computing. AISC, vol. 542, pp. 397–405. Springer, Singapore (2018). https://doi.org/10.1007/978-981-10-3223-3_38
5. Yıldız, B.S., Yıldız, A.R.: Comparison of grey wolf, whale, water cycle, ant lion and sine-cosine algorithms for the optimization of a vehicle engine connecting rod. Mater. Test. **60**(3), 311–315 (2018)
6. Ochoa, P., Castillo, O., Soria, J.: Differential evolution using fuzzy logic and a comparative study with other metaheuristics. In: Melin, P., Castillo, O., Kacprzyk, J. (eds.) Nature-Inspired Design of Hybrid Intelligent Systems. SCI, vol. 667, pp. 257–268. Springer, Cham (2017). https://doi.org/10.1007/978-3-319-47054-2_17
7. Castillo, O., Amador-Angulo, L.: A generalized type-2 fuzzy logic approach for dynamic parameter adaptation in bee colony optimization applied to fuzzy controller design. Inf. Sci. **460–461**, 476–496 (2018)
8. Castillo, O., Neyoy, H., Soria, J., Melin, P., Valdez, F.: A new approach for dynamic fuzzy logic parameter tuning in ant colony optimization and its application in fuzzy control of a mobile robot. Appl. Soft Comput. **28**, 150–159 (2015)
9. González, B., Valdez, F., Melin, P., Prado-Arechiga, G.: Fuzzy logic in the gravitational search algorithm for the optimization of modular neural networks in pattern recognition. Expert Syst. Appl. **42**(14), 5839–5847 (2015)
10. Valdez, F., Melin, P., Castillo, O.: A survey on nature-inspired optimization algorithms with fuzzy logic for dynamic parameter adaptation. Expert Syst. Appl. **41**(14), 6459–6466 (2014)
11. Valdez, F., Melin, P., Castillo, O.: Modular neural networks architecture optimization with a new nature inspired method using a fuzzy combination of particle swarm optimization and genetic algorithms. Inf. Sci. **270**, 143–153 (2014)
12. Ontiveros-Robles, E., Melin, P., Castillo, O.: New methodology to approximate type-reduction based on a continuous root-finding karnik mendel algorithm. Algorithms **10**(3), 77 (2017)
13. Sun, Z., Wang, N., Srinivasan, D., Bi, Y.: Optimal tunning of type-2 fuzzy logic power system stabilizer based on differential evolution algorithm. Int. J. Electr. Power Energy Syst. **62**, 19–28 (2014)
14. Marinaki, M., Marinakis, Y., Stavroulakis, G.E.: A differential evolution algorithm for fuzzy control of smart structures (2012)
15. Bi, Y., Srinivasan, D., Lu, X., Sun, Z., Zeng, W.: Type-2 fuzzy multi-intersection traffic signal control with differential evolution optimization. Expert Syst. Appl. **41**(16), 7338–7349 (2014)

16. Ochoa, P., Castillo, O., Soria, J.: Differential evolution algorithm with interval type-2 fuzzy logic for the optimization of the mutation parameter. In: Castillo, O., Melin, P., Kacprzyk, J. (eds.) Fuzzy Logic Augmentation of Neural and Optimization Algorithms: Theoretical Aspects and Real Applications. SCI, vol. 749, pp. 55–65. Springer, Cham (2018). https://doi.org/10.1007/978-3-319-71008-2_5

17. Kumar, A., Misra, R.K., Singh, D.: Improving the local search capability of effective butterfly optimizer using covariance matrix adapted retreat phase, pp. 1835–1842 (2017)

18. Brest, J., Maucec, M.S., Boskovic, B.: Single objective real-parameter optimization: algorithm jSO, pp. 1311–1318 (2017)

19. Awad, N.H., Ali, M.Z., Suganthan, P.N.: Ensemble sinusoidal differential covariance matrix adaptation with Euclidean neighborhood for solving CEC2017 benchmark problems, pp. 372–379 (2017)

20. Mohamed, A.W., Hadi, A.A., Fattouh, A.M., Jambi, K.M.: LSHADE with semi-parameter adaptation hybrid with CMA-ES for solving CEC 2017 benchmark problems, pp. 145–152 (2017)

21. Ochoa, P., Castillo, O., Soria, J.: Type-2 fuzzy logic dynamic parameter adaptation in a new fuzzy differential evolution method. In: Proceedings of NAFIPS 2016, pp. 1–6 (2016)

22. Leal- Ramírez, C., Castillo, O., Melin, P., Rodríguez, A.: Díaz: simulation of the bird age-structured population growth based on an interval type-2 fuzzy cellular structure. Inf. Sci. 181(3), 519–535 (2011)

23. Melin, P., Amezcua, J., Valdez, F., Castillo, O.: A new neural network model based on the LVQ algorithm for multi-class classification of arrhythmias. Inf. Sci. 279, 483–497 (2014)

24. Castillo, O., Amador-Angulo, L., Castro, J.R., García Valdez, M.: A comparative study of type-1 fuzzy logic systems, interval type-2 fuzzy logic systems and generalized type-2 fuzzy logic systems in control problems. Inf. Sci. 354, 257–274 (2016)

25. Melin, P., Mancilla, A., Lopez, M., Mendoza, O.: A hybrid modular neural network architecture with fuzzy Sugeno integration for time series forecasting. Appl. Soft Comput. 7(4), 1217–1226 (2007)

26. Castillo, O., Melin, P.: Intelligent systems with interval type-2 fuzzy logic. Int. J. Innovative Comput. Inf. Control 4(4), 771–783 (2008)

27. Melin, P., Castillo, O.: Modelling, Simulation and Control of Non-Linear Dynamical Systems: An Intelligent Approach Using Soft Computing and Fractal Theory. CRC Press, Boca Raton (2001)

28. Melin, P., Gonzalez, C.I., Castro, J.R., Mendoza, O., Castillo, O.: Edge-detection method for image processing based on generalized type-2 fuzzy logic. IEEE Trans. Fuzzy Syst. 22(6), 1515–1525 (2014)

29. Melin, P., Castillo, O.: Intelligent control of complex electrochemical systems with a neuro-fuzzy-genetic approach. IEEE Trans. Ind. Electron. 48(5), 951–955 (2001)

30. Mendez, G.M., Castillo, O.: Interval type-2 TSK fuzzy logic systems using hybrid learning algorithm. In: The 14th IEEE International Conference on Fuzzy Systems, FUZZ 2005, pp. 230–235 (2005)

Allocation Centers Problem on Fuzzy Graphs with Largest Vitality Degree

Alexander Bozhenyuk[1][(✉)] ⓘ, Stanislav Belyakov[1] ⓘ,
Margarita Knyazeva[1] ⓘ, and Janusz Kacprzyk[2] ⓘ

[1] Southern Federal University, Nekrasovskiy Street, 44,
347928 Taganrog, Russia
{avb002, beliacov}@yandex.ru,
margarita.knyazeva@gmail.com
[2] Systems Research Institute Polish Academy of Sciences, Newelska 6,
01-447 Warsaw, Poland
janusz.kacprzyk@ibspan.waw.pl

Abstract. The problem of optimal allocation of service centers is considered in this paper. It is supposed that the information received from GIS is presented in the form of second kind fuzzy graphs. Method of optimal allocation as a way to determine fuzzy set of vitality for fuzzy graph is suggested. This method is based on the transition to the complementary fuzzy graph of first kind. The method allows solving not only problem of finding of optimal service centers location but also finding of optimal location for k-centers with the greatest degree and selecting of service center numbers. Based on this method the algorithm searching vitality fuzzy set for second kind fuzzy graphs is considered. The example of finding optimum allocation centers in fuzzy graph is considered as well.

Keywords: Fuzzy graph · Service centers · Vitality fuzzy set

1 Introduction

The worldwide expansion and diversified implementation of geographic information systems (GIS) is largely caused by the need to improve information systems that support decision-making. Application spheres of GIS are huge, thus geoinformation technologies become leaders in information retrieval, display, analytical tools and decision support [1, 2].

However, geographic data are often associated with significant uncertainty. Errors in data that are often used without considering their inherent uncertainties lead to a high probability of obtaining information of doubtful value. Uncertainty presents throughout the process of geographical abstraction: from acquiring data to using it [3, 4].

Data modeling [5–7] is the process of abstraction and generation of real forms of geographic data here. This process provides a conceptual model of the real world. It is doubtful that the geographical complexity can be reduced in models with perfect accuracy. So, the imminent contradiction between the real world and the model stands as inaccuracy and uncertainty can lead to the wrong decision making results.

© Springer Nature Switzerland AG 2018
I. Batyrshin et al. (Eds.): MICAI 2018, LNAI 11288, pp. 379–390, 2018.
https://doi.org/10.1007/978-3-030-04491-6_29

Allocation of centers [8] is the optimization problem that can be effectively solved by GIS. This problem includes optimal allocation of extremely important services, such as hospitals, police stations, fire brigades etc. In some cases the optimality criterion can be considered as distance minimization (travel time) from the service center to the most remote service point, therefore, the problem considers optimization of the "worst case" [9]. At the same time, the information presented in GIS, can be approximate or insufficiently reliable [10].

Information inaccuracy, as well as parameter uncertainty, uncertainty that usually happens in decision-making process, uncertainty caused by environmental influence – are the examples of fuzziness concerning service centers allocation problems.

For example, information inaccuracy can be a result of imprecise measurement of distance between serving facilities, such as distance measure, the measuring accuracy equaling in several centimeters or meters may appear.

Parameter measuring process is connected with the fact that some characteristics are qualitative and don't have numerical equivalents: quality of some part of a route can't be measured but it can be described like "good", "bad", "the worst" and etc. Besides, some objects can change parameters over time, for example, sizes of water reservoir can be changed naturally and its borders are indistinct.

Uncertainty in decision-making process supposes imprecise data about targets and some parameters that can't be determined exactly or it allows variations of the certain range. Implementation of fuzzy theory allows to describe reality more adequate and to find more suitable decision. Uncertainty caused by environmental influence supposes influence of some factors on parameters of objects or relations that changes their values.

Some types of inaccuracy of parameters measurement of objects with complex structure and impact of external factors on parameters that are taking into account for optimization are usually considered. Then location problems are solved under uncertain conditions [11].

We consider that a certain system of transportation ways has n stations. There are k service centres, which may be placed into these stations. Each centre can serve several stations. The degree of a service station by a centre depends on a route which connects them. It is necessary to define the places of best allocation for a given number of centers. In other words, it is necessary to define the places of k centers into n stations so that the «control» of all territory (all stations) is carried out with the greatest possible degree of service.

In works [12–14], this problem has been considered as a problem of allocation centers on fuzzy graphs of the first kind [15, 16], i.e. graphs whose set of vertices is non-fuzzy set, and set of edges is fuzzy set. Sometimes, however, adequate model of such allocation is fuzzy graph is a more general form [17, 18], in which a set of vertices and a set of edges are fuzzy sets. Such a graph in the work [19] was named the fuzzy graph of the second kind.

The approach to finding the best centers allocation on the fuzzy graph of the second kind is considered in this paper.

2 Basic Concepts and Definitions

For choosing of the best location of service centers we can use the notion of conjunctive strength and the reachability degree of vertex. In this case we can compare all of the movement paths among themselves and find routes with a maximal reachability degree.

Let X be a nonempty set. Let ε denote the set of all subsets of X with cardinality 2. Let $U \in \varepsilon$. A graph G is a pair (X,U). The elements of X are thought of as vertices of the graph and the elements of U as the edges.

Definition 1 [17, 18]. A fuzzy graph $\tilde{G} = (X, \sigma, \mu)$ is a triple consisting of a nonempty set X together with a pair of functions $\sigma\colon X \to [0, 1]$ and $\mu\colon U \to [0, 1]$ such that for all $x, y \in X, \mu(xy) \le \sigma(x)\&\sigma(y)$.

Definition 2 [19, 20]. A pair $\tilde{G} = (\tilde{X}, \tilde{U})$ is a second kind fuzzy graph. Here a set $\tilde{X} = \{<\mu_X(x)/x>\}$ is the fuzzy set of vertices, defined on set X, $|X| = n$, and $\tilde{U} = \{<\mu_U(x_i, x_j)/(x_i, x_j)>\}$, $x_i, x_j \in X$ is fuzzy set of directed edges. Here $\mu_X(x) \in [0, 1]$ - membership function for vertex x, $\mu_U(x_i, x_j) \in [0, 1]$ - membership function for edge (x_i, x_j).

We note that in a second kind fuzzy graph there is no relation between the membership functions $\mu_X(x_i)$, $\mu_X(x_j)$, and $\mu_U(x_i, x_j)$.

Definition 3 [15, 18]. A path of fuzzy graph $l(x_i, x_j)$ is called a directed sequence of fuzzy edges from vertex x_i to vertex x_j, in which the final vertex of any edge is the first vertex of the following edge.

A conjunctive strength of second kind fuzzy graph is defined by expression:

$$\mu_l(x_i, x_j) = \underset{<x_k,x_t> \in l(x_i,x_j)}{\&} \mu_U(x_k, x_t)\& \underset{\substack{x_t \ne x_i \\ x_t \ne x_j}}{\mu_X}(x_t).$$

In other words, the conjunctive strength of second kind fuzzy graph is defined by the smallest value of membership functions of vertices and edges that are included in this path, except for the first and last vertices of the path.

Let L is a family of second kind fuzzy graph paths from vertex x_i to vertex x_j. Then the value $\gamma(x_i, x_j) = \max_{l \in L}\{\mu_l(x_i, x_j)\}$ defines the reachability degree of vertex x_j from vertex x_i.

The problem of the best allocation of centers on the fuzzy graph can be limited to the problem of finding a subset of vertices Y, which all the other vertices X/Y of the fuzzy graph achievable from with the greatest reachability degree. There are three strategies of the selection of vertices Y [14].

- We "go" from each vertex of subset X/V, and arrive at a vertex of V;
- We "come out" of any of the vertices of V, and reach all vertices of subset X/V;
- We "come out" of any of the vertices of V, reach all vertices of subset X/V and come back.

In this paper, the third strategy is considered.

Let k is service centers $(k < n)$, placed in the vertices of subset Y, $|Y| = k$, $Y \subset X$, and $\gamma(x_i, x_j)$ is reachability degree of vertex x_j from vertex x_i.

Definition 4. Value

$$V_{\tilde{G}}(Y) = \underset{\forall x_j \in X \backslash Y}{\&} \left(\underset{\forall x_i \in Y}{\vee} \gamma(x_i, x_j) \& \gamma(x_j, x_i) \right)$$

is a *vitality degree of fuzzy graph* $\tilde{G}$ which is served by k-centers from vertex set Y.

Vitality degree $V_{\tilde{G}}(Y) \in [0, 1]$ determines the minimax strong connectivity value between each vertex from set $X \backslash Y$ and a center from set Y.

In other words, one can "leave" the vertex of subset Y, "reach" any vertex of the graph, "serve" it, return to the "initial" vertex while the conjunctive strength of the route will not be less than value $V_{\tilde{G}}(Y)$.

It is clear that value $V_{\tilde{G}}(Y)$ depends either on the number of centers k, or the allocation of the centers on the vertices of graph $\tilde{G}$ (i.e. on the choice of set Y).

Thus, the problem of the allocation of k service centers $(k < n)$ in fuzzy graph $\tilde{G}$ is reduced to determining such a subset of vertices $Y \subset X$, that value of vitality degree $V_{\tilde{G}}(Y)$ reaches its maximum value, that is value $V_{\tilde{G}}(k) = \underset{\substack{\forall Y \subset X \\ |Y| = k}}{\max} \{V_{\tilde{G}}(Y)\}$.

Definition 5. Fuzzy set

$$\tilde{V}_{\tilde{G}} = \{ <V_{\tilde{G}}(1)/1>, <V_{\tilde{G}}(2)/2>, \dots, <V_{\tilde{G}}(n)/n> \},$$

defined on vertex set X, is called a *fuzzy set of vitality* of graph $\tilde{G} = (X, \tilde{U})$. Fuzzy set of vitality $\tilde{V}_{\tilde{G}}$ determines the greatest vitality degrees of graph $\tilde{G}$ if it is served by 1, 2...n centers.

Values $\tilde{V}_{\tilde{G}}(k)$ $(1 \leq k \leq n)$ signify that we can place k-centers in graph $\tilde{G}$ so that there is a route from at least one center to any vertex of graph $\tilde{G}$ and back. The conjunctive strength of the graph will be not less than $\tilde{V}_{\tilde{G}}(k)$.

The fuzzy set of vitality is a fuzzy invariant of a fuzzy graph. It determines the highest degree of the reachability of the vertices for any given number of service centers.

3 Method for Finding of Vitality Fuzzy Set

We will consider the method of finding a family of all service centers with the largest vitality degree for second kind fuzzy graph.

In [14, 15] the method for first kind fuzzy graph was considered. Let Y be a subset of the vertices of fuzzy graph $\tilde{G} = (X, \tilde{U})$ in which the service centers are located and the vitality degree equals to V. One of the two conditions for any vertex $x_i \in X$ can be satisfied:

(a) vertex x_i belongs to set Y;
(b) there is vertex x_j that belongs to set Y and inequalities $\gamma(x_i, x_j) \geq V$ and $\gamma(x_j, x_i) \geq V$ are encountered.

Using the notation quantifier form we can get the truth of the following formula:

$$(\forall x_i \in X)[x_i \in Y \vee (\exists x_j)(x_j \in Y \& \gamma(x_i, x_j) \geq V \& \gamma(x_j, x_i) \geq V)]. \tag{1}$$

For each vertex $x_i \in X$ we assign Boolean variable p_i that takes value 1, if $x_i \in Y$ and 0 otherwise. We assign the fuzzy variable $\xi_{ji} = \gamma(x_j, x_i)$ for the proposition $\gamma(x_j, x_i) \geq V$. Transforming the quantifier form of proposition (1) to the form in terms of logical operations, we obtain a true logical proposition:

$$\Phi_V = \underset{i}{\&}(p_i \vee (\bar{p}_i \rightarrow (\underset{j}{\vee}(p_j \& \gamma_{ji})))).$$

Taking into account the interrelation between the implication operation and disjunction operation, we receive:

$$\Phi_V = \underset{i=\overline{1,n}}{\&} (p_i \vee \underset{j=\overline{1,n}}{\vee}(p_j \& \xi_{ij} \& \xi_{ij})).$$

Suppose $\xi_{ii} = 1$ and consider that the equality $p_i \vee \underset{j}{\vee} p_i \& \xi_{ij} = \underset{j}{\vee} p_j \xi_{ij}$ is true for any vertex x_i, we finally obtain:

$$\Phi_V = \underset{i=\overline{1,n}\, j=\overline{1,n}}{\&}\ \vee\ (\xi_{ij} \& \xi_{ji} \& p_j). \tag{2}$$

We open the parentheses in the expression (2) and reduce the similar terms by following rules (3):

$$a \vee a\&b = a;\ a\&b \vee a\&\bar{b} = a;\ \xi'\&a \vee \xi''\&a\&b. \tag{3}$$

Here, $a, b \in \{0, 1\}$, $\xi' \geq \xi''$, $\xi', \xi'' \in [0, 1]$. Then the expression (2) may be presented as:

$$\Phi_V = \underset{i=\overline{1,l}}{\vee} (p_{1_i} \& p_{2_i} \& \ldots \& p_{k_i} \& V_i). \tag{4}$$

The following property holds: Each disjunctive member in the expression (4) defines the subset of vertices $Y \subseteq X$ with vitality degree V_i of fuzzy graph $\tilde{G} = (X, \tilde{U})$. Here subset Y is minimal, in other words, any subset of Y does not have this property.

Considering method works with reachability vertex matrix of first kind fuzzy graph. So in order to apply the method in case of second kind fuzzy graph $\tilde{G}$ we transform this graph into first kind fuzzy graph $\tilde{G}'$ like that:

- let $y_1 \in X$ is the vertex of initial second kind fuzzy graph $\tilde{G}$, and it is adjacent for t vertices and has a degree μ. Represent vertex x as directed complete t-subgraph of first kind with degree of "inside" edges equals μ.
- connect everyone of the initial "outside" t vertices with one of the vertices of received subgraph.

Example of transformation of second kind fuzzy graph (Fig. 1) to first kind fuzzy graph is presented in the Fig. 2.

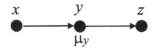

Fig. 1. Initial second kind fuzzy graph.

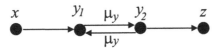

Fig. 2. Transformation to first kind fuzzy graph.

We can prove the next property:

Property 1. Reachability degree of initial second kind fuzzy graph $\tilde{G}$ coincides with reachability degree of first kind fuzzy graph $\tilde{G}'$.

First kind fuzzy graph $\tilde{G}'$ comes out of this transformation. Meanwhile, transition from graph $\tilde{G}$ to graph $\tilde{G}'$ is biunique. Built reachability matrix $N^{(1)}$ of received first kind fuzzy graph $\tilde{G}'$ and pass from it to reachability matrix $N^{(2)}$ of initial second kind fuzzy graph $\tilde{G}$. Apply the foregoing method to the matrix $N^{(2)}$.

Given procedure is considered by example.

Example 1. Let's consider an example of finding of service centers in second kind fuzzy graph $\tilde{G}$ given in the Fig. 3.

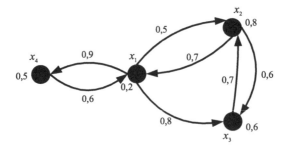

Fig. 3. Second kind fuzzy graph $\tilde{G}$.

Here vertices of graph are objects of transport network and edges are paths that connect the objects.

Let's find centers of graph supposing that it is located in graph vertices. In other words service centers should be located in objects of transport network.

The adjacent matrix of fuzzy graph $\tilde{G}$ is presented as:

$$R_X = \begin{array}{c} \\ x_1 \\ x_2 \\ x_3 \\ x_4 \end{array} \begin{array}{c} \begin{array}{cccc} x_1 & x_2 & x_3 & x_4 \end{array} \\ \left| \begin{array}{cccc} 0,2 & 0,5 & 0,8 & 0,9 \\ 0,7 & 0,8 & 0,6 & 0 \\ 0 & 0,7 & 0,6 & 0 \\ 0,6 & 0 & 0 & 0,5 \end{array} \right| \end{array}.$$

For finding of reachability matrix $N^{(2)}$ of second kind fuzzy graph $\tilde{G}$ we need to construct new complementary fuzzy graph $\tilde{G}'$ of first kind. This graph is presented in the Fig. 4.

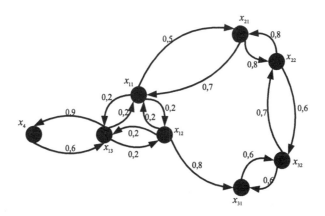

Fig. 4. Complementary fuzzy graph $\tilde{G}'$ of first kind.

The reachability matrix $N^{(1)}$ of fuzzy graph $\tilde{G}'$ can be presented:

$$N^{(1)} = \bigcup_{i=\overline{0,n-1}} R_X^i.$$

Here, R_X^0 - diagonal unitary matrix, R_X^i - i degree of adjacent matrix. Adjacent matrix of first kind fuzzy graph $\tilde{G}'$ takes on form:

$$R_X^1 = \begin{array}{c|cccccccc} & x_{11} & x_{12} & x_{13} & x_{21} & x_{22} & x_{31} & x_{32} & x_4 \\ \hline x_{11} & 0 & 0.2 & 0.2 & 0.5 & 0 & 0 & 0 & 0 \\ x_{12} & 0.2 & 0 & 0.2 & 0 & 0 & 0.8 & 0 & 0 \\ x_{13} & 0.2 & 0.2 & 0 & 0 & 0 & 0 & 0 & 0.9 \\ x_{21} & 0.7 & 0 & 0 & 0 & 0.8 & 0 & 0 & 0 \\ x_{22} & 0 & 0 & 0 & 0.8 & 0 & 0 & 0.6 & 0 \\ x_{31} & 0 & 0 & 0 & 0 & 0 & 0 & 0.6 & 0 \\ x_{32} & 0 & 0 & 0 & 0 & 0.7 & 0.6 & 0 & 0 \\ x_4 & 0 & 0 & 0.6 & 0 & 0 & 0 & 0 & 0 \end{array}$$

Find degree of matrix $R_X^2 = R_X^1 \times R_X^1$, where matrix elements $R_X^2 = \left\| r_{ij}^{(2)} \right\|$, $i,j = \overline{1,n}$ are determined as: $r_{ij}^{(2)} = \bigvee_{k=\overline{1,n}} r_{ik}^{(1)} \& r_{kj}^{(1)}$, where $r_{ik}^{(1)}$ and $r_{kj}^{(1)}$ are elements of matrix R_X^1. Similarly we find $R_X^i = R_X^{i-1} \times R_X^1$, for index $i = 3, 4, 5, 6, 7$. Let's join the matrixes and find reachability matrix $N^{(1)} = \bigcup_{i=\overline{0,7}} R_X^i$:

$$N^{(1)} = \begin{array}{c|cccccccc} & x_{11} & x_{12} & x_{13} & x_{21} & x_{22} & x_{31} & x_{32} & x_4 \\ \hline x_{11} & 1.0 & 0.2 & 0.2| & 0.5 & 0.5| & 0.5 & 0.5| & 0.2 \\ x_{12} & 0.2 & 1.0 & 0.2| & 0.2 & 0.6| & 0.8 & 0.6| & 0.2 \\ x_{13} & 0.2 & 0.2 & 1.0| & 0.2 & 0.2| & 0.2 & 0.2| & 0.9 \\ x_{21} & 0.7 & 0.2 & 0.2| & 1.0 & 0.8| & 0.6 & 0.6| & 0.2 \\ x_{22} & 0.7 & 0.2 & 0.2| & 0.8 & 1.0| & 0.6 & 0.6| & 0.2 \\ x_{31} & 0.6 & 0.2 & 0.2| & 0.6 & 0.6| & 1.0 & 0.6| & 0.2 \\ x_{32} & 0.7 & 0.2 & 0.2| & 0.7 & 0.7| & 0.6 & 1.0| & 0.2 \\ x_4 & 0.2 & 0.2 & 0.6| & 0.2 & 0.2| & 0.2 & 0.2| & 1.0 \end{array}.$$

Each block of complementary matrix conforms to vertex of initial second kind fuzzy graph. So the greatest values are chosen in these blocks. And reachability matrix $N^{(2)}$ of second kind fuzzy graph $\tilde{G}$, given in Fig. 2, will be defined as:

$$N^{(2)} = \begin{array}{c|cccc} & x_1 & x_2 & x_3 & x_4 \\ \hline x_1 & 1 & 0,6 & 0,8 & 0,9 \\ x_2 & 0,7 & 1 & 0,6 & 0,2 \\ x_3 & 0,7 & 0,7 & 1 & 0,2 \\ x_4 & 0,6 & 0,2 & 0,2 & 1 \end{array}.$$

To construct the expression (4) we rewrite expression (2) like this:

$$\Phi_V = \underset{i=\overline{1,n}}{\&} (a_{i1}p_1 \vee a_{i2}p_2 \vee \ldots \vee a_{in}p_n). \tag{5}$$

We convert pair $a_{ij}\&p$ from expression (5) to weighted binary vector $a_{ij}\bar{P}_j$. Here $\bar{P}_j = \|p_i^{(j)}\|$ is a binary vector that has dimension of n. The elements of $\bar{P}_j$ are defined as:

$$p_i^{(j)} = \begin{cases} 1, & \text{if } i = j \\ 0, & \text{if } i \neq j. \end{cases}$$

The conjunction of (a_1p_1) and (a_2p_2) from expression (5) corresponds the conjunction of two weighted binary vectors $a_1\bar{P}_1$ and $a_2\bar{P}_2$, $\bar{P}_1 = ||p_i^{(1)}||$, $\bar{P}_2 = ||p_i^{(2)}||$, $i = \overline{1,n}$, $a_1, a_2 \in [0,1]$. In a vector space the conjunction is defined as $a_1\bar{P}_1 \& a_2\bar{P}_2 = a\bar{P}$, where $a = \min\{a_1, a_2\}$, $\bar{P} = ||p_i||$, $p_i = \max\{p_i^{(1)}, p_i^{(2)}\}$, $i = \overline{1,n}$.

We define the operation $\leq$ "less or equal" between binary vectors. Binary vector $\bar{P}_1$ is less or equal than $\bar{P}_2$ if and only if each element of $\bar{P}_1$ is less or equal than the corresponding element of vector $\bar{P}_2$. Or:

$$(\bar{P}_1 \leq \bar{P}_2) \leftrightarrow (\forall i = \overline{1,n})[p_i^{(1)} \leq p_i^{(2)}].$$

Considering the algebra in space of weighted binary vectors, we can make a rule of absorption:

$$a_1\bar{P}_1 \vee a_2\bar{P}_2 = a_1\bar{P}_1, \text{ if } a_1 \geq a_2 \text{ and } \bar{P}_1 \leq \bar{P}_2. \tag{6}$$

Now we can construct statement (6) using the conjunction operation and the rule of absorption of weighted binary vectors by the following algorithm:

1°. Each element of the first bracketed expression $(j = 1)$ of expression (5) is converted to weighted binary vector. The result is to be written in the first n elements of the buffer vector $\bar{V}_1 = ||v_i^{(1)}||, i = \overline{1,n^2}$.

2°. j incrementing $(j: = j+1)$.

3°. Each element of the bracketed expression j is also converted to weighted binary vectors. The result is to be written in the first n elements of the buffer vector $\bar{V}_2 = ||v_i^{(2)}||, i = \overline{1,n}$.

4°. The next stage consists of the conjunction of two vectors $\bar{V}_1$ and $\bar{V}_2$. The result is placed into the buffer vector $\bar{V}_3 = ||v_i^{(3)}||, i = \overline{1,n^2}$. While placing elements into $\bar{V}_3$, absorption is made using rule (6).

5°. All the elements of vector $\bar{V}_3$ are copied to vector $\bar{V}_1$ $(v_i^{(1)} := v_i^{(3)}, i = \overline{1,n^2})$.

6°. $j: = j+1$.

7°. If $j \leq n$ then goes to 3°, otherwise go to 8°.

8°. Expression (2) is to be built using elements in the vector $\bar{V}_1$. This way we have fuzzy set of vitality of graph $\tilde{G} = (X, \tilde{U})$.

Example 2. The corresponding expression (5) for second kind fuzzy graph $\tilde{G}$, given in the Fig. 2, has the next form:

$$\Phi_V = (1p_1 \vee 0.6p_2 \vee 0.7p_3 \vee 0.6p_4)\&(0.6p_1 \vee 1p_2 \vee 0.6p_3 \vee 0.2p_4)\&$$
$$\&(0.7p_1 \vee 0.6p_2 \vee 1p_3 \vee 0.2p_4)\&(0.6p_1 \vee 0.2p_2 \vee 0.2p_3 \vee 1p_4).$$

Before the first iteration of the algorithm vectors $\bar{V}_1$, $\bar{V}_2$, $\bar{V}_3$ and $\bar{V}_4$ have the following forms:

$$\bar{V}_1 = \begin{pmatrix} 1(1000) \\ 0.6(0100) \\ 0.7(0010) \\ 0.6(0001) \end{pmatrix} \quad \bar{V}_2 = \begin{pmatrix} 0.6(1000) \\ 1(0100) \\ 0.6(0010) \\ 0.2(0001) \end{pmatrix} \quad \bar{V}_3 = \begin{pmatrix} 0.7(1000) \\ 0.6(0100) \\ 1(0010) \\ 0.2(0001) \end{pmatrix} \quad \bar{V}_4 = \begin{pmatrix} 0.6(1000) \\ 0.2(0100) \\ 0.2(0010) \\ 1(0001) \end{pmatrix}.$$

After the first iteration of the algorithm vector $\bar{V}_1$ has the following form:

$$\bar{V}_1 = \begin{vmatrix} 0.6(1000) \\ 1.0(1100) \\ 0.6(0100) \\ 0.7(0110) \\ 0.6(0010) \\ 0.2(0001) \end{vmatrix}.$$

After completing the iterations, finally we have:

$$\bar{V}_1 = \begin{vmatrix} 0.6(1000) \\ 0.7(1101) \\ 1.0(1111) \\ 0.2(0100) \\ 0.6(0101) \\ 0.6(0011) \\ 0.2(0001) \end{vmatrix}.$$

So, the formula (4) for this graph has the form:

$$\Phi_V = 0.6p_1 \vee 0.2p_2 \vee 0.2p_4 \vee 0.6p_2p_4 \vee 0.6p_3p_4 \vee 0.7p_1p_2p_4 \vee 1p_1p_2p_3p_4.$$

It follows from the last equality that the second kind fuzzy graph $\tilde{G}$ has 7 subsets of vertices with the greatest vitality degree, and fuzzy set of vitality is defined as:

$$\tilde{V} = \{ <0.6/1> , <0.6/2> , <0.7/3> , <1/4> \}.$$

The fuzzy set of vitality defines the next optimum allocation of the service centres: If we have 4 service centres then we place these centres into all vertices. The degree of service equals 1 in this case. If we have 3 service centres then we should place these centres into vertices 1, 2, and 4. In this case the degree of service equals 0.7. If we have only one service centre then we should place it into vertex 1. The degree of service equals 0.6 in last case. Thus, we can conclude that there is no preference when placing two centers.

4 Conclusion and Future Work

The problem of defining of the optimum allocation of centers was considered as the problem of the definition of fuzzy vitality set of second kind fuzzy graphs. The algorithm of the definition of fuzzy base set has been proposed. It should be noted; that the considered method makes it possible to define the best service allocations only if the centers are placed in the vertices of a graph (the case of generating new vertices on the edges is not considered). In our future work we are going to examine the problem of the centers' allocation in the temporal fuzzy graphs, i.e. the graphs, edges' membership functions of which may change in discrete time.

Acknowledgments. This work has been supported by the Ministry of Education and Science of the Russian Federation under Project "Methods and means of decision making on base of dynamic geographic information models" (Project part, State task 2.918.2017)

References

1. Slocum, T., McMaster, R., Kessler, F., Howard, H.: Thematic Cartography and Geovisualization, 3rd edn. Pearson Education Limited, London (2014)
2. Fang, Y., Dhandas, V., Arriaga, E.: Spatial Thinking in Planning Practice. Portland State University, Portland (2014)
3. Zhang, J., Goodchild, M.: Uncertainty in Geographical Information. Taylor & Francis Inc., New York (2002)
4. Belyakov, S., Rozenberg, I., Belyakova, M.: Approach to real-time mapping, using a fuzzy information function. In: Bian, F., Xie, Y., Cui, X., Zeng, Y. (eds.) GRMSE 2013. CCIS, vol. 398, pp. 510–521. Springer, Heidelberg (2013). https://doi.org/10.1007/978-3-642-45025-9_50
5. Goodchild, M.: Modelling error in objects and fields. In: Goodchild, M., Gopal, S. (eds.) Accuracy of Spatial Databases, pp. 107–113. Taylor & Francis Inc., Basingstoke (1989)
6. Belyakov, S., Bozhenyuk, A., Rozenberg, I.: Intuitive figurative representation in decision-making by map data. J. Multiple-Valued Logic Soft Comput. **30**, 165–175 (2018)
7. Bozhenyuk, A., Belyakov, S., Rozenberg, I.: The intuitive cartographic representation in decision-making. In: Proceedings of the 12th International FLINS Conference (FLINS 2016), World Scientific Proceeding Series on Computer Engineering and Information Science, 10, pp. 13–18. World Scientific (2016)
8. Kaufmann, A.: Introduction a la Theorie des Sous-Ensemles Flous. Masson, Paris (1977)
9. Christofides, N.: Graph Theory: An Algorithmic Approach. Academic Press, London (1976)
10. Malczewski, J.: GIS and Multicriteria Decision Analysis. Willey, New York (1999)
11. Rozenberg, I., Starostina, T.: Solving of Location Problems Under Fuzzy Data with Using GIS. Nauchniy Mir, Moscow (2006)
12. Bozhenyuk, A., Rozenberg, I.: Allocation of service centers in the GIS with the largest vitality degree. In: Greco, S., Bouchon-Meunier, B., Coletti, G., Fedrizzi, M., Matarazzo, B., Yager, R.R. (eds.) IPMU 2012. CCIS, vol. 298, pp. 98–106. Springer, Heidelberg (2012). https://doi.org/10.1007/978-3-642-31715-6_12

13. Bozheniuk, V., Bozhenyuk, A., Belyakov, S.: Optimum allocation of centers in fuzzy transportation networks with the largest vitality degree. In: Proceedings of the 2015 Conference of the International Fuzzy System Association and the European Society for Fuzzy Logic and Technology, pp. 1006–1011. Atlantis Press (2015)
14. Bozhenyuk, A., Belyakov, S., Gerasimenko, E., Savelyeva, M.: Fuzzy optimal allocation of service centers for sustainable transportation networks service. In: Kahraman, C., Sarı, İ.U. (eds.) Intelligence Systems in Environmental Management: Theory and Applications. ISRL, vol. 113, pp. 415–437. Springer, Cham (2017). https://doi.org/10.1007/978-3-319-42993-9_18
15. Monderson, J., Nair, P.: Fuzzy Graphs and Fuzzy Hypergraphs. Springer, Heidelberg (2000). https://doi.org/10.1007/978-3-7908-1854-3
16. Bershtein, L., Bozhenyuk, A.: Fuzzy graphs and fuzzy hypergraphs. In: Dopico, J., de la Calle, J., Sierra, A. (eds.) Encyclopedia of Artificial Intelligence, Information SCI, pp. 704–709. Hershey, New York (2008)
17. Rosenfeld, A.: Fuzzy graph. In: Zadeh, L.A., Fu, K.S., Shimura, M. (eds.) Fuzzy Sets and Their Applications to Cognitive and Decision Process, pp. 77–95. Academic Press, New York (1975)
18. Mordeson, J., Mathew, S., Malik, D.: Fuzzy Graph Theory with Applications to Human Trafficking. Springer, Cham (2018). https://doi.org/10.1007/978-3-319-76454-2
19. Bozhenyuk, A., Rozenberg, I., Yastrebinskaya, D.: Finding of service centers in GIS described by second kind fuzzy graphs. World Appl. Sci. J. **22**(Special Issue on Techniques and Technologies), 82–86 (2013)
20. Bozhenyuk, A., Belyakov, S., Knyazeva, M., Rozenberg, I.: Optimal allocation centersin second kind fuzzy graphs with the greatest base degree. Adv. Intell. Syst. Comput. **679**, 312–321 (2018)

Fuzzy Design of Nearest Prototype Classifier

Yanela Rodríguez Alvarez[1]([⊠]), Rafael Bello Pérez[2],
Yailé Caballero Mota[1], Yaima Filiberto Cabrera[1],
Yumilka Fernández Hernández[1], and Mabel Frias Dominguez[1]

[1] Departamento de Computación, Universidad de Camagüey,
Circunvalación Norte Km 5 ½, Camagüey, Cuba
{yanela.rodriguez,yaile.caballero,yaima.filiberto,
yumilka.fernandez,mabel.frias}@reduc.edu.cu
[2] Departamento de Ciencias de la Computación,
Universidad Central "Marta Abreu" de las Villas,
Carretera a Camajuaní Km. 5 y ½, Santa Clara, Villa Clara, Cuba
rbellop@uclv.edu.cu

Abstract. In pattern classification problems, many works have been carried out with the aim of designing good classifiers from different perspectives. These works achieve very good results in many domains. However, in general they are very dependent on some crucial parameters involved in the design. An alternative is to use fuzzy relations to eliminate thresholds and make the development of classifiers more flexible. In this paper, a new method for solving data classification problems based on prototypes is proposed. Using fuzzy similarity relations for the granulation of the universe, similarity classes are generated and a prototype is built for each similarity class. In the new approach we replace the relation of similarity between two objects by a binary fuzzy relation, which quantifies the strength of the relationship in a range of [0; 1]. Experimental results show that the performance of our method is superior to other methods.

Keywords: Prototype generation · Similarity relations
Fuzzy-rough sets theory · Classification

1 Introduction

Theories of rough and fuzzy sets are distinct and complementary generalizations of set theory. A fuzzy set allows a membership value other than 0 and 1. A rough set uses three membership functions, a reference set and its lower and upper approximations in an approximation space. There are extensive studies on the relationships between rough and fuzzy sets * [1–4]. Many proposals have been made for the combination of rough and fuzzy sets.

Rough and fuzzy sets are complementary with each other in the sense that fuzzy sets model the ambiguous memberships between elements and classes while rough sets provide a way of approximating indefinable concept with a pair of definable sets within the universe. This observation motivated a lot of researchers to combine fuzzy sets and rough sets together and various kinds of fuzzy rough set models had been proposed in publications. One is the constructive approach that starts with the fuzzy relations on the

© Springer Nature Switzerland AG 2018
I. Batyrshin et al. (Eds.): MICAI 2018, LNAI 11288, pp. 391–400, 2018.
https://doi.org/10.1007/978-3-030-04491-6_30

universe and the lower and upper approximation operators are constructed via these fuzzy relations [5]. The constructive approach of fuzzy rough sets is firstly proposed by Dubois and Prade [6, 7], and Radzikowska and Kerre [8] provide a more general approach of constructing fuzzy rough sets.

A recently proposed competitive scheme termed NPBASIR-CLASS method, concerning to work [9], has been adapted to solve different kinds of problems such as: classification problems on domains with hubness [10] and classification of imbalanced datasets [11]. This method NPBASIR-CLASS to build prototypes use the concepts of Granular Computation [12] and are based on the NPBASIR algorithm [13]. Granulation of a universe is performed using a relation of similarity which generates similarity classes of objects in the universe and for each similarity class one prototype is built. To build the similarity relation the method proposed in [14] is used. A similarity quality measure attempts to establish a correspondence between the granulation induced by the set of input attributes A and that induced by the decision attribute d.

NPBASIR-CLASS make use of crisp similarity relations to build the granulations corresponding to the A and d spaces, respectively. In other words, two objects of the universe will belong to the same similarity class if their degree of indiscernibility (or similarity) goes beyond a user-defined threshold in [0; 1]. Hence, the similarity quality measure depends on two thresholds (one for the input attributes ε_1 and one for the decision attribute ε_2) which have a crucial importance in any subsequent data analysis step that leans upon the similarity classes. Besides, the threshold values will be application dependent, hence calling for a fine-tuning process to maximize the performance of the knowledge discovery process at hand.

Given that NPBASIR-CLASS is fairly sensitive to the values of the ε_1 and ε_2 similarity thresholds, in this paper we tackle this limitation by employing fuzzy sets to categorize their domains through fuzzy binary relations. We show how this facilitates the definition of the similarity relations (as there are fewer parameters to consider) and enhances the system's interpretability without degrading, from a statistical perspective, the efficacy of the subsequent data mining tasks.

The present study examines the combination of rough and fuzzy sets from the perspective of design of Nearest Prototype Classifiers. The key contribution of the proposed work is the development of a novel classification approach that uses Nearest Prototype and then combines the advantages of both rough and fuzzy sets in the classification process. Also, in this case the weights of the similarities between the two objects according to each input attributes have been optimized using Particle Swarm Optimization., which conducts a search over the space of similarity weights for the input attributes and uses the similarity quality measure as the fitness of a candidate weight vector. The performance comparison with state-of-the-art methods proposed in the literature is presented in the experimental part.

2   Related Works

The based method, NPBASIR-CLASS [9], uses a similarity relation R and a set of instances $X = \{X_1, X_2 \ldots X_n\}$, each of which is described by a vector of m descriptive features and belongs to one of k classes $C = \{c_1, c_2, \ldots, c_k\}$. The similarity relation

R is constructed according to the method proposed in [14]; this is based on finding the relation that maximizes the quality of the similarity measure. In this case, the relation R is sought that generates a granulation considering the m descriptive features, as similar as possible to the granulation according to the classes.

Both two granulations are built using the crisp binary relations R_1 and R_2 defined in Eqs. (1) and (2):

$$xR_1y \iff F_1(x, y) \geq \varepsilon_1 \tag{1}$$

$$xR_2y \iff F_2(x, y) \geq \varepsilon_2 \tag{2}$$

where x and y are two objects in U, F_1 is a similarity function over the input attributes, F_2 is a similarity function over the decision attribute and ε_1, ε_2 are their respective similarity thresholds. The similarity function F_1 and F_2 usually takes the form in Eqs. (3) and (4):

$$F_1(x, y) = \sum_{i=1}^{N} w_i * \delta_i(x_i, y_i) \tag{3}$$

$$F_2(x, y) = \delta_d(x_d, y_d) \tag{4}$$

where $N = |A|$ is the number of input attributes, $\delta i(\cdot, \cdot)$ is the similarity function of two objects x and y with respect to the i-th attribute and w_i its associated weight. The goal then, is to find the relations R_1 and R_2 such that $R_1(x, y)$ and $R_2(x, y)$ are as similar as possible. To do so, the following sets are defined:

$$N_1(x) = y \in U : xR_1y \tag{5}$$

$$N_2(x) = y \in U : xR_2y \tag{6}$$

where $N_1(x)$ and $N_2(x)$ contain the objects that are similar enough to x according to the input attribute set A and the decision attribute d, respectively. The similarity degree $0 \leq \phi(x) \leq 1$ between both sets for a given object x is hence quantified as shown in Eq. (7):

$$\phi(x) = \frac{|N_1(x) \cap N_2(x)|}{0.5 * |N_1(x)| + 0.5*|N_2(x)|} = \frac{2|N_1(x) \cap N_2(x)|}{|N_1(x)| + |N_2(x)|} \tag{7}$$

The similarity quality measure $\Theta(\cdot)$ of a decision system DS of $M = |DS|$ objects is then defined in Eq. (8):

$$\Theta(SD) = \frac{\sum_{\forall x \in U} \phi(x)}{|U|} \tag{8}$$

Filiberto et al. [15] review different applications of the similarity quality measure in machine learning, e.g. improving classifiers like Multi-Layer Perceptron [14, 16] and

K-nearest neighbors (KNN) [14, 17], in prototype construction [18] and the induction of classification rules [13].

To determine the weight vector $w = (w_1, \ldots, w_N)$ in the similarity function F_1 of Eq. (3), the authors in [16] employ Particle Swan Optimization PSO to navigate the numerical space of all candidate weight vectors. Each particle is represented by a position vector p and a velocity vector v. The position encodes a candidate weight vector w whose fitness value is precisely the similarity quality measure $\Theta(DS)$ in Eq. (8) induced by w over the decision system DS. The attributewise similarity is defined as shown in Eq. (9).

$$\partial_i(x_i, y_i) = \begin{cases} 1 - \frac{|(x_i - y_i)|}{Max(n_i) - Min(n_i)} & \text{if } A_i \text{ is numerical} \\ 1 & \text{if } A_i \text{ is nominal} \wedge x_i = y_i \\ 0 & \text{if } A_i \text{ is nominal} \wedge x_i \neq y_i \end{cases} \tag{9}$$

where A_i is the i_th attribute of the decision system DS and x_i, y_i are the values of Ai in objects x and y, respectively. The particles update their velocities and positions as per the standard rules in Eqs. (10) and (11).

$$v_i(t+1) = w \cdot v_i(t) + c_1(t) \cdot r_1(t) \cdot (pbest(t) - p_i(t)) + c_2(t) \cdot r_2(t) \cdot (gbest(t) - p_i(t)) \tag{10}$$

$$p_i(t+1) = p_i(t) + v_i(t+1) \tag{11}$$

where t is the current iteration number, p_i and v_i are the particle's position and velocity in the $i = 1..N$ dimension, w is the inertia weight, c_1 and c_2 are two acceleration constants, r_1 and r_2 are two uniformly distributed random numbers in $[0; 1]$, $pbest(t)$ and $gbest(t)$ are the respective best solutions found by the particle itself and the swarm as a whole. This iterative update process takes place until the stop criterion is met, e.g., reaching a maximum number of iterations. At the end of the PSO search, the $gbest$ particle encodes in its position the best weight vector $w*$ found, which is then plugged into the similarity function F_1 in Eq. (3), thus allowing the calculation of the similarity relation R_1 depicted in Eq. (1).

3   Our Proposed

The impact of the similarity thresholds ε_1 and ε_2 upon the NPBASIR-CLASS approach was assessed and discussed in previously sections as well. In general, this scheme was found to be fairly sensitive to small variations of these thresholds, thus causing the PSO metaheuristic to converge to different local optima in the weight space, hence affecting the stability of this technique. In an attempt to overcome this limitation and increase the overall system interpretability without degrading its performance, this paper employs fuzzy sets to characterize the domain of both similarity thresholds.

Taking into account [19], we propose the use of fuzzy binary relations in lieu of crisp binary relations. Zadeh [18] defined a fuzzy binary relation R between $X \subseteq U$ and $Y \subseteq U$ as a fuzzy subset of $X \times Y$, i.e. the similarity $\mu_R(x, y)$ between any two objects of the universe can take values in the interval [0; 1]. The value of $\mu_R(x, y)$ represents the strength of the relationship between x and y [19]. The use of fuzzy relations makes the computational methods more tolerant and flexible to imprecision, especially in the case of mixed data (numerical and nominal variables). Taking these criteria into account, the use of fuzzy relations in the NPBASIR-CLASS algorithm is proposed in this paper.

Using a fuzzy approach to build the similarity quality measure requires the rewriting of the binary relations R_1 and R_2, defined in Eqs. (1) and (2), as fuzzy binary relations defined in Eqs. (12) and (13):

$$xR_1^F y \Leftrightarrow F_1(x, y) \text{ is High\_1} \tag{12}$$

$$xR_2^F y \Leftrightarrow F_2(x, y) \text{ is High\_2} \tag{13}$$

where R_1^F and R_2^F, are the fuzzy binary relations for the input attributes A and the decision attribute d, respectively and *High_1* and *High_2* are fuzzy sets that quantify the similarity between objects x and y in their respective attribute spaces. Both fuzzy sets are modeled after sigmoidal membership functions whose analytical expressions are given in Eqs. (14) and (15) and schematic depictions provided in Figs. 1(a) and (b).

$$\mu_{High\_1}(x) = \begin{cases} 0 & \text{if } x \le 0.70 \\ \frac{2(x-0.70)^2}{1+2(x-0.70)^2} & \text{otherwise} \end{cases} \tag{14}$$

$$\mu_{High\_2}(x) = \begin{cases} 0 & \text{if } x < 0.75 \\ 2\left(\frac{x-0.75}{0.90-0.75}\right)^2 & \text{if } 0.75 \le x \ge 0.85 \\ 1 - 2\left(\frac{x-0.75}{0.90-0.75}\right)^2 & \text{if } 0.85 \le x \ge 0.90 \\ 1 & \text{otherwise} \end{cases} \tag{15}$$

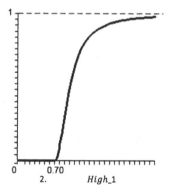
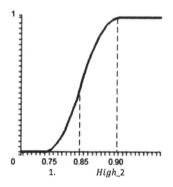

2.   *High_1*

1.   *High_2*

Fig. 1. The sigmoidal membership functions for the *High_1* and *High_2* fuzzy sets

From the *High_1* and *High_2* fuzzy sets we can build the fuzzy versions of $N_1(x)$ and $N_2(x)$ by replacing Eqs. (5) and (6) with Eqs. (16) and (17).

$$N_1^F(x) = \{(y, \mu_{High\_1}(F_1(x, y))) \ \forall y \in U\} \tag{16}$$

$$N_2^F(x) = \{(y, \mu_{High\_2}(F_2(x, y))) \ \forall y \in U\} \tag{17}$$

The fuzzy similarity degree between both sets [20] for an object $\vec{x}$ is calculated as the similarity between the fuzzy sets $N_1^F(x)$ and $N_2^F(x)$ according to Eq. (18):

$$\phi^F(x) = \frac{\sum_{i=1}^{M}(1 - |\mu_{High\_1}(x_i) - \mu_{High\_2}(x_i)|)}{M} \tag{18}$$

With the fuzzy similarity degree in Eq. (18), we can formalize the fuzzy similarity quality measure of a decision system DS of $M = |DS|$ objects as shown in Eq. (19):

$$\Theta^F(DS) = \frac{\sum_{i=1}^{M} \phi^F(x)}{M} \tag{19}$$

With the advent of fuzzy logic into the similarity-quality-based framework, numerical thresholds are no longer required as part of the definition of the similarity relation. Therefore, we reduce the number of degrees of freedom (algorithm parameters) under consideration, increase system interpretability (since the *High_1* and *High_2* fuzzy sets are intuitive for any user).

4   Experimental Results

In this section, we present some experimental facts regarding the proposed NPBASIR-FUZZY classification methodology and compare it with other state-of-the-art classification algorithms in terms of accuracy. The proposed NPBASIR-FUZZY classification methodology has been deployed on 17 datasets from UCI Machine Learning repository [21]. The 17 datasets and their description with number of attributes, instances, continuous attributes, discrete attributes, classes and missing values are shown below in the Table 1. These data sets include a wide range of domains and a variety of data characteristics such as number of classes, instances, and attributes. The prediction accuracy has been measured by applying a 10-fold cross-validation where each dataset is randomly partitioned into 10 approximately equally sized subsets (or folds or tests). The induction algorithm is executed 10 times; in each time it is trained on the data that is outside one of the subsets and the generated classifier is tested on that subset. The estimated accuracy for each cross-validation fold is a random variable that depends on the random partitioning of the data. So, for each dataset, we repeated 10-fold cross-validation 10 times. The estimated accuracy is the average over the ten 10-fold cross-validations.

Table 1. Description of data sets

Datasets	#Att.	#Inst,	#Cont.	#Disc.	#Class	Missing values
Breast-cancer	10	286	0	10	2	9/0.3%
Breast-w	9	683	9	0	2	16/2%
Diabetes	8	768	8	0	2	None
Flags	30	194	10	20	8	None
Haberman	4	306	3	1	2	None
Heart-Statlog	13	270	13	0	2	None
Ionosphere	35	351	34	1	2	None
Iris	4	150	4	0	3	None
Mfeat-fourier	76	2000	76	0	10	None
Optdigits	64	5620	64	0	10	None
Pima	8	768	8	0	2	None
Postoperative-patient-data	9	90	0	9	3	3/3%
Primary-tumor	18	339	0	18	21	224/66%
Solar-flare_1	13	323	0	13	2	None
Solar-flare_2	13	1066	0	13	2	None
Sponge	46	76	3	43	3	22/29%
Waveform	40	5000	40	0	3	None

This time we chose of the state-of-the-art classification algorithms which have been included within the KEEL [22] software tool, an open source software for multi-stage analysis in data mining. Specifically, 4 algorithms were selected from those that appear within the family of the Fuzzy Instance Based Learning: FRKNNA-C [23], FRNN-C [24], FRNN_FRS-C [25], FRNN_VQRS-C [25]. These algorithms following the fuzzy-rough nearest-neighbor approach based on the fuzzy-rough sets theory, like as our proposed. Table 2 shows, for each method, the classification accuracy (%) achieved by the state-of-the-art classification algorithms and our proposed, tested on 17 UCI datasets.

To statistically validate the experimental results, nonparametric tests for multiple comparisons with a control method were used. In Table 3 we show the average ranks obtained by each method in the Friedman test. The P-value computed by Iman and Daveport Test is equal 0.000007910702. The resulting p-value = 0.004989 577133 < α = 0.05, this indicates that there are indeed significant performance differences in the group. Besides, Table 4 show the Holm post hoc comparison (Friedman) where p-values are obtained in by applying post hoc methods over the results of Friedman procedure. The results reported in Table 4 reject all null hypotheses, hence confirming the superiority of the control method.

From the above analysis we can conclude that by replacing the similarity thresholds ε_1, ε_2 in the definition of the crisp binary relations in Eqs. (1) and (2) with the *High*_1 and *High*_2 fuzzy sets in the fuzzy binary relations shown in Eqs. (11) and (12), the accuracy of the classification is not statistically deteriorated. Additionally, the number of parameters in the automatic similarity learning procedure is reduced and the overall system interpretability is enhanced.

Table 2. Classification accuracy (%)

Datasets	FRKNNA-C	FRNN-C	FRNN_FRS-C	FRNN_VQRS-C	NPBASIR-FUZZY
Breast-cancer	65.78	71.7	66.13	64.74	74.51
Breast-w	97.07	88.14	95.47	95.32	97.22
Diabetes	72.52	66.41	70.06	70.06	74.61
Flags	51.05	54.13	29.89	26.79	60.21
Haberman	71.55	73.53	64.04	64.04	73.84
Heart-Statlog	76.67	81.85	75.93	74.44	82.22
Ionosphere	64.1	0	76.07	64.1	80.33
Iris	92.67	88.67	95.33	95.33	96
Mfeat-fourier	78.35	77.5	78.8	78.65	79
Optdigits	9.86	0	9.86	9.54	89.57
Pima	71.11	66.67	70.6	70.6	73.83
Postoperative-patient-data	62.22	71.11	63.33	60	71.11
Primary-tumor	36.9	43.34	33.34	15.65	40.99
Solar-flare_1	97.84	97.84	97.22	94.73	97.84
Solar-flare_2	99.53	99.53	99.34	99.53	99.34
Sponge	92.5	95	55	55.18	92.5
Waveform	72.94	83.6	72.2	72.28	81.46

Table 3. Average rankings of Friedman test

Algorithm	Ranking
NPBASIR-FUZZY	1.5
FRKNNA-C	2.9118
FRNN-C	3.0294
FRNN-FRS-C	3.5294
FRNN-VQRS-C	4.0294

Table 4. Post hoc comparison table with NPBASIR-FUZZY as the control method for $\alpha = 0.05$

i	Algorithm	z	p	Holm	Hypothesis
4	FRNN-VQRS-C	4.664005	0.000003	0.0125	Rejected
3	FRNN-FRS-C	3.74205	0.000183	0.016667	Rejected
2	FRNN-C	2.820096	0.004801	0.025	Rejected
1	FRKNNA-C	2.603165	0.009237	0.05	Rejected

5   Conclusions

This paper introduces a fuzzy modeling of one of the Nearest Prototype Classifiers, in the base method NPBASIR-CLASS we change the numerical thresholds by a fuzzy approach in the definition of the binary relations that underlie any similarity-based inference process. This modification improving the interpretability of the system and reducing the number of parameters that will need to be fine-tuned. This will, in turn, shorten the time necessary to test, validate and deploy the system. The experimental study carried out on several datasets, four learning algorithms based on the fuzzy-rough sets theory and non-parametric statistical methods have been used to compare and analyze the accuracy of the algorithms. The previously study permit concludes that the proposed method NPBASIR-FUZZY outperforms the rest of the algorithms with regard to the accuracy results on test data. NPBASIR-FUZZY is a compact and interpretable Nearest Prototype Classifier.

Future research may focus on simulating this approach on prototypes selection based on similarity relations for classification problems, taking NPBASIR SEL-CLASS [10] as base method, and no limit the simulation for fuzzy-rough NN approach only.

Acknowledgment. This research has been partially sponsored by VLIR-UOS Network University Cooperation Programme - Cuba.

References

1. Feng, F., et al.: Soft sets combined with fuzzy sets and rough sets: a tentative approach. Soft. Comput. **14**(9), 899–911 (2010)
2. Lin, T.Y., Cercone, N.: Rough Sets and Data Mining: Analysis of Imprecise Data. Springer, New York (2012). https://doi.org/10.1007/978-1-4613-1461-5
3. Ziarko, W.P., Sets, R.: Fuzzy Sets and Knowledge Discovery. Springer, London (2012)
4. Yao, Y.: Combination of rough and fuzzy sets based on α-level sets. In: Lin, T.Y., Cercone, N. (eds.) Rough sets and Data Mining, pp. 301–321. Springer, Boston (1997). https://doi.org/10.1007/978-1-4613-1461-5_15
5. Tsang, E.C., et al.: Hybridization of fuzzy and rough sets: present and future. In: Bustince, H., Herrera, F., Montero, J. (eds.) Fuzzy Sets and Their Extensions: Representation Aggregation and Models, pp. 45–64. Springer, Heidelberg (2008). https://doi.org/10.1007/978-3-540-73723-0_3
6. Dubois, D., Prade, H.: Rough fuzzy sets and fuzzy rough sets. Int. J. Gener. Syst. **17**(2–3), 191–209 (1990)
7. Dubois, D., Prade, H.: Putting rough sets and fuzzy sets together. In: Słowiński, R. (ed.) Intelligent Decision Support, pp. 203–232. Springer, Dordrecht (1992). https://doi.org/10.1007/978-94-015-7975-9_14
8. Radzikowska, A.M., Kerre, E.E.: A comparative study of fuzzy rough sets. Fuzzy Sets Syst. **126**(2), 137–155 (2002)
9. Fernández Hernández, Y.B., et al.: An approach for prototype generation based on similarity relations for problems of classification. Computación y Sistemas **19**(1), 109–118 (2015)
10. Rodríguez, Y., et al., An Approach to solve classification problems on domains with hubness using rough sets and Nearest Prototype. In: 16th Mexican International Conference on Artificial Intelligence. Springer, Ensenada (2017)

11. Rodríguez, Y., et al.: Similar prototype methods for class imbalanced data classification. In: 2nd International Symposium on Fuzzy and Rough Sets (ISFUROS 2017), Varadero, Cuba, 24–26 October 2017. Springer, Heidelberg (2017)
12. Yao, Y.: Granular computing: basic issues and possible solutions. In: Proceedings of the 5th Joint Conference on Information Sciences. Citeseer (2000)
13. Filiberto, Y., et al.: Algoritmo para el aprendizaje de reglas de clasificación basado en la teoría de los conjuntos aproximados extendida. Dyna **78**(169), 62–70 (2011)
14. Filiberto, Y., et al.: A method to build similarity relations into extended Rough Set Theory. In: 2010 10th International Conference on Intelligent Systems Design and Applications (ISDA). IEEE (2010)
15. Filiberto, Y., et al.: An analysis about the measure quality of similarity and its applications in machine learning. In: Fourth International Workshop on Knowledge Discovery, Knowledge Management and Decision Support. Atlantis Press (2013)
16. Filiberto Cabrera, Y., Bello Pérez, R., Mota, Y.C., Jimenez, G.R.: Improving the MLP learning by using a method to calculate the initial weights of the network based on the quality of similarity measure. In: Batyrshin, I., Sidorov, G. (eds.) MICAI 2011. LNCS (LNAI), vol. 7095, pp. 351–362. Springer, Heidelberg (2011). https://doi.org/10.1007/978-3-642-25330-0_31
17. Filiberto, Y., et al.: Using PSO and RST to predict the resistant capacity of connections in composite structures. In: González, J.R., Pelta, D.A., Cruz, C., Terrazas, G., Krasnogor, N. (eds.) NICSO 2010. SCI, vol. 284, pp. 359–370. Springer, Heidelberg (2010). https://doi.org/10.1007/978-3-642-12538-6_30
18. Bello-García, M., García-Lorenzo, M.M., Bello, R.: A method for building prototypes in the nearest prototype approach based on similarity relations for problems of function approximation. In: Batyrshin, I., González Mendoza, M. (eds.) MICAI 2012. LNCS (LNAI), vol. 7629, pp. 39–50. Springer, Heidelberg (2013). https://doi.org/10.1007/978-3-642-37807-2_4
19. Fernandez, Y., et al.: Learning similarity measures from data with fuzzy sets and particle swarms. In: 2014 11th International Conference on Electrical Engineering, Computing Science and Automatic Control (CCE). IEEE (2014)
20. Wang, W.-J.: New similarity measures on fuzzy sets and on elements. Fuzzy Sets Syst. **85**(3), 305–309 (1997)
21. Blake, C.L., Merz, C.J.: UCI repository of machine learning databases (1998) http://www.ics.uci.edu/~mlearn/MLRepository.html. Accessed 18 Mar 2018
22. Triguero, I., et al.: KEEL 3.0: an open source software for multi-stage analysis in data mining. Int. J. Comput. Intell. Syst. **10**(1), 1238–1249 (2017)
23. Bian, H., Mazlack, L.: Fuzzy-rough nearest-neighbor classification approach. In: In: NAFIPS 2003, 22nd International Conference of the North American Fuzzy Information Processing Society. IEEE (2003)
24. Sarkar, M.: Fuzzy-rough nearest neighbor algorithms in classification. Fuzzy Sets Syst. **158**(19), 2134–2152 (2007)
25. Jensen, R., Cornelis, C.: Fuzzy-rough nearest neighbour classification. In: Peters, J.F., Skowron, A., Chan, C.C., Grzymala-Busse, J.W., Ziarko, W.P. (eds.) Transactions on Rough Sets XIII. LNCS, vol. 6499, pp. 56–72. Springer, Heidelberg (2011). https://doi.org/10.1007/978-3-642-18302-7_4

A Fuzzy Harmony Search Algorithm for the Optimization of a Benchmark Set of Functions

Cinthia Peraza, Fevrier Valdez, Oscar Castillo[(✉)], and Patricia Melin

Division of Graduate Studies, Tijuana Institute of Technology, Tijuana, Mexico
cinthia.peraza18@tectijuana.edu.mx,
{fevrier,ocastillo,pmelin}@tectijuana.mx

Abstract. A fuzzy harmony search algorithm (FHS) is presented in this paper. This method uses a fuzzy system for dynamic adaptation of the harmony memory accepting (HMR) parameter along the iterations, and in this way achieving control of the intensification and diversification of the search space. This method was previously applied to classic benchmark mathematical functions with different number of dimensions. However, in this case we decided to apply the proposed FHS to benchmark mathematical problems provided by the CEC 2015 competition, which are unimodal, multimodal, hybrid and composite functions to check the efficiency for the proposed method. A comparison is presented to verify the results obtained with respect to the original harmony search algorithm and fuzzy harmony search algorithm.

Keywords: Harmony search algorithm
Optimization of benchmark mathematical functions
Fuzzy logic and dynamic parameter adaptation

1 Introduction

This paper focuses on optimizing benchmark mathematical functions of the set given in the CEC 2015 competition. In previous works other simple functions are considered using the FHS method [1–5]. The proposed method is a modification of the original harmony search algorithm (HS), which uses the improvisation process of jazz musicians [22, 23]. This method has been used to solve optimization problems, and there are also hybrid methods such as in [6–8]. The main difference between the variants and the existing methods in literature is that this method is based on the original harmony search algorithm and uses fuzzy logic to dynamically adjust the HMR parameter, to achieve a control of the search space applied to mathematical functions. This paper focuses on complex mathematical functions to verify the stability for the proposed method using fuzzy logic to dynamically adapt the algorithm parameters as the iterations are progressing, and this technique has been applied in other metaheuristics methods which have achieved good results [9–13]. Fuzzy logic is used in order to control the diversification and intensification, processes in the search space and enable finding the global optimum, avoiding stagnation and premature convergence. Different

© Springer Nature Switzerland AG 2018
I. Batyrshin et al. (Eds.): MICAI 2018, LNAI 11288, pp. 401–412, 2018.
https://doi.org/10.1007/978-3-030-04491-6_31

types of mathematical functions are considered in this article, as they are unimodal, multimodal, composite and hybrid, and these benchmark mathematical functions have been used by different methods [14, 15]. Fuzzy logic has also recently been used in existing metaheuristic methods because it uses linguistic variables and rules with which fuzzy models make decisions [16], and has been used in areas of evolutionary algorithms with fuzzy logic [17, 18], engineering problems [19, 20] among others.

This paper describes the FHS method that can be applied for solving global optimization problems. CEC each year proposes new mathematical functions to be tested and challenge different methods [15, 21]. The proposed method uses fuzzy logic to achieve an efficient adjustment of parameters in harmony search and the CEC 2015 benchmark functions are evaluated with 10 and 30 dimensions.

This paper is organized as follows. The original harmony search algorithm is presented in Sect. 2. The proposed method with fuzzy logic is shown in Sect. 3. The benchmark mathematical functions are summarized in Sect. 4. The comparison between the original harmony search algorithm (HS) and FHS is presented in Sect. 5. Finally, Conclusions are offered in Sect. 6.

2 Original Harmony Search Algorithm

The harmony search (HS) algorithm is a new population based metaheuristic search technique that mimics the process of music improvisation [10, 17]. In this method there are five basic steps, which are:

Step 1: Initialize the problem and parameters

$$\text{Minimize} f(x) s.t. x(j) \in [LB(j), UB(j)\}, j = 1, 2, \ldots, n] \tag{1}$$

Step 2: Initialize the Harmony memory (HM)

$$HM = \begin{bmatrix} x_1^1 & x_2^1 & \cdots & x_N^1 & f(x^1) \\ x_1^2 & x_2^2 & \cdots & x_N^2 & f(x^2) \\ \vdots & \vdots & \vdots & \vdots & \vdots \\ x_1^{HMS} & x_2^{HMS} & \cdots & x_N^{HMS} & f(x^{HMS}) \end{bmatrix} \tag{2}$$

Step 3: Improvise New Harmony

$$X_{new}(j) = X_{new}(j) \pm r \times BW \tag{3}$$

Step 4: Update Harmony Memory
To update the HM with a new solution vector, x_{new}, the objective function will be used to evaluate them. A comparison is made is the new vector solution is better than the worst historical vector solution the worst historical is excluded and substituted with a new one.

Step 5: Check the stopping criteria

The process is repeated until the number of improvisations (NI) is satisfied, otherwise the process repeats steps 3 and 4. Finally the best solution is achieved and considered as the best result to the problem. The HMR parameter represents the intensification or exploitation, the *PArate* and randomization parameters represent the diversification or exploration of the algorithm.

3 Proposed Method with Fuzzy Logic

This section shows the methodology of the proposed method, as Sect. 2 describes the detailed operation of the original harmony search algorithm. The proposed method is based on the HS algorithm, in previous works the behavior of the original harmony search algorithm was studied [1–5], and tests were performed with the *PArate* and *HMR* parameters and making adjustments to the *HMR* parameter achieves better results. The value of the *HMR* parameter is constant during the process of improvisation of the algorithm, based on this it is decided to make it a fuzzy parameter. The main contribution is to dynamically change the *HMR* parameter using a fuzzy system as the number of iterations advances and with this to achieve more diversity of solutions. The pseudocode of the proposed method is illustrated in Fig. 1:

1. *Define objective function* $f(x), x = (X_1, \dots, X_n)^T$
2. *Initial generate harmonics (matrix of real numbers)*
3. *Define Pitch adjustment rate (PArate) and limits of tone*
4. *Define harmony memory accepting (HMR)*
5. *While (t<Maximum number of iterations)*
6. **Calculate iteration using Equation (5)**
7. **Calculate new HMR parameter using a fuzzy system using Equation (4)**
8. *Generate a new harmony and accept the best system*
9. *Setting the tone for new harmonies (solutions)*
10. *If (rand > HMR)*
11. *Choose an existing harmony randomly*
12. *Else if (rand > Parate)*
13. *Setting the tone at random within a bandwidth*
14. *Else*
15. *Generate a new harmony through a randomization*
16. *End if*
17. *Accepting new harmonies (solutions) best*
18. *End while*
19. *To find the best solutions*

Fig. 1. Pseudocode of the FHS

3.1   Design of the Fuzzy Harmony Search System (FHS)

In this section the graphical representation of the fuzzy system is illustrated by Fig. 2, using one input, which is the iteration, and one output, which is the *HMR* parameter and the system if of Mamdani type.

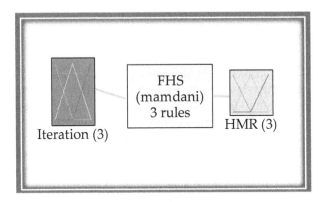

Fig. 2. Scheme of the proposed method (FHS)

The FHS method uses one output that is the *HMR* parameter and the input is the *iteration* variable as presented in Fig. 3.

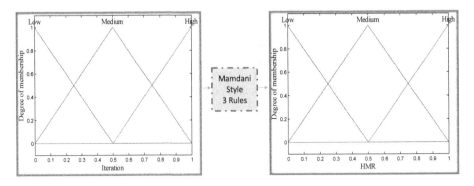

Fig. 3. Input and Output of the fuzzy system.

To represent this idea this parameter is converted into a fuzzy parameter, in this method, this value is considered to be fuzzy as it is updated within the FHS progress, and is determined by Eq. (4), where *HMR* is changing values in the [0, 1] range.

$$HMR = \frac{\sum_{i=1}^{r_{hmr}} \mu_i^{hmr}(hmr_{1i})}{\sum_{i=1}^{r_{hmr}} \mu_i^{hmr}} \qquad (4)$$

Where HMR, is the memory considerations; r_{hmr}, is the number of rules of the fuzzy system corresponding to hmr; hmr_{1i}, is the output result for rule i corresponding to hmr; μ_i^{hmr}, is the membership function of rule i corresponding to hmr.

The representation of the input and the output variables in the fuzzy system is presented in Fig. 3, the input is known as "iteration", and is granulated into three triangular membership functions and they are *Low*, *Medium* and *High*. The output called "HMR" uses values from 0 to 1, and the output is granulated into three triangular membership functions and they are *Low*, *Medium* and *High*. The fuzzy system is of Mamdani type and contains three fuzzy rules.

The rules are designed based on the study of parameters of the algorithm, so that in the initial iterations it will explore and by the final iterations will exploit the search space, and in this case the rules are on the increase. Rules are summarized in Table 1.

Table 1. Rules for the fuzzy system (FHS)

Iteration	HMR		
	Low	Medium	High
Low	Low	–	–
Medium	–	Medium	–
High	–	–	High

Table 1 represents the idea of increasing the output of the rules as iterations are progressing. To represent the input of the rules the following equation is used:

$$Iteration = \frac{Current\ Iteration}{Maximun\ of\ iterations} \qquad (5)$$

This phase considers a percentage of elapsed iterations to find the values of HMR. It initializes with low values of HMR so that the algorithm has diversification and then achieve intensification.

4 Experimental Setting

For the experiments we used the 14 mathematical functions presented in Table 2, provided by the CEC 2015 competition. In this case the problem is finding the global minimum for each mathematical function. These functions are tested with 10 and 30 dimensions.

Table 2 presents the equations used to perform the experiments, functions 1 and 2 are unimodal functions, 3 to 5 are simple multimodal functions, 6 to 8 are hybrid functions, 9 to 15 are composition functions, the search range used for all functions are [−100,100] and each experiments ran 51 times and during each run, the function was evaluated for 10000 * D times. The parameter D denoted the dimension of the function. Table 3 presents the global minimum for each mathematical function.

Table 2. Mathematical functions for the experiments

No.	Equations
1	$f_1(x) = \sum_{i=1}^{D} (10^6)^{\frac{i-1}{D-1}} x_i^2$
2	$f_2(x) = x_1^2 + 10^6 \sum_{i}^{D} x_i^2$
3	$f_3(x) = 10^6 x_1^2 + \sum_{i=2}^{D} x_i^2$
4	$f_4(x) = \sum_{i=1}^{D-1} \left(100 (x_i^2 - x_{i+1})^2 + (x_i - 1)^2 \right)$
5	$f_5(x) = -20 exp \left(-0.2 \sqrt{\frac{1}{D} \sum_{i=1}^{D} x_i^2} \right) - exp \left(\frac{1}{D} \sum_{i=1}^{D} \cos(2\pi x_i) \right) + 20 + e$
6	$f_6(x) = \sum_{i=1}^{D} \left(\sum_{k=0}^{kmax} [a^k \cos(2\pi b^k (x_i + 0.5))] \right) - D \sum_{i=1}^{kmax} [a^k \cos 2\pi b^k \cdot 0.5]$ $a = 0.5, b = 3, kmax = 20$
7	$f_7(x) = \sum_{i=1}^{D} \frac{x_i^2}{4000} - \prod_{i=1}^{D} \cos \left(\frac{x_i}{\sqrt{i}} \right) + 1$
8	$f_8(x) = \sum_{i=1}^{D} (x_i^2 - 10 \cos(2\pi x_i) + 10)$
9	$f_9(x) = 418.9829 \times D - \sum_{i=1}^{D} g(z_i), \quad z_i$ $\quad = x_i + 4.209687462275036e + 002$ $g(z_i) = z_i \sin \left(|z_i|^{\frac{1}{2}} \right) \qquad if\ |z_i| \le 500$ $g(z_i) = (500 - mod(z_i, 500)) \sin(\sqrt{|500 - mod(z_i, 500)|}) - \frac{(z_i - 500)^2}{10000D}\ if\ z_i > 500$ $g(z_i) = (mod(|z_i|, 500) - 500) \sin(\sqrt{|mod(|z_i|, 500) - 500|}) - \frac{(z_i - 500)^2}{10000D}\ if\ z_i$
10	$f_{10}(x) = \frac{10}{D^2} \prod_{i=1}^{D} \left(1 + i \sum_{j=1}^{32} \frac{|2^j x_i - round(2^j x_i)|}{2^j} \right)^{\frac{10}{D^{1.2}}} - \frac{10}{D^2}$
11	$f_{11}(x) = \left| \sum_{i=1}^{D} x_i^2 - D \right|^{1/4} + \frac{\left(0.5 \sum_{i=1}^{D} x_i^2 + \sum_{i=1}^{D} x_i \right)}{D + 0.5}$
12	$f_{12}(x) = \left| \left(\sum_{i=1}^{D} x_i^2 \right)^2 - \left(\sum_{i=1}^{D} x_i \right)^2 \right|^{\frac{1}{2}} + \frac{\left(0.5 \sum_{i=1}^{D} x_i^2 + \sum_{i=1}^{D} x_i \right)}{D + 0.5}$
13	$f_{13}(x) = f_7(f_4(x_1, x_2)) + f_7(f_4(x_3, x_4)) + \ldots + f_7(f_4(x_{D-1}, x_D)) + f_7(f_4(x_D, x_1))$
14	Scaffer's F6 Function: $g(x, y) = 0.5 + \frac{(\sin^2(\sqrt{x^2 + y^2}) - 0.5)}{(1 + 0.001(x^2 + y^2))^2} f_{14}(x) = g(x_1, x_2) + g(x_2, x_3) + \ldots + g(x_{D-1}, x_D) + g(x_D, x_1)$

Table 3 represents the local optimum, which is obtained using $F_i' = F_i - F_i*$ where F_i is the result obtained for function and F_i* is the local optimum for each function and are used as g_i. In this way, the function values of global optima of g_i are equal to 0 for all composition functions. A value of 100 was added to function 1, 200 to function 2, ..., and a value of 1500 was added to function 15, all of which were subtracted from the calculated mean errors so that the best value for each function remains zero. With hybrid functions as the basic functions, the composition function can have different properties for different variables subcomponents.

Table 3. Global minimum for each function

No.	Functions	$F_i* = F_i(x^*)$
F1	Rotated high conditioned elliptic function	100
F2	Rotated cigar function	200
F3	Shifted and rotated Ackley's function	300
F4	Shifted and rotated Rastrigin's function	400
F5	Shifted and rotated Schwefel's function	500
F6	Hybrid function 1 ($N = 3$)	600
F7	Hybrid function 2 ($N = 4$)	700
F8	Hybrid function 3($N = 5$)	800
F9	Composition function 1 ($N = 3$)	900
F10	Composition function 2 ($N = 3$)	1000
F11	Composition function 3 ($N = 5$)	1100
F12	Composition function 4 ($N = 5$)	1200
F13	Composition function 5 ($N = 5$)	1300
F14	Composition function 6 ($N = 7$)	1400
F15	Composition function 7 ($N = 10$)	1500

5 Experimental Results

This section discusses the obtained results and a comparison between the HS and the FHS. For the tests we used the values indicated in Table 4.

Table 4. Parameter values used in the methods

Parameter	Simple HS	Fuzzy HS
Harmonies	100	100
Dimensions	10, 30	10, 30
Iterations	100000, 300000	100000, 300000
HMR	0.95	Dynamic
PArate	0.75	0.75

Table 4 shows the parameters used in each method, where we can find each value used as the number of harmonies, the dimensions and the maximum number of iterations for each dimension number and value of the *HMR* parameter used in each method. Tables 5 to 6 are presenting the results of the experiments using the original harmony search algorithm with the parameters indicated in Table 4.

The results obtained in Tables 5 to 6 are presenting, the best, worst, mean, median and standard deviations as a result of 51 runs for each mathematical function using the original harmony search algorithm. The following tables show the results obtained by applying the 15 benchmark functions using the proposed FHS method. Tables 7 to 8 indicate the values obtained with 10 dimensions.

Table 5. Results obtained using HS with 10 dimensions

10 dimension					
Function	Best	Worst	Median	Mean	Std
F1	$2.01E+03$	$1.21E+06$	$4.08E+04$	$1.91E+04$	$2.47E+05$
F2	$2.00E+02$	$3.13E+04$	$1.68E+03$	$7.10E+02$	$9.31E+03$
F3	$3.00E+02$	$3.20E+02$	$3.20E+02$	$3.20E+02$	$2.84E+00$
F4	$4.01E+02$	$4.20E+02$	$4.08E+02$	$4.08E+02$	$3.82E+00$
F5	$5.17E+02$	$1.27E+03$	$7.46E+02$	$7.25E+02$	$1.79E+02$
F6	$6.30E+02$	$4.90E+03$	$1.50E+03$	$1.32E+03$	$1.07E+03$
F7	$7.00E+02$	$7.03E+02$	$7.01E+02$	$7.01E+02$	$6.39E-01$
F8	$8.02E+02$	$6.77E+03$	$1.44E+03$	$1.44E+03$	$1.69E+03$
F9	$1.03E+03$	$1.17E+03$	$1.07E+03$	$1.08E+03$	$2.98E+01$
F10	$2.59E+04$	$1.51E+07$	$2.51E+06$	$5.73E+05$	$2.94E+06$
F11	$1.48E+03$	$1.97E+03$	$1.72E+03$	$1.70E+03$	$1.32E+02$
F12	$1.33E+03$	$1.38E+03$	$1.36E+03$	$1.36E+03$	$1.05E+01$
F13	$1.35E+03$	$1.63E+03$	$1.38E+03$	$1.39E+03$	$5.02E+01$
F14	$1.12E+04$	$2.74E+04$	$2.02E+04$	$1.98E+04$	$2.97E+03$
F15	$1.73E+03$	$1.16E+04$	$3.88E+03$	$3.53E+03$	$2.62E+03$

Table 6. Results obtained using HS with 30 dimensions

30 dimension					
Function	Best	Worst	Median	Mean	Std
F1	$6.77E+05$	$3.00E+07$	$3.56E+06$	$2.83E+06$	$5.26E+06$
F2	$2.03E+02$	$1.45E+06$	$6.09E+03$	$2.05E+03$	$2.02E+05$
F3	$3.21E+02$	$3.21E+02$	$3.21E+02$	$3.21E+02$	$8.57E-02$
F4	$4.31E+02$	$4.90E+02$	$4.54E+02$	$4.55E+02$	$1.32E+01$
F5	$1.88E+03$	$6.31E+03$	$3.21E+03$	$3.12E+03$	$7.65E+02$
F6	$2.20E+04$	$2.34E+06$	$2.47E+05$	$1.87E+05$	$4.02E+05$
F7	$7.06E+02$	$7.15E+02$	$7.10E+02$	$7.10E+02$	$2.02E+00$
F8	$1.48E+04$	$4.78E+05$	$1.10E+05$	$8.00E+04$	$9.19E+04$
F9	$1.37E+03$	$1.97E+03$	$1.61E+03$	$1.61E+03$	$1.50E+02$
F10	$2.39E+07$	$1.00E+09$	$3.15E+08$	$1.75E+08$	$2.29E+08$
F11	$2.91E+03$	$3.99E+03$	$3.29E+03$	$3.31E+03$	$2.45E+02$
F12	$1.40E+03$	$1.53E+03$	$1.47E+03$	$1.47E+03$	$2.88E+01$
F13	$1.62E+03$	$5.57E+03$	$2.39E+03$	$2.47E+03$	$8.84E+02$
F14	$1.06E+05$	$2.58E+05$	$1.62E+05$	$1.57E+05$	$3.85E+04$
F15	$9.12E+04$	$1.15E+06$	$5.12E+05$	$3.74E+05$	$2.56E+05$

The results in Tables 7 to 8 are presenting, the best, worst, mean, median and standard deviations as a result of 51 runs for each mathematical with different dimension function using the proposed fuzzy harmony search algorithm. Table 9 shows the comparison between the results obtained when applying the 15 functions of the CEC2015 using the HS and FHS method.

Table 7. Results obtained using FHS with 10 dimensions

10 dimensions					
Function	Best	Worst	Median	Mean	Std
F1	$2.60E-03$	$5.96E-02$	$1.81E-02$	$2.62E-02$	$1.58E-02$
F2	$1.13E+00$	$3.05E+00$	$1.17E+00$	$1.51E+00$	$4.63E-01$
F3	$0.00E+00$	$2.13E+01$	$2.09E+01$	$2.05E+01$	$2.94E+00$
F4	$0.00E+00$	$1.55E+02$	$1.29E+02$	$1.24E+02$	$2.52E+01$
F5	$0.00E+00$	$3.01E+03$	$2.48E+03$	$2.43E+03$	$4.24E+02$
F6	$0.00E+00$	$1.18E-02$	$5.50E-03$	$4.05E-03$	$2.83E-03$
F7	$0.00E+00$	$1.20E+02$	$5.16E+01$	$5.70E+01$	$2.70E+01$
F8	$0.00E+00$	$1.40E-03$	$5.00E-04$	$3.61E-04$	$2.57E-04$
F9	$1.00E+02$	$1.00E+02$	$1.00E+02$	$1.00E+02$	$4.68E-02$
F10	$2.25E+02$	$4.13E+03$	$5.98E+02$	$1.11E+03.$	$1.05E+03$
F11	$2.54E+00$	$4.53E+02$	$3.00E+02$	$2.63E+02$	$1.06E+02$
F12	$1.01E+02$	$1.04E+02$	$1.02E+02$	$1.02E+02$	$6.19E-01$
F13	$2.65E+01$	$4.02E+01$	$3.23E+01$	$3.20E+01$	$2.53E+00$
F14	$1.00E+02$	$1.04E+04$	$7.14E+03$	$5.44E+03$	$2.81E+03$
F15	$1.00E+02$	$1.00E+02$	$1.00E+02$	$1.00E+02$	$0.00E+00$

Table 8. Results obtained using FHS with 30 dimensions

30 dimension					
Function	Best	Worst	Median	Mean	Std
F1	$1.58E-02$	$8.23E-02$	$4.54E-02$	$4.52E-02$	$1.46E-02$
F2	$1.02E+00$	$1.84E+00$	$1.44E+00$	$1.45E+00$	$1.87E-01$
F3	$0.00E+00$	$2.15E+01$	$2.13E+01$	$2.09E+01$	$2.99E+00$
F4	$0.00E+00$	$7.65E+02$	$6.51E+02$	$6.38E+02$	$1.05E+02$
F5	$8.42E+03$	$1.00E+04$	$9.42E+03$	$9.36E+03$	$3.79E+02$
F6	$9.00E-04$	$1.04E-02$	$3.90E-03$	$4.32E-03$	$2.62E-03$
F7	$0.00E+00$	$2.20E+03$	$1.37E+03$	$1.34E+03$	$4.33E+02$
F8	$1.00E-04$	$4.10E-03$	$1.30E-03$	$1.31E-03$	$8.59E-04$
F9	$1.03E+02$	$2.54E+02$	$1.03E+02$	$1.09E+02$	$2.73E+01$
F10	$3.80E+03$	$3.14E+05$	$6.55E+04$	$1.05E+05$	$8.73E+04$
F11	$3.05E+02$	$8.95E+02$	$6.78E+02$	$6.17E+02$	$1.82E+02$
F12	$1.05E+02$	$1.11E+02$	$1.08E+02$	$1.08E+02$	$1.20E+00$
F13	$1.03E+02$	$1.29E+02$	$1.18E+02$	$1.18E+02$	$5.99E+00$
F14	$3.19E+04$	$3.87E+04$	$3.53E+04$	$3.48E+04$	$1.92E+03$
F15	$1.00E+02$	$1.00E+02$	$1.00E+02$	$1.00E+02$	$1.41E-11$

Table 9. Summary of results with 10 and 30 dimensions with HS and FHS methods

Function	Dimension			
	10		30	
	HS	FHS	HS	FHS
F1	$1.91E+04$	$2.62E-02$	$2.83E+06$	$4.52E-02$
F2	$7.10E+02$	$1.51E+00$	$2.05E+03$	$1.45E+00$
F3	$3.20E+02$	$2.05E+01$	$3.21E+02$	$2.09E+01$
F4	$4.08E+02$	$1.24E+02$	$4.55E+02$	$6.38E+02$
F5	$7.25E+02$	$2.43E+03$	$3.12E+03$	$9.36E+03$
F6	$1.32E+03$	$4.05E-03$	$1.87E+05$	$4.32E-03$
F7	$7.01E+02$	$5.70E+01$	$7.10E+02$	$1.34E+03$
F8	$1.44E+03$	$3.61E-04$	$8.00E+04$	$1.31E-03$
F9	$1.08E+03$	$1.00E+02$	$1.61E+03$	$1.09E+02$
F10	$5.73E+05$	$1.11E+03$	$1.75E+08$	$1.05E+05$
F11	$1.70E+03$	$2.63E+02$	$3.31E+03$	$6.17E+02$
F12	$1.36E+03$	$1.02E+02$	$1.47E+03$	$1.08E+02$
F13	$1.39E+03$	$3.20E+01$	$2.47E+03$	$1.18E+02$
F14	$1.98E+04$	$5.44E+03$	$1.57E+05$	$3.48E+04$
F15	$3.53E+03$	$1.00E+02$	$3.74E+05$	$1.00E+02$

6   Conclusions

In this paper 15 benchmark functions provided by the CEC 2015, which are uni-modal, multimodal, hybrid and composite to give more complexity to these benchmark problems, were considered by the FHS and HS methods to test the effectiveness of these methods. When performing the experiments with these mathematical functions it was verified that when using the FHS method a better control of the search space is obtained and better results are obtained unlike the original method. It is possible to demonstrate that when using fuzzy logic for the dynamic parameter adaptation of the algorithm, in this case the HMR parameter changes as the number of iterations advances, the function of this parameter is to achieve control of the exploitation of the search space, in early iterations the algorithm explores and in last iterations exploits the search space to obtain the diversity of solutions, and this indicates that the FHS method has the potential to find global minima of complex functions. As future work we can apply the proposed optimization method in different applications, like in [24–30].

References

1. Peraza, C., Valdez, F., Castillo, O.: An adaptive fuzzy control based on harmony search and its application to optimization. In: Melin, P., Castillo, O., Kacprzyk, J. (eds.) Nature-Inspired Design of Hybrid Intelligent Systems. SCI, vol. 667, pp. 269–283. Springer, Cham (2017). https://doi.org/10.1007/978-3-319-47054-2_18

2. Peraza, C., Valdez, F., Castillo, O.: Comparative study of type-1 and interval type-2 fuzzy systems in the fuzzy harmony search algorithm applied to Benchmark functions. In: Kacprzyk, J., Szmidt, E., Zadrożny, S., Atanassov, Krassimir T., Krawczak, M. (eds.) IWIFSGN/EUSFLAT -2017. AISC, vol. 643, pp. 162–170. Springer, Cham (2018). https://doi.org/10.1007/978-3-319-66827-7_15

3. Peraza, C., Valdez, F., Garcia, M., Melin, P., Castillo, O.: A new fuzzy harmony search algorithm using fuzzy logic for dynamic parameter adaptation. Algorithms **9**(4), 69 (2016)

4. Peraza, C., Valdez, F., Melin, P.: Optimization of Intelligent controllers using a type-1 and interval type-2 fuzzy harmony search algorithm. Algorithms **10**(3), 82 (2017)

5. Peraza, C., Valdez, F., Castro, J.R., Castillo, O.: Fuzzy dynamic parameter adaptation in the harmony search algorithm for the optimization of the ball and beam controller. Adv. Oper. Res. **2018**, 1–16 (2018)

6. Ameli, K., Alfi, A., Aghaebrahimi, M.: A fuzzy discrete harmony search algorithm applied to annual cost reduction in radial distribution systems. Eng. Optim. **48**(9), 1529–1549 (2016)

7. Arabshahi, P., Choi, J.J., Marks, R.J., Caudell, T.P.: Fuzzy parameter adaptation in optimization: some neural net training examples. IEEE Comput. Sci. Eng. **3**(1), 57–65 (1996)

8. Patnaik, S.: Recent Developments in Intelligent Nature-Inspired Computing: IGI Global (2017)

9. Bernal, E., Castillo, O., Soria, J.: Imperialist Competitive Algorithm with Fuzzy Logic for Parameter Adaptation: A Parameter Variation Study. In: Atanassov, K.T., et al. (eds.) Novel Developments in Uncertainty Representation and Processing. AISC, vol. 401, pp. 277–289. Springer, Cham (2016). https://doi.org/10.1007/978-3-319-26211-6_24

10. Caraveo, C., Valdez, F., Castillo, O.: A new optimization meta-heuristic algorithm based on self-defense mechanism of the plants with three reproduction operators. Soft. Comput. **22**(15), 4907–4920 (2018)

11. Amador-Angulo, L., Castillo, O.: A new fuzzy bee colony optimization with dynamic adaptation of parameters using interval type-2 fuzzy logic for tuning fuzzy controllers. Soft. Comput. **22**(2), 571–594 (2018)

12. Valdez, F., Vazquez, J.C., Melin, P., Castillo, O.: Comparative study of the use of fuzzy logic in improving particle swarm optimization variants for mathematical functions using co-evolution. Appl. Soft Comput. **52**, 1070–1083 (2017)

13. Perez, J., Valdez, F., Castillo, O., Melin, P., Gonzalez, C., Martinez, G.: Interval type-2 fuzzy logic for dynamic parameter adaptation in the bat algorithm. Soft. Comput. **21**(3), 667–685 (2017)

14. Awad, N., Ali, M.Z., Reynolds, R.G.: A differential evolution algorithm with success-based parameter adaptation for CEC 2015 learning-based optimization. In: 2015 IEEE Congress on Evolutionary Computation (CEC), Sendai, Japan, pp. 1098–1105 (2015)

15. Rueda, J.L., Erlich, I.: Testing MVMO on learning-based real-parameter single objective benchmark optimization problems. In: 2015 IEEE Congress on Evolutionary Computation (CEC), Sendai, Japan, pp. 1025–1032 (2015)

16. Tan, W.W., Chua, T.W.: Uncertain rule-based fuzzy logic systems: introduction and new directions (Mendel, J.M.; 2001) [book review]. IEEE Comput. Intell. Mag. **2**(1), 72–73 (2007)
17. Castillo, O., Ochoa, P., Soria, J.: Differential evolution with fuzzy logic for dynamic adaptation of parameters in mathematical function optimization. In: Angelov, P., Sotirov, S. (eds.) Imprecision and Uncertainty in Information Representation and Processing. SFSC, vol. 332, pp. 361–374. Springer, Cham (2016). https://doi.org/10.1007/978-3-319-26302-1_21
18. Valdez, F., Melin, P., Castillo, O.: Evolutionary method combining particle swarm optimization and genetic algorithms using fuzzy logic for decision making, pp. 2114–2119 (2009)
19. Gao, K.Z., Suganthan, P.N., Pan, Q.K., Tasgetiren, M.F.: An effective discrete harmony search algorithm for flexible job shop scheduling problem with fuzzy processing time. Int. J. Prod. Res. **53**, 5896–5911 (2015)
20. Geem, Z.W., Kim, J.H., Loganathan, G.V.: A New Heuristic Optimization Algorithm: Harmony Search. Simulation **76**(2), 60–68 (2001)
21. Caraveo, C., Valdez, F., Castillo, O.: A new optimization metaheuristic based on the self-defense techniques of natural plants applied to the CEC 2015 Benchmark functions. In: Kacprzyk, J., Szmidt, E., Zadrożny, S., Atanassov, Krassimir T., Krawczak, M. (eds.) IWIFSGN/EUSFLAT -2017. AISC, vol. 641, pp. 380–388. Springer, Cham (2018). https://doi.org/10.1007/978-3-319-66830-7_34
22. Yang, X.: Nature Inspired Metaheuristic Algorithms University of Cambridge, United Kingdom, pp 73–76, Luniver Press (2010)
23. Yang, X.S., Geem, Z.W.: Music-inspired Harmony Search Algorithm: Theory And Applications. Springer, Heidelberg (2009). https://doi.org/10.1007/978-3-642-00185-7
24. Leal- Ramírez, C., Castillo, O., Melin, P., Rodríguez Díaz, A.: Simulation of the bird age-structured population growth based on an interval type-2 fuzzy cellular structure. Inf. Sci. **181**(3), 519–535 (2011)
25. Melin, P., Amezcua, J., Valdez, F., Castillo, O.: A new neural network model based on the LVQ algorithm for multi-class classification of arrhythmias. Inf. Sci. **279**, 483–497 (2014)
26. Castillo, O., Amador-Angulo, L., Castro, J.R., García Valdez, M.: A comparative study of type-1 fuzzy logic systems, interval type-2 fuzzy logic systems and generalized type-2 fuzzy logic systems in control problems. Inf. Sci. **354**, 257–274 (2016)
27. Melin, P., Mancilla, A., Lopez, M., Mendoza, O.: A hybrid modular neural network architecture with fuzzy Sugeno integration for time series forecasting. Appl. Soft Comput. **7**(4), 1217–1226 (2007)
28. Castillo, O., Melin, P.: Intelligent systems with interval type-2 fuzzy logic. Int. J. Innovative Comput. Inf. Control **4**(4), 771–783 (2008)
29. Melin, P., Castillo, O.: Modelling, simulation and control of non-linear dynamical systems: an intelligent approach using soft computing and fractal theory. CRC Press, Boca Raton (2001)
30. Melin, P., Castillo, O.: Intelligent control of complex electrochemical systems with a neuro-fuzzy-genetic approach. IEEE Trans. Ind. Electron. **48**(5), 951–955

An Innovative and Improved Mamdani Inference (IMI) Method

Hamid Jamalinia[1,2], Zahra Alizadeh[1], Samad Nejatian[2,3],
Hamid Parvin[1,4], and Vahideh Rezaie[2,5(✉)]

[1] Department of Computer Engineering, Yasooj Branch,
Islamic Azad University, Yasooj, Iran
[2] Young Researchers and Elite Club, Yasooj Branch,
Islamic Azad University, Yasooj, Iran
v.rezaie@iauyasooj.ac.ir
[3] Department of Electrical Engineering, Yasooj Branch,
Islamic Azad University, Yasooj, Iran
[4] Young Researchers and Elite Club, Nourabad Mamasani Branch,
Islamic Azad University, Nourabad Mamasani, Iran
[5] Department of Mathematic, Yasooj Branch,
Islamic Azad University, Yasooj, Iran

Abstract. For a fuzzy system, inputs can be considered as crisp ones or fuzzy ones or a combination of them. Generally, the inputs are of crisp type; but sometimes they are of fuzzy type. For fuzzy inputs, the min max method for measuring the amount of matching is used. The min max method is studied in the paper and its weaknesses will be discovered in the current paper. We propose an alternative approach which is called an *innovative and improved mamdani inference method* (IIMI). We will show that all weaknesses of the previous min max method have been managed in the proposed inference method.

Keywords: Max min inference method · Inference method · Fuzzy inference

1 Introduction

A *fuzzy implication* (FI) rule is a crucial element in any *fuzzy inference engine* (FIE) [1, 2]. FIE can takes fuzzy inputs or crisp inputs. In this paper, we only talk about FIE with fuzzy inputs. Therefore, a FIE maps a set of one or more input *fuzzy sets* (FS), $A_i \subset U_i$, and an output FS, $B \subset V$ [3]. A FI rule can be considered as a *fuzzy relation* (FR) in the input/output product space, $U_1 \times U_2 \times \ldots \times U_n \times V$ [4] where n is the number of input variables. If n is equal to one, the FS $B' \subset V$ for an input FS $A'_1 \subset U_i$ can be determined by *generalized modus ponens* (GMP) [4–6]. If the *fuzzy rule-base* (FRB) contains over one rule, i.e. n is greater than one, then inference can be taken according to a FIE [7–9].

If FRB contains over one rule, 2 approaches can be employed to infer from FRB: (a) *composition*-based inference (CBI) and (b) *individual-rule*-based inference (IRBI) [10–12]. CBI is out of the paper discussion. IRBI contains 3 parts: (b.I) determining the

© Springer Nature Switzerland AG 2018
I. Batyrshin et al. (Eds.): MICAI 2018, LNAI 11288, pp. 413–422, 2018.
https://doi.org/10.1007/978-3-030-04491-6_32

amounts which the input matches in antecedents of all rules, (b.II) obtaining the consequences of all rules, and (b.III) aggregating the obtained consequences into a FS control action [13, 14]. The step b.I can be taken by "Max Min" (MM) method [10, 14, 15]. MM method will be explained in next section. Its drawbacks are also presented in that section.

In the section three, an alternative to MM method has been introduced. This method is inspired by the approach proposed in [23, 24]. The method proposed there was "Surface Matching Degree" (SMD). The proposed method is an extension of SMD and is named "Consolidated SMD" (CSMD). CSMD-based IRBI is a general FIE suitable for and applicable to all fuzzy systems specially to oscillatory ones. Finally, the paper is concluded in Sect. 4.

2 Traditional Fuzzy Inference

Let's assume $A_1(u)$ is the sole antecedent in one of fuzzy rules in FRB. Also assume we have an input as $A_1'(u) = \{(u_0, 1)\}$, i.e. the input is crisp. The matching degree (α) in MM method can be computed according to following equation.

$$\alpha = \min\left(1, \mu_{A_1}(u_0)\right) = \mu_{A_1}(u_0) \tag{1}$$

If $A_1'(u)$ is not a crisp input, the matching degree in MM method may be reformulated as follows [16–18]:

$$\alpha = \max_{u'} \min_{u'} \left(\mu_{A_1}(u'), \mu_{A_1'}(u')\right) \tag{2}$$

Suppose that the fuzzy term $(A_1(u) = max\{min\{0.25 \times u - 2.5, 4.5 - 0.25 \times u\}, 0\})$ is the antecedent of the rule, and the five different inputs are given. The inputs are as follows: $A_1^0(u) = \{(16, 1)\}$, $A_1^1(u) = max\{min\{u - 15.5, 17.5 - u\}, 0\}$, $A_1^2(u) = max\{min\{0.5 \times u - 7.5, 9.5 - 0.5 \times u\}, 0\}$, $A_1^3(u) = max\{min\{0.25 \times u - 3.5, 5.5 - 0.25 \times u\}, 0\}$, and $A_1^4(u) = max\{min\{\frac{u-10}{12}, \frac{34-u}{12}\}, 0\}$. The drawbacks of MM approach are as follows:

The point c, where the matching degree is computed by MM method, is located in $u = 16$ for all inputs. While the input membership functions are different, the point α is fixed for all of them. The membership value for the crisp input, i.e. $A_1^0(u)$ when $u = 16$, is equal to 1 and for fuzzy input terms 1 to 4, $A_1^1(u)$, $A_1^2(u)$, $A_1^3(u)$ and $A_1^4(u)$ when $u = 16$, is equal to 0.5. It has been observed that the matching degree is fixed by changing in input membership values. It is an unexpected phenomenon.

Suppose that the fuzzy input term is an isosceles. The center of this term is its center of gravity. As it is obvious, the inputs $A_1^3(u)$ and $A_1^4(u)$ cut the fuzzy term $A_1(u)$, in the fixed point α. The input $A_1^4(u)$ is wider than $A_1^3(u)$ and therefore, it is fuzzier. Also, its center of gravity is more far away than $A_1^3(u)$. In the same situation, the fuzziness of the input increases, when the gravity center of input gets away. It appears to be logical that, the value of matching degree decreases, whatever the input term be more far from than the point c. However, the input $A_1^4(u)$ is wider and fuzzier than $A_1^3(u)$, the matching degrees are the same. It is equal to 0.5 for all of those inputs. In reality, in this approach, the distance between center of input and point α, does not influence in the determination of matching degree. It is the second challenge of common approach to determine the matching degree.

The highlighted area in Fig. 1, named θ, is the intersection area between the input $A_1'(u)$ and the fuzzy term $A_1(u)$. This area θ is a section of input that is true part of the rule antecedent $A_1(u)$. The ratio of θ area to the total input area $A_1(u)$ is called R. R shows how much from the input is true in the rule antecedent. In the same situation, it can be meaningful that the greater value of R leads to greater matching degree. However, R can be negligible; the matching degree obtained by MM method is significant. It does not take into account the value of input that is true in the rule antecedent. It was the third shortcoming of above mentioned approach. However, this ratio for the input $A_1^4(u)$ is only 0.17 and for input $A_1^1(u)$ is 0.58, the matching degree is 0.5 for those inputs by MM method. Thus, the MM approach does not consider R.

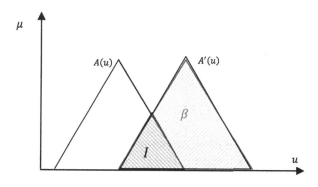

Fig. 1. The ratio of areas (α)

In this section, the weak points of the traditional MM method (or Mamdani Inference Method (MIM)) for measuring the matching degree were mentioned. It has been proven that the aforementioned MM method operate illogically on some cases.

3 CSMD: Consolidated Surface Matching Degree

The underlying idea of the proposed method is to use the intersection area between fuzzy input term and the antecedent of the rule. This idea is consolidated form of SMD [23, 24]. The goal of CSMD is to propose FIE which solves the weak points of MM method. We use R, as another parameter in determination of matching degree. Thus, the matching degrees will be changed, dynamically. It depends on input shape, amount of fuzziness of input term and the value of R.

Depending on β and I being shown in Fig. 1, the ratio of areas (α) is defined according to Eq. 3:

$$\alpha = \frac{\beta}{I} \tag{3}$$

where I is the intersection area between $A_1(u)$ and $A_1'(u)$ and β is the total input area $A_1'(u)$. The α will be faded, if fuzzy inputs is wider and fuzzier. Therefore, it can be as a good parameter to model the fuzziness of input and the distance from point α. This ratio, R, from input fuzzy terms 4 to 1 and in limit state for crisp input 0 are respectively equal to 0.17, 0.25, 0.37, 0.58, and 0.75. It can be observed that this parameter makes different values for different fuzzy inputs with the same fixed point α.

In other words, by getting the center of fuzzy input away from α, the changes of R, tend to fade. In fact, this is what the Eq. 3 is lacked. A very good feature of this parameter is adaptation. It is able to adapt itself to input transformation. By using this parameter with its adaptability feature in definition of matching degree, we will have an adaptive matching degree. Also, we can use R as another parameter beside with that parameter α (derived from Eq. 3). This combination can yield a dynamic and adaptive result for determination of matching degree. It will have the advantages of both Max-Min and the proposed approaches, simultaneously.

Our proposition to define CSMD is to combine the parameters, normalized R and α, as Eq. 5:

$$CSMD(R, \alpha) = \frac{w_1 f(R_n) + w_2 g(\alpha)}{w_1 + w_2} \tag{4}$$

Where R_n is the normalized ratio of surfaces, α is the value that comes from Eq. 3. f and g are two functions in range [0, 1] that can be chosen depending on specific problem. Also, w_1 and w_2 are two weights in [0, 1]. By defining the Eq. 5, we can profit from both approaches by weights w_1 and w_2.

3.1   Computation of R

In this section, the ratio R is computed in three states between isosceles fuzzy terms. These three possible states between fuzzy input term and rule antecedent are shown in Fig. 2.

State 1 shows that the input A' and rule antecedent A cut each other at just one point. The junction point in this state is called μ. State 2 shows that the input A' and rule antecedent A cut each other at two points, when A' is smaller than A. State 3 shows that the input A' and rule antecedent (A) cut each other at two points, when A' be larger than A. In two recent states, the larger and smaller points are called μ and $\mu2$, respectively.

As shown in Fig. 3, suppose that the half bases of terms A' and A are respectively I and r. Then to compute the ratio of areas in state 1, R1, we have:

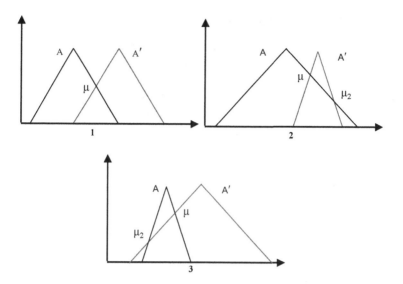

Fig. 2. Three possible states between fuzzy input term and rule antecedent.

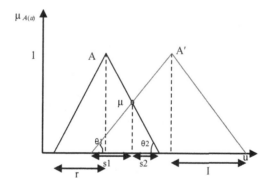

Fig. 3. Computation of R_n

$$\cot\theta1 = \frac{I}{1} = \frac{S1}{\mu} \Rightarrow S1 = I\mu$$

$$\cot\theta2 = \frac{r}{1} = \frac{S2}{\mu} \Rightarrow S2 = r\mu \tag{5}$$

Then:

$$S1 + S2 = (I + r)\mu \tag{6}$$

By substituting Eq. 6 in Eq. 3, we have:

$$R_1 = \frac{A_1}{A_2} = \frac{0.5 * (S1 + S2)\mu}{0.5 * 2 * I} = \frac{(r+I)\mu * \mu}{2 * I} \tag{7}$$

The Eq. 8 will be obtained by summarizing of Eq. 7:

$$R_1 = \frac{(r+I)\mu^2}{2I} \tag{8}$$

Similarly, it has been proved that R2 and R3 for states 2 and 3 obtained from the Eqs. 9 and 10 [22]:

$$R_2 = \mu + \frac{(r-I)(\mu - \mu_2)\mu_2}{2I} \tag{9}$$

$$R_3 = \frac{r}{I}\mu - \frac{(r-I)(\mu - \mu_2)\mu_2}{2I} \tag{10}$$

3.2   Normalization of R by Fixed μ

It is apparent from Eq. 1, that the matching degree for the crisp input is the membership value of antecedent of rule at junction point. Since this value is determined by human expert [19–21], the value of R must be normalized such it is μ for crisp input. Therefore, when a fuzzy input tends to crisp input (crispness), the value of R has to tend to μ.

As shown in Fig. 4, if we go from input 2 toward narrower input 1, the R1 will be replaced with R2. With a fixed point α (here, as μ), whatever we go from input 1

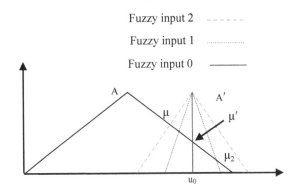

Fig. 4. Computation of μ′

toward 0 ($I \to 0$), the ratio of R goes from R2 to its. In the other hand, whatever the fuzziness of input decreases, R goes to its maximum limit value.

It has been shown that this value occurs in state 2 (R2) and its value trends to $2\mu' - \mu'2$ [22]. So we have:

$$\lim_{I \to 0} R = 2\mu' - \mu'^2 \tag{11}$$

Since this value is obtained from limit state of fuzzy input, it must be such normalized that it becomes equal to μ.

$$R = 2\mu' - \mu'^2 \Rightarrow \mu' = \frac{R}{2 - \mu'} \tag{12}$$

So, the normalized ratio of areas (Rn) has been defined as:

$$R_n = \frac{R}{2 - \mu'} \tag{13}$$

To change the equations in terms of μ, we have to derive μ from μ'. It has been shown that μ' will obtain from the equation below:

$$\mu' = \mu - \frac{I(1 - \mu)}{r} \tag{14}$$

By replacing this equation in the Eq. 13 we will have:

$$R_n = \frac{R}{2 - \mu'} = \frac{rR}{2r - r\mu + I(1 - \mu)} \tag{15}$$

This equation is a normalized equation that returns the true value for crisp input. By applying this equation in Eq. 5, we can reach to a parametric and adaptive one. We can select the functions f and g, depending on problem.

3.3   Choice of Parameters

By choosing the functions f and g as identical functions and $w1 = w2 = 0.5$ from Eq. 5, we will have:

$$CSMD(R_n, \alpha) = \frac{R_n + \mu}{2} \tag{16}$$

Replacing Rn as Eq. 15 in Eq. 16 CSMD will be:

$$CSMD(R_n, \alpha) = \frac{rR + \mu^2(r - I) + \mu(2r + I)}{2(2r - r\mu + I(1 - \mu))} \tag{17}$$

Equation 17 performs an averaging between Max-Min and CSMD methods. That means to benefit from both common and proposed methods simultaneously, we can average over those.

3.4   Example and Comparison

By presenting two examples and tables in this section, we try to make a comparison between the common approach (α) and CSMD.

Figures 5 and 6 show that whatever the input fuzzy term be wider and its center of gravity be more than far away, CSMD will decrease.

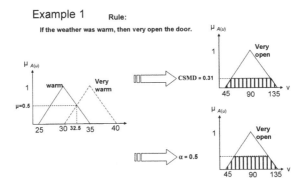

Fig. 5. Comparison between α and CSMD for fuzzy input "very warm" by R = 0.25, μ = 0.5.

In Table 1, CSMD is computed for input terms of Fig. 1. Table 1 shows that, α is 0.5 for all of those inputs, however CSMD is adaptively different, depending on input shape.

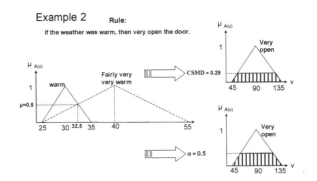

Fig. 6. Comparison between α and CSMD for fuzzy input "fairly very very warm" by R = 0.17, μ = 0.5.

Table 1. A comparison between α and CSMD for fuzzy input terms in Fig. 1.

	Input 4	Input 3	Input 2	Input 1	Crisp input 0
α	0.5	0.5	0.5	0.5	0.5
CSMD	**0.29**	**0.31**	**0.36**	**0.43**	**0.5**

Table 2, shows the values of matching degree obtained from common and CSMD method, in limit states.

Table 2. A comparison between α and CSMD for fuzzy input terms in Fig. 1.

	The limit value for crisp input	The limit value for fuzzy input by infinite amount of fuzziness
Common approach (α)	$\alpha = \mu'$	$\alpha = \mu'$
CSMD	$CSMD = \mu'$	$CSMD = \frac{\mu}{2}$

Two points in Tables 1 and 2 is considerable: first that however, the values obtained from common approach are fixed for every state, a different value is obtained from CSMD for each input. Second, the CSMD values with any parameter choices are trended to μ′, in limit state.

4 Conclusion and Future Work

In this paper, three shortcomings of common approach "Max-Min" for determining of matching degree is investigated, and an adaptive method called "Centralized Surface Matching Degree (CSMD)" is proposed. We can change and adapt the parameters of CSMD depends on the kind of inference engine and its application. It can help the inference engine in different applications. However, the values obtained from Max-Min method are fixed for every state of inputs that have a same μ; different values are obtained from CSMD, adaptively. It depends on input shape and the amount of input term that is true of the rule antecedent.

In addition, CSMD has the following benefits: first it seems that using CSMD can be more logical than common approach. Second, emphasizing to distance can influence in system stability. It can cause to faster convergence by decreasing fuzziness in feedback systems.

References

1. Zadeh, L.A.: Outline of a new approach to the analysis of complex systems and decision processes. IEEE Trans. Syst. Man Cybern. **SMC-3**, 28–44 (1973)
2. Zadeh, L.A., Fu, K.S., Tanaka, K., Shimura, M. (eds.): Calculus of fuzzy restrictions, Journal, Fuzzy Sets and Their Applications to Cognitive and Decision Processes. Academic, New York (1975)

3. Lee, C.C.: Fuzzy logic in control systems: fuzzy logic controller-part I. IEEE Trans. Syst. Man Cybern. **20**(2), 404–418 (1990)
4. Wang, L.-X.: A Course in Fuzzy Systems and Control. Prentice-Hall International Inc, Upper Saddle River (1997)
5. Dubois, D., Prade, H.: Fuzzy logics and the generalized modus ponens revisited. Cybern. Syst. **15**, 3–4 (1984)
6. Gupta, M.M., Kandel, A., Bandler, W., Kiszka, J.B.: The generalized modus ponens under sup-min composition_a theoretical study. In: Approximate Reasoning in Expert Systems, Amsterdam, North-Holland, pp. 217–232 (1985)
7. Fukami, S., Mizumoto, M., Tanaka, K.: Some considerations of fuzzy conditional inference. Fuzzy Sets Syst. **4**, 243–273 (1980)
8. Baldwin, J., Guild, N.: Modeling controllers using fuzzy relations. Kybernetes **9**, 223–229 (1980)
9. Baldwin, J.F., Pilsworth, B.W.: Axiomatic approach to implication for approximate reasoning with fuzzy logic. Fuzzy Sets Syst. **3**, 193–219 (1980)
10. Mamdani, E.H., Assilian, S.: An experiment in linguistic synthesis with a fuzzy logic controller. Int. J. Man-Mach. Stud. **7**, 1–13 (1974)
11. Sugeno, M.: An introductory survey of fuzzy control. Inform. Sci. **36**, 59–83 (1985)
12. Lee, C.C.: Fuzzy logic in control systems: fuzzy logic controller, part II. IEEE Trans. Syst. Man Cybern. **20**(2), 404–418 (1990)
13. Zimmermann, H.J.: Fuzzy Set Theory and Its Applications, 3rd edn. Kluwer Academic Publishers, New York (1996)
14. Mamdani, E.H.: Application of fuzzy logic to approximate reasoning using linguistic synthesis. IEEE Trans. Comput. **26**, 1182–1191 (1977)
15. Zadeh, L.A., Hayes, J.E., Michie, D., Kulich, L.I. (eds.): A theory of approximate reasoning. In: Machine Intelligence, New York, vol. 9, pp. 149–194 (1979)
16. Tsukamoto, Y., Gupta, W., (eds.) An approach to fuzzy reasoning method. Adv. Fuzzy Set Theor. Appl. 137–149 (1979). North-Holland, Amsterdam
17. Sugeno, M., Takagi, T.: Multidimensional fuzzy reasoning. Fuzzy Sets Syst. **9**, 313–325 (1983)
18. Wangming, W.: Equivalence of some methods on fuzzy reasoning. IEEE (1990)
19. Mizumotom, M., Zimmermann, H.: Comparison of fuzzy reasoning methods. Fuzzy Sets Syst. **8**, 253–283 (1982)
20. Mizumoto, M.: Comparison of various fuzzy reasoning methods. In: Proceedings 2nd IFSA Congress, Tokyo, Japan, pp. 2–7, July 1987
21. Mamdani, E.H.: Advances in the linguistic synthesis of fuzzy controllers. Int. J. Man-Mach. Stud. **8**, 669–678 (1976)
22. Alizadeh, H.: Adaptive matching degree, Technical report of Fuzzy Course, Iran University of Science and Technology (2007). (in Persian)
23. Alizadeh, H., Mozayani, N.: A new approach for determination of matching degree in fuzzy inference. In: Proceedings of the 3rd International Conference on Information and Knowledge Technology (IKT07), Faculty of Engineering, Ferdowsi University of Mashad, Mashad, Iran, 27–29 November 2007. (in Persian)
24. Alizadeh, H., Mozayani, N., Minaei, B.B.: Adaptive matching for improvement of fuzzy inference engine. In: Proceedings of the 13th National CSI Computer Conference (CSICC08), Kish Island, Persian Gulf, Iran, 9–11 March 2008. (in Persian)

A General Method for Consistency Improving in Decision-Making Under Uncertainty

Virgilio López-Morales$^{(\boxtimes)}$, Joel Suárez-Cansino, Ruslan Gabbasov, and Anilu Franco Arcega

Centro de Investigación en Tecnologías de Información y Sistemas, Univ. Aut. del Edo. de Hidalgo, Carr. Pachuca - Tulancingo km. 4.5, 42184 Pachuca, Mexico
{virgilio,jsuarez,ruslan-gabbasov,afranco}@uaeh.edu.mx

Abstract. In order to ascertain and solve a particular Multiple Criteria Decision Making (MCDM) problem, frequently a diverse group of experts must share their knowledge and expertise, and thus uncertainty arises from several sources. In those cases, the Multiplicative Preference Relation (MPR) approach can be a useful technique. An MPR is composed of judgements between any two criteria components which are declared within a crisp rank and to express decision maker(s) (DM) preferences. Consistency of an MPR is obtained when each expert has her/his information and, consequently, her/his judgments free of contradictions. Since inconsistencies may lead to incoherent results, individual Consistency should be sought after in order to make rational choices. In this paper, based on the Hadamard's dissimilarity operator, a methodology to derive intervals for MPRs satisfying a consistency index is introduced. Our method is proposed through a combination of a numerical and a nonlinear optimization algorithms. As soon as the synthesis of an interval MPR is achieved, the DM can use these acceptably consistent intervals to express flexibility in the manner of her/his preferences, while accomplishing some a priori decision targets, rules and advice given by her/his current framework. Thus, the proposed methodology provides reliable and acceptably consistent Interval MPR, which can be quantified in terms of Row Geometric Mean Method (RGMM) or the Eigenvalue Method (EM). Finally, some examples are solved through the proposed method in order to illustrate our results and compare them with other methodologies.

Keywords: Decision-making support systems
Multiple criteria decision-making · Analytic hierarchy process
Consistency · Multiplicative preference relations
Uncertain decision-making

© Springer Nature Switzerland AG 2018
I. Batyrshin et al. (Eds.): MICAI 2018, LNAI 11288, pp. 423–434, 2018.
https://doi.org/10.1007/978-3-030-04491-6_33

1   Introduction

Group Decision Making (GDM) is a process where multiple decision makers (or experts) act collectively, analyze problems, evaluate options according to a set of criteria, and select a solution from a collection of alternatives [3]. The importance of solving these topics is paramount as applications can be found in management science, operational research, industrial and chemical engineering, among others Cf. [4,8–11,18].

As noted in [7], when organizations gather specialized groups o larger groups, and the number of alternatives increases, unanimity may be difficult to attain, particularly in diversified groups. For this reason flexible or milder benchmarks (definitions) of consensus and consistency have been employed. Consensus has to do with group cooperation since the alternative, option, or course of action to be attained is the best representative for the entire group. On the other hand, consistency is obtained when each expert has her/his information and, consequently, her/his judgments free of contradictions. Since inconsistencies may lead to incoherent results, Individual Consistency should be sought after in order to make rational choices [5,15], and is then related with the management of human subjectivity, imprecision, hesitation or uncertainty along the decision-making process.

For instance, in the AHP method, an MPR or a pairwise comparison matrix is composed of judgements between any two criteria components which are declared within a crisp rank, called Saaty's Scale ($SS \in [\frac{1}{9}\ 9]$). An MPR is also called a subjective judgement matrix and is adopted to express decision maker(s) (DM) preferences.

Thus, AHP method is a very common method for multi-criteria decision making and still remains an easy and a reliable approach, as many real applications have demonstrated. Nevertheless, dealing with expert's uncertainties is still an open problem.

In this paper these problems are addressed and solved through the Hadamard product to measure the dissimilarity of two matrices and an algorithm based on a Non Linear Optimization Approach (NLOA). The main contribution is a couple of algorithms for giving reliable intervals for a group of DM whom have proposed a set of MPRs possibly inconsistent. After some iterations, these algorithms return a set of reliable intervals MPRs which are now consistent within an arbitrary threshold. Thus, the set of DM can now confidently pick up an MPR or the entire Interval MPR from their corresponding reliable Interval MPR, in order to re-express her/his final judgment decision. The main goal is to develop a system that, once a set of DM has proposed a set of MPRs, then a set of reliable intervals[1] can be generated by the system. In this manner, DM could be more confident, by re-expressing their judgments within these reliable intervals; despite their inherent uncertainty and imprecision, due to incomplete information or the evolving problem complexity.

---

[1] Where individually consistency holds.

The paper is organized as follows: In Sect. 2, some preliminaries are given to support a basis for the main methodologies and techniques previously described above. In Sect. 3, the methodology applicable for obtaining reliable intervals given by MPRs is introduced, and the main role in the GDM framework is enlightened. Then in Sect. 4, the GDM implementation is described in detail via some numerical examples. Finally, in Sect. 5 some concluding remarks and discussions, about the main advantages of the methodology and future research recommendations are provided.

2 Preliminaries

In the following, some necessary concepts and properties are introduced to support our contribution. Further details can be found in [6,12,17].

Consider a GDM problem and let $D = \{d_1, d_2, \cdots, d_m\}$ be the set of DM, and $C = \{c_1, c_2, \cdots, c_n\}$ be a finite set of alternatives, where c_i denotes the ith alternative. With an MPR, a DM provides judgments for every pair of alternatives which reflect her/his degree of preference of the first alternative over the second. Thus, an MPR, for instance $A = (a_{ij})_{n \times n}$ is a positive reciprocal $n \times n$ matrix, $a_{ij} > 0$, such that $a_{ji} = 1/a_{ij}, \forall i, j \in N$, and consequently $a_{ii} = 1$; where $i \in N$. Note that a_{ij} belongs precisely to the Saaty's scale and is interpreted as the ratio of the preference intensity of alternative c_i to that of c_j.

An MPR $n \times n$ matrix is called a completely consistent matrix (cf. [12]) if

$$a_{ij} = a_{il}a_{lj}, \ \forall \ i, j, l \in N. \tag{1}$$

Thus, a completely consistent matrix $K = (k_{ij})_{n \times n}$ can be constructed from (1) as follows,

$$k_{ij} = \prod_{r=1}^{n} (a_{ir}a_{rj})^{1/n}. \tag{2}$$

Some Lemmata are given below in order to obtain our main results.

Lemma 1. *Suppose $a > 0$, $\epsilon > 1$, then*

$$1.0 < a + \frac{1}{a} \leq a^\epsilon + \left(\frac{1}{a}\right)^\epsilon \tag{3}$$

where equality holds if and only if $a = 1$.

Proof: It is straightforward since the addition of inequalities and reciprocal values (Cf. [2], pp. 29–31).

Lemma 2. *Suppose $a > 0$, $\epsilon < 1$, then*

$$1.0 < a^\epsilon + \left(\frac{1}{a}\right)^\epsilon \leq a + \frac{1}{a} \tag{4}$$

where equality holds if and only if $a = 1$.

426 V. López-Morales et al.

Proof: It is straightforward. Idem.

Lemma 3. *Let us suppose* $a > 1$, $\bar{\epsilon} \leq 1$, $\overset{+}{\epsilon} \geq 1$, *then, without loss of generality, any given interval^2 in* $\left[\frac{1}{9} \quad 9\right]$ *can be obtained by*

$$\begin{bmatrix} \bar{a} & \overset{+}{a} \end{bmatrix} = \begin{bmatrix} a^{\bar{\epsilon}} & a^{\overset{+}{\epsilon}} \end{bmatrix}$$

where $\bar{a} \leq \overset{+}{a}$ *and equality holds if and only if* $\bar{\epsilon} = \overset{+}{\epsilon} = 1$.

Proof: It is straightforward (Cf. [2], pp. 657–658).

2.1 Measuring the Consistency of an MPR

The Hadamard product is a useful operator to measure the degree of deviation between two MPRs, where given $A = (a_{ij})_{n \times n}$ and $B = (b_{ij})_{n \times n}$ is defined by

$$C = (c_{ij})_{n \times n} = A \circ B = a_{ij}b_{ij}. \tag{5}$$

The degree of dissimilarity of A and B is given by $d(A, B) = \frac{1}{n^2}e^T A \circ B^T e$. I.e.,

$$d(A, B) = \frac{1}{n^2}\Sigma_{i=1}^n \Sigma_{j=1}^n a_{ij}b_{ji} = \frac{1}{n}\left[\frac{1}{n}\Sigma_{i=1}^{n-1}\Sigma_{j=i+1}^n (a_{ij}b_{ji} + a_{ji}b_{ij}) + 1\right]. \tag{6}$$

where $e = (1, 1, \cdots, 1)_{n \times 1}^T$.

Note that $d(A, B) \geq 1$ where $d(A, B) = 1$ if and only if $A = B$ and $d(A, B) = d(B, A)$.

Thus, through the Hadamard product, the consistency index of A is defined as $CI_K(A) = d(A, K)$, where K is the corresponding completely consistent matrix obtained from A.

Then if

$$d(A, K) = CI_K(A) \leq \overline{CI}, \tag{7}$$

where $\overline{CI}$ is an acceptable threshold value, then we call matrix A as an MPR with an acceptable consistency.

An MPR A is *completely* consistent if and only if $CI_K(A) = 1$. Thus, a threshold used to test the compatibility of two MPRs up to an acceptable level of consistency, was suggested in [13,16] as $\overline{CI} = 1.1$.

On the other hand, trough the EM, the consistency ratio (CR) was defined in [12] as $CR = CI/RI$, where the Random Index (RI) is the average value of the CI for random matrices using the Saaty's Scale. Moreover, an MPR is only accepted as a consistent matrix if and only if its $\overline{CR} \leq 0.1$.

When it comes to measuring the degree of dissimilarity between two matrices, both methods yield the same results. Nevertheless, as previously stated in [13], the Hadamard product is more reliable to measure the dissimilarity of two matrices constructed by ratio scales.

---

^{2} For our practical purposes.

2.2 Prioritization Method

The process of deriving a priority vector $w = (w_1, w_2, \cdots, w_n)^T$ from an MPR, is called a prioritization method, where $w_l \geq 0$ and $\sum_{l=1}^{n} w_l = 1$. Two prioritization methods are commonly used:

(1) The eigenvalue method (EM), (proposed by [12] and [13]), where the principal right eigenvector of A, called λ_{max}, is the desired priority vector w, which can be obtained by solving

$$Aw = \lambda w, \quad e^T w = 1. \tag{8}$$

(2) Row geometric mean method (RGMM) or logarithmic least square method: The RGMM uses the L^2 metric by defining an objective function of the following optimization problem:

$$\begin{cases} \min \sum_{i=1}^{n} \sum_{j>i} [\ln(a_{ij}) - (\ln(w_i) - \ln(w_j))]^2 \\ s.t. \ w_i \geq 0, \ \sum_{i=1}^{n} w_i = 1 \end{cases}, \tag{9}$$

where, a unique solution exists with the geometric means of the rows of matrix A:

$$w_i = \frac{(\prod_{j=1}^{n} a_{ij})^{1/n}}{\sum_{i=1}^{n} (\prod_{j=1}^{n} a_{ij})^{1/n}}. \tag{10}$$

As both methods (EM and RGMM) generate similar results (Cf. [14]) and since the group of DM is assumed acting together as a unit, AIJ and RGMM become appropriate methods to give reliable intervals in assessment of consistency model in GDM. Thus, AIJ and RGMM methods are used in the remainder of this paper.

2.3 Individual Consistency Improving Algorithm

Algorithm 1
   Input: The individual multiplicative preference relations $A_l = (a_{ij})_{n \times n}$, $l = 1, 2, \cdots, m$, the Consistency parameter $\theta \in (0, 1)$, the maximum number of iterative times $h_{\max} \geq 1$, the Consistency threshold $\overline{CI}^3$.
   Output: the adjusted multiplicative preference relation $\overline{A_l}$ and the Consistency index $CI_H(\overline{A_l})$.
   Step 1. Set $A_{l,0} = (a_{ij,0}^l)_{n \times n} = A_l = (a_{ij}^l)_{n \times n}$ and $h = 0$.
   Step 2. Compute K_l by (2) and the Consistency index $CI_H(A_l)$, where

$$CI_H = d(A_l, K_l). \tag{11}$$

   Step 3. If $CI_H(A_l) \leq \overline{CI}$, then go to Step 5; otherwise, go to the next step.
   Step 4. Apply the following strategy to update the last matrix $A_l = (a_{ij,l})_{n \times n}$.

---

[3] Usually $\overline{CI} = 1.1$ however it can be selected by the project designer.

$$A_{l+1} = (A_l)^\theta \circ (K_l)^{1-\theta}. \tag{12}$$

where $\theta \in (0,1)$. Let $l = l + 1$, and return to **Step 2**.

Step 5. Let $\overline{A} = A_l$. Output $\overline{A}_l$ and $CI_H(\overline{A}_l)$.

Step 6. End.

Based on the Algorithm 1, Eq. (2) and Definition given by Eq. (7):

Theorem 1. *For each iteration r, the consistency of the MPR under analysis is improved.*

I.e., $CI_K^{r+1}(\overset{*}{A}_{r+1}) < CI_K^r(\overset{*}{A}_r)$ and $\lim_{r \to \infty} CI_K^r(\overset{*}{A}_r) \le \beta, \forall \ \beta > 1$.

Proof: From step 4 of the Algorithm 1 and Eq. (2), it follows:

$$a_{ij,r+1} = (a_{ij,r})^\theta \cdot (k_{ij,r})^{1-\theta},$$
$$k_{ij,r+1} = \prod_{l=1}^n (a_{il,r+1} \cdot a_{lj,r+1})^{\frac{1}{n}} = \prod_{l=1}^n \left[(a_{il,r} \cdot a_{lj,r})^\theta \cdot (k_{il,r} \cdot k_{lj,r})^{1-\theta} \right]^{\frac{1}{n}}$$
$$= k_{ij,r}. \tag{13}$$

Then,

$$a_{ij,r+1} k_{ji,r+1} = (a_{ij,r})^\theta \cdot (k_{ij,r})^{1-\theta} k_{ji,r} = (a_{ij,r})^\theta \cdot (k_{ji,r})^{\theta-1} k_{ji,r} = (a_{ij,r} k_{ji,r})^\theta. \tag{14}$$

By using Lemma 2, one obtains:

$$a_{ij,r+1} k_{ji,r+1} + a_{ji,r+1} k_{ij,r+1} = (a_{ij,r} k_{ji,r})^\theta + (a_{ji,r} k_{ij,r})^\theta \le a_{ij,r} k_{ji,r} + a_{ji,r} k_{ij,r}. \tag{15}$$

Let a pair (i, j) be chosen such that the inequality strictly holds. I.e., since $A^{(r)} \ne K^{(r)}$, then there exists at least one pair such that:

$$\frac{\frac{1}{n^2} \sum_{i=1}^{n-1} \sum_{j=i+1}^n (a_{ij,r+1} k_{ji,r+1} + a_{ji,r+1} k_{ij,r+1}) + \frac{1}{n}}{\frac{1}{n^2} \sum_{i=1}^{n-1} \sum_{j=i+1}^n (a_{ij,r} k_{ji,r} + a_{ji,r} k_{ij,r}) + \frac{1}{n}} < 1. \tag{16}$$

It implies, $CI_K(\overset{*}{A}_{r+1}) < CI_K(\overset{*}{A}_r)$. Thus, we have $CI_K(\overset{*}{A}_r) \ge 1, \quad \forall r$, and consequently, the sequence $\{CI_K(\overset{*}{A}_r)\}$ is monotone decreasing and has a lower bound. Apply the limit existence theorem for a sequence where $\lim_{r \to \infty} CI_K(\overset{*}{A}_r)$ exists, thus $\overset{*}{A}_\infty = \lim_{r \to \infty} (\overset{*}{A}_r)$. Assume that $CI_K(\overset{*}{A}_\infty) = \lim_{r \to \infty} CI_K(\overset{*}{A}_r) = \inf\{CI_K(\overset{*}{A}_r)\}$.

By contradiction, it can be shown that the $\lim_{r \to \infty} CI_K(\overset{*}{A}_r) \le \beta$.

Since $\overset{*}{A}_r$ is calculated iteratively in the Algorithm 1, $CI_K(\overset{*}{A}_r)$ decreases as the number of iterations increases. $\qquad \square$

3   Reliable Intervals Programming Method

In order to have an assessment of individual consistency (CI_H), one can measure the compatibility of A_l with respect to (w.r.t.) its own completely consistent matrix K given by Eq. (2). Thus,

$$CI_H(A_l) = d(A_l, K) \leq \overline{CI}, \tag{17}$$

where $\overline{CI} = 1.1$, A_l, $l = 1, 2, \cdots, m$ is an individual MPR.
    From Eqs. (17), (2) and (6), for CI_H it follows $d(A, K) \leq \overline{CI} \Rightarrow$

$$1.0 \leq \tfrac{1}{n} \left[\tfrac{1}{n} \sum_{i=1}^{n-1} \sum_{j=i+1}^{n} \left(a_{ij} \prod_{r=1}^{n} (a_{jr}a_{ri})^{1/n} + a_{ji} \prod_{r=1}^{n} (a_{ir}a_{rj})^{1/n} \right) + 1 \right] \leq \overline{CI}. \tag{18}$$

3.1   Reliable Intervals for Individual Consistency

By applying the Algorithm 1 to a matrix A_l, and when the modifier parameter θ is close to 1, i.e. $0 << \theta < 1$, a new matrix defined as A_{max}^O is obtained. This matrix has slight modifications[4] from the original A_l under analysis. Naturally, A_{max}^O verifies individual Consistency by Eq. (17). On the other hand, when θ is close to zero, i.e as $0 < \theta << 1$, a new matrix, defined as A_{min}^O is obtained.
    Based on this fact, an interval matrix A_I^O can be defined as follows,

$$A_M^O \equiv \left[a_{ij,min}^O - a_{ij,max}^O \right]_{n \times n}, \quad i, j \in N, \ M = 1, 2, \cdots, m. \tag{19}$$

    Naturally, for each one of the A_M^O interval matrices synthesized from A_l, $l = 1, 2, \cdots, m$; they couldn't verify, for every combination of values chosen within the intervals, the inequality given in Eq. (18). Then a Nonlinear Minimization Approach (NLOA) can solve the inequalities from Eq. (18) for consistency evaluation, and the bounds imposed by Eq. (19).
    From Eq. (19), following inequalities can be stated in terms of optimization variables x_i, $i = 1, 2, \cdots, n$ as follows:

$$a_{12,min}^O \leq x_1 \leq a_{12,max}^O, \cdots, a_{(n-1)n,min}^O \leq x_{\frac{n^2-n}{2}} \leq a_{(n-1)n,max}^O, \tag{20}$$

and the inequalities given from Eq. (18) are $1.0 \leq \overline{CI} \leq 1.1$.
    Finally, the initialization point x_0 used in the NLOA can be set at any point within the corresponding interval given by Eq. (20).
    In the following, an algorithm based on a NLOA is used to obtain reliable intervals for assessment of based decision models such as those given by Algorithm 1 for index $\overline{CI}$.
    The following method obtains reliable and acceptably Consistent intervals matrices. Thus, we can find an interval matrix verifying individual consistency in the whole interval.

---

[4] Once the Algorithm 1 of the Sect. 2.3 has converged.

In order to use a NLOA, a Sequential Quadratic Programming (SQP) algorithm can be found in [1]. In the following, our algorithm is described in detail.

Algorithm 2

Input: $A_M^O = (a_{ij,min}^O - a_{ij,max}^O)_{n \times n}$, the initial interval matrix; x_0: the initial value for the nonlinear optimization; which is to be defined within the corresponding interval; $\overline{CI}$ for the individual Consistency assessment.

Output: $\overline{A}_M^O$: the consistency interval matrix computed and verifying interval conditions given by Eq. (18).

Step 1: Get the function for assessment of individual Consistency given by Eq. (18)

Step 2: Define for the nonlinear optimization algorithm

$$a_{ij,min}^O \le x_i \le a_{ij,max}^O; j > i, \ i,j = 1,2,\cdots,n. \tag{21}$$

Thus, assign the linear inequality constraints as follows:

$$c(2k+1) = x_i - a_{ij,max}^O; \ c(2(k+1)) = x_i - a_{ij,min}^O; \\ j > i, \ i,j = 1,2,\cdots,n, \ k = 0,1,2,\cdots, \frac{n^2-n}{2}. \tag{22}$$

Step 3: Obtain the acceptable index of individual consistency, $1.0 \le CI \le \overline{CI}$.

Thus, based on matrix K given by Eq. (2), the nonlinear inequality constraints imposed is given by (18).

Step 4: Solve the former nonlinear optimization problem using an algorithm to minimize it and obtain the matrix $\overline{A}_{Mmin}^O = (a_{Mij,min}^O)_{n \times n}$. Solve again the same nonlinear optimization problem but this time in order to maximize it. Obtain $\overline{A}_{Mmax}^O = (a_{Mij,max}^O)_{n \times n}$.

Step 5: Compose the Consistency Interval Matrix $\overline{A}_M^O$ as follows:

$$\overline{A}_M^O = (\overline{a}_{Mij,min}^O - \overline{a}_{Mij,max}^O)_{n \times n} \tag{23}$$

where $(\overline{a}_{Mij,min}^O - \overline{a}_{Mij,max}^O)$, stands for the interval obtained.

Step 6: end.

A scheme of the Algorithm implementation is depicted in Fig. 1.

4   The Complete Process of Improving Consistency for an MPR

In Fig. 1 is shown a complete support model for a Decision Making problem based on Reliable Intervals.

4.1   Numerical Examples

In the following, Algorithms 1 and 2 are applied to numerical examples, built and used by several authors [6,17], and tested in order to illustrate their performance.

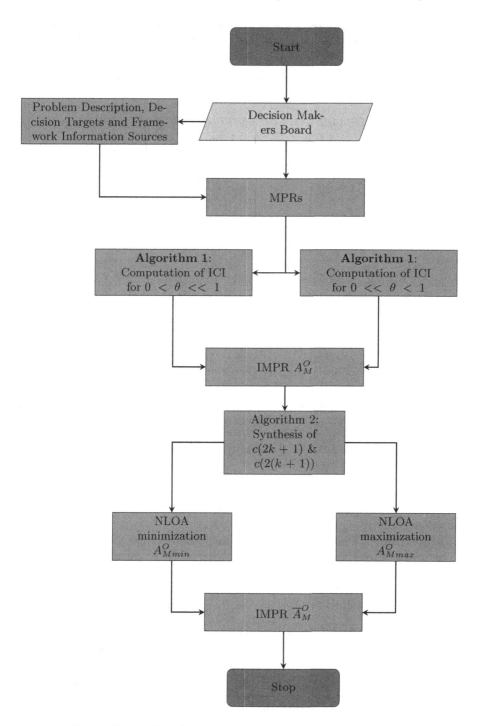

Fig. 1. Process flowchart for improving individual consistency.

Example 1: Let us suppose a set of five DM providing the following judgment matrices $\{A_1, \cdots, A_5\}$ on a set of four alternatives C_1, C_2, C_3, and C_4, which need to be ranked from best to the worst. Let $w^{(k)} = (w_1^{(k)}, \cdots, w_4^{(k)})^T$ be the individual priority vector derived from judgment matrix A_k using RGMM or the eigenvector method. A_k and $w^{(k)}$, $(k = 1, 2, \cdots, 5)$ are

$$
A_1 = \begin{pmatrix} 1\,4\,6\,7 \\ 1\,3\,4 \\ *\,1\,2 \\ *\,*\,1 \end{pmatrix}, \quad A_2 = \begin{pmatrix} 1\,5\,7\,9 \\ 1\,4\,6 \\ *\,1\,2 \\ *\,*\,1 \end{pmatrix}, \quad A_3 = \begin{pmatrix} 1\,3\,5\,8 \\ 1\,4\,5 \\ *\,1\,2 \\ *\,*\,1 \end{pmatrix},
$$

$$
A_4 = \begin{pmatrix} 1\,6\,7\,8 \\ 1\,5\,5 \\ *\,1\,4 \\ *\,*\,1 \end{pmatrix}, \quad A_5 = \begin{pmatrix} 1\,1/2\,1\,2 \\ 1\ \ 2\,3 \\ *\ \ 1\,4 \\ *\ \ *\,1 \end{pmatrix},
$$
(24)

$$
\begin{aligned}
w^{(1)} &= (0.6145, 0.2246, 0.0985, 0.0624), \quad w^{(2)} = (0.6461, 0.2270, 0.0793, 0.0476), \\
w^{(3)} &= (0.5393, 0.2764, 0.0967, 0.0575), \quad w^{(4)} = (0.6514, 0.2174, 0.0885, 0.0428), \\
w^{(5)} &= (0.2221, 0.4134, 0.2641, 0.1004).
\end{aligned}
$$
(25)

where $*$ stands for the corresponding inverted terms of symmetric entries.

Thus, the corresponding priority vector $w^c = (w_1^c, w_2^c, w_3^c, w_4^c, w_5^c)^T$, is calculated and listed below.

By applying the Algorithm 1 the Consistency assessment is addressed with $\overline{CI} = 1.1$ for each $A_i, i = 1, 2, \cdots, 5$, one obtains:

$$
\begin{aligned}
CI_H(A_1) &= 1.0255, CI_H(A_2) = 1.0450, CI_H(A_3) = 1.0226 \\
CI_H(A_4) &= 1.1194, CI_H(A_5) = 1.0241, CI_H(A^c) = 1.0324.
\end{aligned}
$$
(26)

MPRs A_1, A_2, A_3, A_5 are of acceptable consistency, however A_4 does not. Then, in the following we apply Algorithm 1 in order to obtain reliable intervals where consistency holds.

Algorithm 1: Processing Consistency in the intervals

Step 1: Take $0 < \theta << 1$, for instance $\theta = 0.01$ and apply Algorithm 1. Then, re-execute Algorithm 1 for $0 << \theta < 1$, for instance $\theta = 0.99$. After 1 step for $\theta = 0.01$ and 9 steps for $\theta = 0.99$, a Consistency Interval MPR, defined by Eq. (19) can be obtained as:

$$
A_4 = \begin{pmatrix} (1,1) & (3.0175, 5.6504) & (7.0307, 7.3598) & (8.4579, 15.1293) \\ & (1,1) & (2.4748, 4.7021) & (5.0070, 5.0805) \\ & * & (1,1) & (2.0816, 3.7782) \\ & * & * & (1,1) \end{pmatrix},
$$
(27)

Note: Eq. (27) is not reliable, since it does not verify conditions imposed on the Saaty's scale.

By applying the Algorithm 2, a reliable consistency interval $\overline{A}_4^O$ is obtained, and if the fourth DM decides, for example, that the midpoints are the better evaluation, a final MPR follows:

$$\overline{A}_4 = \begin{pmatrix} 1 & 4.33395 & 7.19525 & 8.72895 \\ & 1 & 3.58845 & 5.04375 \\ * & & 1 & 2.9299 \\ * & * & & 1 \end{pmatrix}, \tag{28}$$

The final ranking of the alternatives is $C_1 > C_2 > C_3 > C_4$, which coincides with [6,17] where MPR A_4 is slightly different but evaluated in a similar ratio. This result indicates that C_1 is the best option, nevertheless the strategy to pick up a suitable point in Consistent reliable intervals is revealed to be useful when, given a particular situation the DM has to observe some constraints imposed by the(ir) framework or express the(ir) uncertainties.

5   Concluding Remarks and Future Work

In order to provide a flexible tool for DM when they are required to derive a suitable Multiplicative Preference Relation, this paper demonstrates the utilization of a methodology to synthesize reliable intervals where consistency constraints hold. Once decision makers have proposed their MPRs, our algorithm can solve for intervals from well-known decision support models. One advantage of our algorithm is that DM can re-express their preferences within an interval where usually, they have to observe some constraints based on decision targets, framework rules and advice. Depending on the analyzed problem, a certain level of flexibility can be found.

Another advantage of our approach when DM pick up the complete Interval Multiplicative Preference Relation is to have a degree of certainty and at the same time they can observe the constraints imposed by their framework. In our approach, reliable interval MPRs provide a distinct advantage in interpretation of hesitancy and uncertainty about the final consistency.

Our approach is proposed based on some numerical algorithms where a nonlinear optimization algorithm is concurrently applied.

References

1. Beale, E.M.L.: Numerical methods. In: Abadie, J. (ed.) Nonlinear Programming. North-Holland, Amsterdam (1967)
2. Bronshtein, I.N., Semendyayev, K.A., Musiol, G., Mühlig, H.: Handbook of Mathematics. Springer, Heidelberg (2015). https://doi.org/10.1007/978-3-662-46221-8
3. Cabrerizo, F.J., Herrera-Viedma, E., Pedrycz, W.: A method based on PSO and granular computing of linguistic information to solve group decision making problems defined in heterogeneous contexts. Eur. J. Oper. Res. **230**(3), 624–633 (2013). https://doi.org/10.1016/j.ejor.2013.04.046

4. Campanella, G., Ribeiro, R.A.: A framework for dynamic multiple-criteria decision making. Decis. Support Syst. **52**, 52–60 (2011)
5. Chiclana, F., Mata, F., Martínez, L., Herrera-Viedma, E., Alonso, S.: Integration of a consistency control module within a consensus decision making model. Int. J. Uncertainty Fuzziness Knowl.-Based Syst. **16**(01), 35–53 (2008)
6. Dong, Y., Zhang, G., Hong, W.C., Xu, Y.: Consensus models for AHP group decision making under row geometric mean prioritization method. Decis. Support Syst. **49**(3), 281–289 (2010)
7. Herrera-Viedma, E., Cabrerizo, F.J., Kacprzyk, J., Pedrycz, W.: A review of soft consensus models in a fuzzy environment. Inf. Fusion **17**, 4–13 (2014). https://doi.org/10.1016/j.inffus.2013.04.002
8. Huang, C.C., Lin, S.H.: Sharing knowledge in a supply chain using the semantic web. Expert Syst. Appl. **37**(4), 3145–3161 (2010)
9. IEOM: International conference in Dhaka, Bangladesh. In: Proceedings of the 2010 International Conference on Industrial Engineering and Operations Management, pp. 48–83, January 2010
10. Ma, L., Li, H.: Using Gower plots and decision balls to rank alternatives involving inconsistent preferences. Decis. Support Syst. **51**, 712–719 (2011)
11. Ribeiro, R., Moreira, A., van den Broek, P., Pimentel, A.: Hybrid assessment method for software engineering decisions. Decis. Support Syst. **51**, 208–219 (2011)
12. Saaty, T.: The Analytic Hierarchy Process. McGraw-Hill, New York (1980)
13. Saaty, T.: A ratio scale metric and the compatibility of ratio scales: the possibility of arrow's impossibility theorem. Appl. Math. Lett. **7**(6), 45–49 (1994)
14. Srdjevic, B.: Linking analytic hierarchy process and social choice methods to support group decision-making in water management. Decis. Support Syst. **42**, 2261–2273 (2007)
15. Urena, R., Chiclana, F., Morente-Molinera, J.A., Herrera-Viedma, E.: Managing incomplete preference relations in decision making: a review and future trends. Inf. Sci. **302**, 14–32 (2015)
16. Wang, L.: Compatibility and group decision making. Syst. Eng. Theory Pract. **20**, 92–96 (2002)
17. Wu, Z., Xu, J.: A consistency and consensus based decision support model for group decision making with multiplicative preference relations. Decis. Support Syst. **52**(3), 757–767 (2012)
18. Yu, L., Lai, K.: A distance-based group decision-making methodology for multiperson multi-criteria emergency decision support. Decis. Support Syst. **51**, 307–315 (2011)

Author Index

Printed in the United States
By Bookmasters